EDITORIAL COPY

The
Bunyaviridae

THE VIRUSES

Series Editors

HEINZ FRAENKEL-CONRAT, *University of California*
Berkeley, California

ROBERT R. WAGNER, *University of Virginia School of Medicine*
Charlottesville, Virginia

THE VIRUSES: Catalogue, Characterization, and Classification
Heinz Fraenkel-Conrat

Other volumes in the series:

THE BACTERIOPHAGES
Volumes 1 and 2 • Edited by Richard Calendar

THE BUNYAVIRIDAE
Edited by Richard M. Elliott

THE CORONAVIRIDAE
Edited by Stuart G. Siddell

THE INFLUENZA VIRUSES
Edited by Robert M. Krug

THE PAPOVAVIRIDAE
Volume 1 • Edited by Norman P. Salzman
Volume 2 • Edited by Norman P. Salzman and Peter M. Howley

THE PARAMYXOVIRUSES
Edited by David W. Kingsbury

THE PARVOVIRUSES
Edited by Kenneth I. Berns

THE PLANT VIRUSES
Volume 1 • Edited by R. I. B. Francki
Volume 2 • Edited by M. H. V. Van Regenmortel and Heinz Fraenkel-Conrat
Volume 3 • Edited by Renate Koenig
Volume 4 • Edited by R. G. Milne

THE REOVIRIDAE
Edited by Wolfgang K. Joklik

THE RETROVIRIDAE
Volumes 1–4 • Edited by Jay A. Levy

THE RHABDOVIRUSES
Edited by Robert R. Wagner

THE TOGAVIRIDAE AND FLAVIVIRIDAE
Edited by Sondra Schlesinger and Milton J. Schlesinger

THE VIROIDS
Edited by T. O. Diener

A complete listing of volumes in this series appears at the back of this volume.

The Bunyaviridae

Edited by

RICHARD M. ELLIOTT

Institute of Virology
University of Glasgow
Glasgow, Scotland

PLENUM PRESS • NEW YORK AND LONDON

Library of Congress Cataloging-in-Publication Data

The bunyaviridae / edited by Richard M. Elliott.
       p.   cm. -- (The Viruses)
     Includes bibliographical references and index.
     ISBN 0-306-45178-6
     1. Bunyaviruses.   I. Elliott, Richard M.   II. Series.
   QR398.7.B86 1996
   576'.6484--dc20   96-4224
                                                          CIP

ISBN 0-306-45178-6

To my daughters, Katie and Emma,
and to the memory of their mother, Margaret,
for their love and support.

Contributors

David H. L. Bishop, NERC Institute of Virology and Environmental Microbiology, Oxford OX1 3SR, England

Keith Bupp, Departments of Neurology and Microbiology, University of Pennsylvania Medical Center, Philadelphia, Pennsylvania 19104-6146

Charles H. Calisher, Arthropod-borne Infectious Diseases Laboratory, Colorado State University, Fort Collins, Colorado 80523

Richard M. Elliott, Institute of Virology, University of Glasgow, Glasgow G11 5JR, Scotland

Colomba Giorgi, Laboratory of Virology, Istituto Superiore di Sanità, 00161 Rome, Italy

Rob Goldbach, Department of Virology, Agricultural University, 6709 PD Wageningen, The Netherlands

Francisco González-Scarano, Departments of Neurology and Microbiology, University of Pennsylvania Medical Center, Philadelphia, Pennsylvania 19104-6146

Thomas G. Ksiazek, Special Pathogens Branch, Division of Viral and Rickettsial Diseases, National Center for Infectious Diseases, Centers for Disease Control and Prevention, Atlanta, Georgia 30333

Ho Wang Lee, Asan Institute for Life Sciences, Asan Medical Center, Seoul, Korea

D. J. McClain, U.S. Army Medical Research Institute of Infectious Diseases, Fort Detrick, Frederick, Maryland 21702-5011

A. C. Marriott, NERC Institute of Virology and Environmental Microbiology, Oxford OX1 3SR, England

Lars Melin, Ludwig Institute for Cancer Research, Stockholm Branch, S-17177 Stockholm, Sweden

J. C. Morrill, U.S. Army Medical Research Institute of Infectious Diseases, Fort Detrick, Frederick, Maryland 21702-5011

Neal Nathanson, Departments of Neurology and Microbiology, University of Pennsylvania Medical Center, Philadelphia, Pennsylvania 19104-6146

Stuart T. Nichol, Special Pathogens Branch, Division of Viral and Rickettsial Diseases, National Center for Infectious Diseases, Centers for Disease Control and Prevention, Atlanta, Georgia 30333

P. A. Nuttall, NERC Institute of Virology and Environmental Microbiology, Oxford OX1 3SR, England

C. J. Peters, Special Pathogens Branch, Division of Viral and Rickettsial Diseases, National Center for Infectious Diseases, Centers for Disease Control and Prevention, Atlanta, Georgia 30333

Dick Peters, Department of Virology, Agricultural University, 6709 PD Wageningen, The Netherlands

Ralf F. Pettersson, Ludwig Institute for Cancer Research, Stockholm Branch, S-17177 Stockholm, Sweden

C. R. Pringle, Biological Sciences Department, University of Warwick, Coventry CV4 7AL, United Kingdom

Pierre E. Rollin, Special Pathogens Branch, Division of Viral and Rickettsial Diseases, National Center for Infectious Diseases, Centers for Disease Control and Prevention, Atlanta, Georgia 30333

Connie S. Schmaljohn, Virology Division, United States Army Medical Research Institute of Infectious Diseases, Fort Detrick, Frederick, Maryland 21702

Preface

In a review of the *Bunyaviridae* in 1990 I described these viruses as the Cinderellas of animal virology. A glance at the content of the present volume reveals this no longer to be the case. Much detail on the molecular biology of this large and diverse group of viruses is now known, at least one representative of each of the five genera has been fully sequenced, and we are on the verge of being able to manipulate *Bunyaviridae* genomes.

The *Bunyaviridae* is a fascinating family of viruses whose members are able to infect vertebrates, invertebrates, and plants. They display intriguing coding strategies and modes of replication, and their maturation at the Golgi provides a useful model to study intracellular protein transport. Interest in the *Bunyaviridae* goes beyond mere academic entertainment, however, as their involvement in disease, of humans, our animals and plants, is increasingly being recognized. Further, in an era of emerging (or reemerging) infectious diseases the *Bunyaviridae* do not disappoint: outbreaks of Rift Valley fever throughout Africa, possibly associated with changes in virulence or pathogenicity of the virus, and the identification of new hantaviruses associated with severe respiratory infection in the United States, are excellent examples. As most members of the family have small mammals as their natural reservoir, changing environmental conditions, both natural and induced, could well lead to increased contact with humans, resulting in new disease syndromes and problems emerging in the future. Hence, this family of viruses warrants continued study.

The diversity of viruses within the family does lead to problems in terminology, though, as one has to distinguish familial from generic characteristics. Hence, the rather clumsy term "*Bunyaviridae*" is used adjectively in this book when talking about the family whereas "bunyavirus" is reserved for discussion of the *Bunyavirus* genus. Perhaps renaming the *Bunyavirus* genus as "*Orthobunyavirus*" may overcome this problem and we can revert to the currently misused descriptor "bunyavirus" when considering the family as a whole.

I am grateful to the contributors to this book who undertook the task of writing the various chapters, generally with enthusiasm, which provide a wide coverage of the *Bunyaviridae*. An overview of *Bunyaviridae* taxonomy is given in Chapter 1 and the next five chapters deal with the molecular biology of the five genera: *Bunyavirus, Hantavirus, Nairovirus, Phlebovirus,* and *Tospovirus.* These are followed by general chapters on *Bunyaviridae* assembly and intracellular protein transport and *Bunyaviridae* genetics, and then four chapters dealing with *Bunyaviridae* diseases, including the newly described hantavirus pulmonary syndrome. In the final chapter a synopsis of comparative features of the family and some thoughts on evolution are given. Inevitably there is some overlap among the different chapters, but this has been kept to a minimum while still enabling each chapter to be read in isolation rather than reading from cover to cover.

Finally, I would like to acknowledge the members of my own laboratory for their continued support, and to thank Ms. Mary Phillips Born of Plenum Press for her patience during the extended gestation period needed to produce this volume.

Richard M. Elliott

Glasgow, Scotland

Contents

Chapter 3

Molecular Biology of Hantaviruses

Connie S. Schmaljohn

Chapter 4

Molecular Biology of Nairoviruses

A. C. Marriott and P. A. Nuttall

Chapter 5

Molecular Biology of Phleboviruses

Colomba Giorgi

Chapter 6

Molecular and Biological Aspects of Tospoviruses

Rob Goldbach and Dick Peters

Chapter 7

Synthesis, Assembly, and Intracellular Transport of *Bunyaviridae* Membrane Proteins

Ralf F. Pettersson and Lars Melin

    IV. G1 and G2 Accumulate in the Golgi Complex 175
    V. Mapping the Golgi-Retention Signal in G1/G2 177
        A. Bunyamwera Virus 178
        B. Hantaan Virus 179
        C. Punta Toro Virus 179
        D. Uukuniemi Virus 180
    VI. Conclusions and Perspectives 181
    VII. References 183

Chapter 8

Genetics and Genome Segment Reassortment

C. R. Pringle

Chapter 9

Pathogenesis of Diseases Caused by Viruses of the *Bunyavirus* Genus

Francisco González-Scarano, Keith Bupp, and Neal Nathanson

Chapter 10

Epidemiology and Pathogenesis of Hemorrhagic Fever with Renal Syndrome

Ho Wang Lee

Chapter 11

Hantavirus Pulmonary Syndrome and Newly Described Hantaviruses in the United States

Stuart T. Nichol, Thomas G. Ksiazek, Pierre E. Rollin, and C. J. Peters

Chapter 12

Epidemiology and Pathogenesis of Rift Valley Fever and Other Phleboviruses

J. C. Morrill and D. J. McClain

Chapter 13

The *Bunyaviridae*: Concluding Remarks and Future Prospects

Richard M. Elliott

The
Bunyaviridae

CHAPTER 1

History, Classification, and Taxonomy of Viruses in the Family *Bunyaviridae*

CHARLES H. CALISHER

I. INTRODUCTION

Viruses in the family *Bunyaviridae* are found worldwide. Most are arboviruses (*arthropod-borne animal viruses*), maintained in nature by a biological (propagative) transmission cycle between susceptible vertebrate hosts and hematophagous arthropods, such as mosquitoes, biting flies, and ticks. Vertebrate hosts of these viruses include rodents and other small mammals, primates, birds, and ungulates. Arboviruses usually occur in silent, sylvatic transmission cycles; however, more than 60 members of the family *Bunyaviridae* have been reported to cause disease in humans or livestock and some cause disease in sea birds or plants (Table I).

II. DISCUSSION

When Smithburn and his colleagues made the initial isolation of Bunyamwera virus from *Aedes* species mosquitoes during studies of yellow fever in Uganda in 1943 (Smithburn *et al.*, 1946), they could scarcely have imagined that this apparently insignificant virus would become the type species

CHARLES H. CALISHER • Arthropod-borne Infectious Diseases Laboratory, Colorado State University, Fort Collins, Colorado 80523.

The Bunyaviridae, edited by Richard M. Elliott, Plenum Press, New York, 1996.

TABLE I. 313 Viruses of the Family *Bunyaviridae,*
Listed by Geographic Distribution and Whether Associated with Illness[a]

	Geographic distribution[b]	Associated illness[c]	Principal arthropod vector
Genus *Bunyavirus* (Bunyamwera supergroup) (172)			
Anopheles A group (15)			
Anopheles A	SA	—	Mosquitoes
Arumateua	SA	—	Mosquitoes
Caraipé	SA	—	Mosquitoes
CoAr 1071	SA	—	Mosquitoes
CoAr 3624	SA	—	Mosquitoes
CoAr 3627	SA	—	Mosquitoes
ColAn 57389	SA	—	Mosquitoes
H 32580	SA	—	Mosquitoes
Las Maloyas	SA	—	Mosquitoes
Lukuni	SA	—	Mosquitoes
SPAr 2317	SA	—	Mosquitoes
Tacaiuma	SA	Human	Mosquitoes
Trombetas	SA	—	Mosquitoes
Tucuruí	SA	—	Mosquitoes
Virgin River	NA	—	Mosquitoes
Anopheles B group (2)			
Anopheles B	SA	—	Mosquitoes
Boraceia	SA	—	Mosquitoes
Bakau group (5)			
Bakau	Asia	—	Mosquitoes
Ketapang	Asia	—	Mosquitoes
Nola	Africa	—	Mosquitoes
Tanjong Rabok	Asia	—	?
Telok Forest	Asia	—	?
Bunyamwera group (32)			
AG83 1746	SA	—	Mosquitoes
Anhembi	SA	—	Mosquitoes
Batai (Calovo)	Asia, Europe	—	Mosquitoes
BeAr 328208	SA	—	Mosquitoes
Birao	Africa	—	Mosquitoes
Bozo	Africa	—	Mosquitoes
Bunyamwera	Africa	Human	Mosquitoes
Cache Valley	NA	Sheep, cattle	Mosquitoes
CbaAr 426	SA	—	Mosquitoes
Fort Sherman	SA	Human	Mosquitoes
Germiston	Africa	Human	Mosquitoes
Iaco	SA	—	Mosquitoes
Ilesha	Africa	Human	Mosquitoes
Kairi	SA	Equine	Mosquitoes
Lokern	NA	—	Mosquitoes, culicoid flies
Macaua	SA	—	Mosquitoes
Maguari	SA	—	Mosquitoes
Main Drain	NA	Equine	Mosquitoes, culicoid flies
Mboke	Africa	—	Mosquitoes
Ngari	Africa	—	Mosquitoes

(*continued*)

TABLE I. (*Continued*)

	Geographic distribution[b]	Associated illness[c]	Principal arthropod vector
Bunyamwera group (32) (*cont.*)			
Northway	NA	—	Mosquitoes
Playas	SA	—	Mosquitoes
Potosi	NA	—	Mosquitoes
Santa Rosa	NA	—	Mosquitoes
Shokwe	Africa	Human	Mosquitoes
Sororoca	SA	—	Mosquitoes
Taiassui	SA	—	Mosquitoes
Tensaw	NA	—	Mosquitoes
Tlacotalpan	NA	—	Mosquitoes
Tucunduba	SA	—	Mosquitoes
Wyeomyia	SA	Human	Mosquitoes
Xingu	SA	Human	Mosquitoes
Bwamba group (2)			
Bwamba	Africa	Human	Mosquitoes
Pongola	Africa	Human	Mosquitoes
Group C (14)			
63U 11	SA	—	Mosquitoes
Apeu	SA	Human	Mosquitoes
Bruconha	SA	—	Mosquitoes
Caraparu	SA, NA	Human	Mosquitoes
Gumbo Limbo	NA	—	Mosquitoes
Itaqui	SA	Human	Mosquitoes
Madrid	NA	Human	Mosquitoes
Marituba	SA	Human	Mosquitoes
Murutucu	SA	Human	Mosquitoes
Nepuyo	SA, NA	Human	Mosquitoes
Oriboca	SA	Human	Mosquitoes
Ossa	NA	Human	Mosquitoes
Restan	SA	Human	Mosquitoes
Vinces	SA	—	Mosquitoes
California group (14)			
AG83 497	SA	—	Mosquitoes
California encephalitis	NA	Human	Mosquitoes
Guaroa	SA, NA	Human	Mosquitoes
Inkoo	Europe	Human	Mosquitoes
Jamestown Canyon	NA	Human	Mosquitoes
Keystone	NA	—	Mosquitoes
La Crosse	NA	Human	Mosquitoes
Melao	SA	—	Mosquitoes
San Angelo	NA	—	Mosquitoes
Serra do Navio	SA	—	Mosquitoes
Snowshoe hare	NA	Human	Mosquitoes
South River	NA	—	Mosquitoes
Tahyña (Lumbo)	Europe (Africa)	Human	Mosquitoes
Trivittatus	NA	—	Mosquitoes

(*continued*)

TABLE I. (*Continued*)

	Geographic distribution[b]	Associated illness[c]	Principal arthropod vector
Capim group (10)			
Acara	SA, NA	—	Mosquitoes
Benevides	SA	—	Mosquitoes
Benfica	SA	—	Mosquitoes
Bushbush	SA	—	Mosquitoes
Capim	SA	—	Mosquitoes
Guajara	SA, NA	—	Mosquitoes
GU71U 344	NA	—	Mosquitoes
GU71U 350	NA	—	Mosquitoes
Juan Diaz	NA	—	?
Moriche	SA	—	Mosquitoes
Gamboa group (8)			
75V 2374	SA	—	Mosquitoes
75V 2621	SA	—	Mosquitoes
78V 2441	SA	—	Mosquitoes
Alajuela	NA	—	Mosquitoes
Brus Laguna	NA	—	Mosquitoes
Gamboa	NA	—	Mosquitoes
Pueblo Viejo	SA	—	Mosquitoes
San Juan	SA	—	Mosquitoes
Guama group (12)			
Ananindeua	SA	—	Mosquitoes
Bertioga	SA	—	?
Bimiti	SA	—	Mosquitoes
Cananeia	SA	—	Mosquitoes
Catu	SA	Human	Mosquitoes
Guama	SA, NA	Human	Mosquitoes
Guaratuba	SA	—	Mosquitoes
Itimirim	SA	—	?
Mahogany Hammock	NA	—	Mosquitoes
Mirim	SA	—	Mosquitoes
Moju	SA	—	Mosquitoes
Timboteua	SA	—	Mosquitoes
Koongol group (2)			
Koongol	Australia	—	Mosquitoes
Wongal	Australia	—	Mosquitoes
Minatitlan group (2)			
Minatitlan	NA	—	?
Palestina	SA	—	Mosquitoes
Nyando group (2)			
Eret 147	Africa	—	Mosquitoes
Nyando	Africa	Human	Mosquitoes
Olifantsvlei group (5)			
Bobia	Africa	—	Mosquitoes
Botambi	Africa	—	Mosquitoes
Dabakala	Africa	—	Mosquitoes
Olifantsvlei	Africa	—	Mosquitoes
Oubi	Africa	—	Mosquitoes

(continued)

TABLE I. (*Continued*)

	Geographic distribution[b]	Associated illness[c]	Principal arthropod vector
Patois group (7)			
Abras	SA	—	Mosquitoes
Babahoyo	SA	—	Mosquitoes
Estero Real	NA	—	Ticks
Pahayokee	NA	—	Mosquitoes
Patois	NA	—	Mosquitoes
Shark River	NA	—	Mosquitoes
Zegla	NA	—	?
Simbu group (25)			
Aino	Asia, Australia	—	Mosquitoes, culicoid flies
Akabane	Africa, Asia, Australia	Cattle	Mosquitoes, culicoid flies
Buttonwillow	NA	—	Culicoid flies
Douglas	Australia	—	Culicoid flies
Facey's Paddock	Australia	—	?
Ingwavuma	Africa, Asia	Pigs	Mosquitoes
Inini	SA	—	?
Jatobal	SA	—	?
Kaikalur	Asia	—	Mosquitoes
Manzanilla	SA	—	?
Mermet	NA	—	Mosquitoes
Oropouche	SA	Human	Culicoid flies, mosquitoes
Para	SA	—	Mosquitoes
Peaton	Australia	—	Culicoid flies
Sabo	Africa	—	Culicoid flies
Sango	Africa	—	Culicoid flies, mosquitoes
Sathuperi	Africa, Asia	—	Culicoid flies, mosquitoes
Shamonda	Africa	—	Culicoid flies
Shuni	Africa	—	Culicoid flies, mosquitoes
Simbu	Africa	—	Mosquitoes
Thimiri	Africa, Asia	—	?
Tinaroo	Australia	—	Culicoid flies
Utinga	SA	—	?
Utive	SA	—	?
Yaba 7	Africa	—	?
Tete group (6)			
Bahig	Africa, Europe	—	Ticks
Batama	Africa	—	?
Matruh	Africa, Europe	—	Ticks
Tete	Africa	—	?
Tsuruse	Asia	—	?
Weldona	NA	—	Culicoid flies

(*continued*)

TABLE I. (*Continued*)

	Geographic distribution[b]	Associated illness[c]	Principal arthropod vector
Turlock group (5)			
Lednice	Europe	—	Mosquitoes
M'Poko	Africa	—	Mosquitoes
Turlock	NA, SA	—	Mosquitoes
Umbre	Asia	—	Mosquitoes
Yaba-1	Africa	—	Mosquitoes
Ungrouped in genus *Bunyavirus* (4)			
Kaeng Khoi	Asia	—	Nest bugs
Leanyer	Australia	—	Mosquitoes
Mojui dos Campos	SA	—	?
Termeil	Asia	—	Mosquitoes
Genus *Hantavirus* (10)			
Hantaan group			
Black Creek Canal	NA	Human	None
Convict Creek	NA	Human	None
Dobrava-Belgrade	Europe	Human	None
Hantaan	Asia, Europe	Human	None
El Moro	NA	?	None
Leaky	NA	?human?	None
Prospect Hill	NA	—	None
Puumala	Europe, Asia	Human	None
Seoul	Asia, Europe, NA, SA	Human	None
Sin Nombre	NA	Human	None
Thailand	Asia	—	None
Thottapalayam	Asia	—	None
Genus *Nairovirus* (34)			
Crimean-Congo hemorrhagic fever (CCHF) group (3)			
CCHF	Africa, Asia, Europe	Human	Ticks, culicoid flies
Hazara	Asia	—	Ticks
Khasan	Asia	—	Ticks
Dera Ghazi Khan group (6)			
Abu Hammad	Africa, Asia	—	Ticks
Abu Mina	Africa	—	?
Dera Ghazi Khan	Asia	—	Ticks
Kao Shuan	Asia, Australia	—	Ticks
Pathum Thani	Asia	—	Ticks
Pretoria	Africa	—	Ticks
Hughes group (10)			
Farallon	NA	—	Ticks
Fraser Point	Europe	—	Ticks
Great Saltee	Europe	—	Ticks
Hughes	NA, SA	Seabirds	Ticks

(*continued*)

TABLE I. (*Continued*)

	Geographic distribution[b]	Associated illness[c]	Principal arthropod vector
Hughes group (10) (*cont.*)			
Puffin Island	Europe	—	Ticks
Punta Salinas	SA	—	Ticks
Raza	NA	—	Ticks
Sapphire II	NA	—	Ticks
Soldado	Africa, Europe, SA	—	Ticks
Zirqa	Asia	—	Ticks
Nairobi sheep disease group (2)			
Dugbe	Africa	Human, cattle	Ticks
Nairobi sheep disease (= Ganjam)	Africa, Asia	Human, cattle	Ticks, culicoid flies, mosquitoes
Qalyub group (4)			
Bakel	Africa	—	Ticks
Bandia	Africa	—	Ticks
Omo	Africa	—	?
Qalyub	Africa	—	Ticks
Sakhalin group (7)			
Avalon	NA, Europe	—	Ticks
Clo Mor	Europe	—	Ticks
Kachemak Bay	NA	—	Ticks
Paramushir	Asia	—	Ticks
Sakhalin	Asia	—	Ticks
Taggert	Australia	—	Ticks
Tillamook	NA	—	Ticks
Thiafora group (2)			
Erve	Europe	—	?
Thiafora	Africa	—	?
Genus *Phlebovirus* (51)			
Sandfly fever group (23)			
Bujaru complex			
Bujaru	SA	—	?
Munguba	SA	—	Phlebotomine flies
Candiru complex			
Alenquer	SA	Human	?
Candiru	SA	Human	?
Itaituba	SA	—	?
Nique	NA	—	Phlebotomine flies
Oriximina	SA	—	Phlebotomine flies
Turuna	SA	—	Phelbotomine flies
Chilibre complex			
Cacao	NA	—	Phlebotomine flies
Chilibre	NA	—	Phlebotomine flies
Frijoles complex			
Frijoles	NA	—	Phlebotomine flies
Joa	SA	—	?

(*continued*)

TABLE I. (*Continued*)

	Geographic distribution[b]	Associated illness[c]	Principal arthropod vector
Sandfly fever group (23) (*cont.*)			
Punta Toro complex			
Buenaventura	NA, SA	—	Phlebotomine flies
Punta Toro	NA, SA	Human	Phlebotomine flies
Rift Valley fever complex			
Belterra	SA	—	?
Icoaraci	SA	—	Phlebotomine flies, mosquitoes
Rift Valley fever	Africa	Human, cattle	Mosquitoes
Salehabad complex			
Arbia	Europe	—	Phlebotomine flies
Salehabad	Asia	—	Phlebotomine flies
Sandfly fever Naples complex			
Karimabad	Asia	—	Phlebotomine flies
Sandfly fever Naples	Europe, Africa, Asia	Human	Phlebotomine flies
Tehran	Asia	—	Phlebotomine flies
Toscana	Europe	Human	Phlebotomine flies
No complex assigned in sandfly fever group (16)			
Aguacate	NA	—	Phlebotomine flies
Anhanga	SA	—	?
Arboledas	SA	—	Phlebotomine flies
Arumowot	Africa	—	Mosquitoes
Caimito	NA	—	Phlebotomine flies
Chagres	NA	Human	Phlebotomine flies, mosquitoes
Corfou	Europe	—	Phlebotomine flies
Gabek Forest	Africa	—	?
Gordil	Africa	—	?
Itaporanga	SA	—	Mosquitoes
Odrenisrou	Africa	—	Mosquitoes
Pacui	SA	—	Phlebotomine flies
Rio Grande	NA	—	?
Saint-Floris	Africa	—	?
Sandfly fever Sicilian	Europe, Africa, Asia	Human	Phlebotomine flies
Urucuri	SA	—	?
Uukuniemi (12)			
EgAn 1825-61	Africa	—	?
Fin V 707	Europe	—	?
Grand Arbaud	Europe	—	Ticks
Manawa	Asia	—	Ticks
Murre	Europe	—	?
Oceanside	NA	—	Ticks
Ponteves	Europe	—	Ticks
Precarious Point	Australia	—	Ticks

(continued)

TABLE I. (*Continued*)

	Geographic distribution[b]	Associated illness[c]	Principal arthropod vector
Uukuniemi (12) (*cont.*)			
RML 105355	NA	—	Ticks
St. Abbs Head	Europe	Seabirds	Ticks
Uukuniemi	Europe	—	Ticks
Zaliv Terpeniya	Asia	—	Ticks
Genus *Tospovirus* (4)			
Tomato spotted wilt (3)	Worldwide	Plants	Thrips[c]
Impatiens necrotic spot (1)		Plants	
Grouped but unassigned viruses of family *Bunyaviridae* (19)			
Bhanja group (3)			
Bhanja	Africa, Asia, Europe	Human	Ticks
Forecariah	Africa	—	Ticks
Kismayo	Africa	—	Ticks
Kaisodi group (3)			
Kaisodi	Asia	—	Ticks
Lanjan	Asia	—	Ticks
Silverwater	NA	—	Ticks
Mapputta group (4)			
Gan Gan	Australia	—	Mosquitoes
Mapputta	Australia	—	Mosquitoes
Maprik	Australia	—	Mosquitoes
Trubanaman	Australia	—	Mosquitoes
Tanga group (2)			
Okola	Africa	—	Mosquitoes
Tanga	Africa	Human	Mosquitoes
Resistencia group (3)			
Antequera	SA	—	Mosquitoes
Barranqueras	SA	—	Mosquitoes
Resistencia	SA	—	Mosquitoes
Upolu group (2)			
Aransas Bay	NA	—	Ticks
Upolu	Australia	—	Ticks
Yogue group (2)			
Kasokero	Africa	Human	?
Yogue	Africa	Human	?
Ungrouped and unassigned viruses of the family *Bunyaviridae* (23) (*cont.*)			
Bangui	Africa	Human	Mosquitoes
Batken	Asia	—	Mosquitoes, ticks
Belém	SA	—	?
Belmont	Australia	—	Mosquitoes
Bobaya	Africa	—	?
Caddo Canyon	NA	—	Ticks
Chim	Asia	—	Ticks
Enseada	SA	—	Mosquitoes
Issyk-Kul (Keterah)	Asia	Human	Mosquitoes, ticks
Jos	Africa	—	Ticks
Kowanyama	Australia	—	Mosquitoes

(*continued*)

TABLE I. (*Continued*)

	Geographic distribution[b]	Associated illness[c]	Principal arthropod vector
Ungrouped and unassigned viruses of the family *Bunyaviridae* (23) (*cont.*)			
Lone Star	NA	—	Ticks
Pacora	NA	—	Mosquitoes
Razdan	Asia	—	Ticks
Santarem	SA	—	?
Sedlec	Europe	—	Mosquitoes
Sunday Canyon	NA	—	Ticks
Tai	Africa	—	Mosquitoes
Tamdy	Asia	Human	Ticks
Tataguine	Africa	Human	Mosquitoes
Wanowrie	Africa, Asia	Human	Mosquitoes, ticks
Witwatersrand	Africa	—	Mosquitoes
Yacaaba	Australia	—	Mosquitoes

[a]Mention of a virus name is not meant to constitute priority of publication.
[b]SA, South America; NA, North America.
[c]Whereas there is evidence for infections of humans and other animals with many of these viruses, only those that are known to cause illnesses are listed here.

of both the genus *Bunyavirus* and the virus family *Bunyaviridae*. Thus, it was largely by historical accident that this African virus that causes only minor illness in humans has become a linchpin of modern molecular, taxonomic, and diagnostic procedures.

By 1960, it was known that eastern equine encephalitis and western equine encephalitis and certain other viruses were related to each other and had been placed in what was termed the "Group A arboviruses" and that Japanese encephalitis, yellow fever, St. Louis encephalitis, West Nile, and certain other viruses were related to each other and had been placed in what was termed the "Group B arboviruses." Casals and Whitman (1960), having demonstrated that Bunyamwera, Wyeomyia (Colombia), Cache Valley (Utah), and Kairi (Trinidad) viruses were related antigenically, suggested the name "Bunyamwera group" for these viruses. Relationships between these viruses initially were based on studies of their complement-fixing (CF) antigens, but further studies of these viruses, as well as of Germiston (South Africa), Batai (Malaysia), Guaroa (Colombia), and related viruses from Brazil, revealed that whereas CF tests showed them to be closely related, neutralization tests showed them to be clearly distinct.

In the next decades a multitude of newly recognized arboviruses were isolated and workers at the Rockefeller Foundation Laboratories and elsewhere detected relationships between and among many of them. For every two or more viruses shown to be antigenically related but distinct from each other by quantitative serologic criteria in one or more tests, but related to other viruses, a serogroup was established. In current practice, the first discovered virus of a newly recognized serogroup lends its name to that serogroup. Individual members, referred to as "viruses" or "types," are antigenically related but distinct. "Subtypes" and "varieties" are isolates separable at lower levels of differentiation, by the use of special tests or reagents such as kinetic assays and monoclonal antibodies (Calisher *et al.*, 1980).

Casals and his co-workers were working in the World Health Organization's Collaborating Centre for Arbovirus Reference and Research, first at the Rockefeller Institute, then at Yale University, and had available to them the world's largest collection of arboviruses. They made the most of this resource, carrying out well-conceived and meticulous studies of the interrelationships of many of these viruses. Within about a 10-year period, not only had the Bunyamwera serogroup been recognized, but serogroups C, Guama, California, Capim, Anopheles A, Simbu, Bwamba, Patois, Koongol, Tete, and others had been distinguished (Casals and Whitman, 1961; Hammond and Reeves, 1952; Theiler and Downs, 1973; Whitman and Casals, 1961; Whitman and Shope, 1962). Remarkably, at least one virus in each of these groups reacted serologically with at least one virus (or antibody to it) of another group, suggesting to Casals that these viruses all were interrelated, albeit distantly and enigmatically. These cross-reactions between one or another member of the various serogroups often were weak and some appeared tenuous, but all were repeatable. Therefore, because the data indicated that these viruses somehow were interrelated, Casals (1963) suggested placing them in a "Bunyamwera Supergroup."

Electron microscopic studies supported and extended the observation that viruses of this supergroup were not clearly distinguishable by size, morphology, or morphogenesis in infected cells (Holmes, 1971; Murphy *et al.*, 1973). Subsequent studies of these characteristics and of the structure–function relationships, replicative mechanisms, biology, and ecology of viruses of the family *Bunyaviridae* have been reviewed (Bishop, 1990; Calisher and Thompson, 1983; Elliott, 1990; Gonzalez-Scarano and Nathanson, 1990; Grimstad, 1988; Le Duc, 1979; Schmaljohn and Patterson, 1990).

Casals had suggested that if two or more viruses shared common antigens, they must be related. This simple but effective premise has never been disproved, but as additional arboviruses were studied it became clear that some shared size, morphology, and morphogenetic and other characteristics but could not be shown to be antigenically related to viruses of the Bunyamwera Supergroup. Further analyses demonstrated that many of these viruses were related to each other but not to other known viruses and therefore were placed in their own serogroups, for example, a large number of viruses related to those that cause sandfly fever (sandfly fever serogroup) and another large number of viruses related to Uukuniemi virus (Uukuniemi serogroup). Whereas sandfly fever serogroup viruses were isolated most often from biting flies, Uukuniemi serogroup viruses were isolated most often from ticks. Many viruses from ticks and other sources, including those causing Nairobi sheep disease and Crimean hemorrhagic fever, were shown to be interrelated and it was suggested that they form another Supergroup (Casals and Tignor,

1980). Clearly, the natural cycles of these viruses were reflective of other differences, differences not recognized at that time.

All members of the family *Bunyaviridae* studied thus far have three segments (large, L; medium, M; and small, S) of single-stranded RNA; intact virions are spherical or oval, enveloped, and 90 to 100 nm in diameter; virions have glycoprotein surface projections; lipid comprises 20 to 30% of the virion by weight and forms part of the lipoprotein envelope, which is cell-derived. Carbohydrate comprises 2 to 7% of the weight of the virion and is incorporated as a component of the glycoproteins and glycolipids (Schmaljohn and Patterson, 1990). All members of the family so far tested are labile to acid (pH 3), lipid solvent (ether, chloroform), detergent (sodium deoxycholate), heat (56°C/15–30 min), formalin, 70% ethanol, 5% iodine, and ultraviolet irradiation.

Virions consist of a unit membrane envelope with spikes surrounding a somewhat unstructured interior from which a helical, 2.5-nm-wide nucleocapsid can be extracted. Constituent synthesis takes place in the cytoplasm, and morphogenesis occurs without prior core formation by budding directly into the Golgi complex and vesicles of infected cells (Murphy *et al.*, 1973; see Chapter 7). Host RNA sequences prime viral mRNA translation, involving a process similar to cap-snatching in orthomyxoviruses (Kolakofsky and Hacker, 1991). No enzymatic function has been associated with the envelope glycoproteins.

Biochemical and molecular similarities between viruses of the Bunyamwera Supergroup were sufficient to prompt the International Committee for Taxonomy of Viruses (Bishop *et al.*, 1980) to recognize them as being members of the genus *Bunyavirus* (vernacular term "bunyaviruses") in the family *Bunyaviridae*. At the same time, viruses belonging to sandfly fever, Uukuniemi, Nairobi sheep disease, and Hantaan serogroups, which morphologically resemble the Bunyamwera Supergroup members but which do not share antigens with them, were placed in other genera within the family, i.e., *Phlebovirus* ("phleboviruses"), *Uukuvirus* ("uukuviruses"), *Nairovirus* ("nairoviruses"), and *Hantavirus* ("hantaviruses"), respectively. Because the phleboviruses and uukuviruses are even more closely related to each other than they are to viruses of other genera within this family, the International Committee for Taxonomy of Viruses more recently has approved their integration in a single genus, *Phlebovirus* (Francki *et al.*, 1991), which includes sandfly fever Sicilian, sandfly fever Naples, and Rift Valley fever viruses.

The three negative-sense RNA species of bunyaviruses are L RNA: 2.7 to 3.1×10^6 ($\approx$7000 bases); M RNA: 1.8 to 2.3×10^6 (4450 to 4540 bases); S RNA: 0.28 to 0.50×10^6 (850 to 990 bases); the RNA comprises 1 to 2% of the total virion by weight. Differences exist between terminal nucleotide sequences of gene segments of viruses of different genera within the family.

Virions usually contain four structural proteins: two external glycoproteins (G1, G2), a nucleocapsid protein (N), and a large protein (L), which is presumed to be a transcriptase. A single open reading frame in the M RNA encodes the glycoproteins, which are cotranslationally cleaved to G1 and G2; the hemagglutinin activity is associated with the glycoproteins. The N protein and a nonstructural protein are coded in overlapping reading frames by the S RNA; CF antigens are associated principally with the nucleocapsid protein. The L protein is coded by the L RNA (Endres *et al.*, 1989).

Antibodies against selected epitopes on the G1 protein have neutralizing activity (Gentsch *et al.*, 1980; Gonzalez-Scarano *et al.*, 1982), and also protect mice from virus challenge (N. Nathanson *et al.*, unpublished observations). Because most neutralizing monoclonal antibodies inhibit hemagglutination, the G1 protein likely is involved in the attachment of virus to cell receptors. The G1 protein is an important determinant of virulence in vertebrate hosts and of infectivity for mosquitoes; it also appears to be the most important protein in protective immunity (Gonzalez-Scarano and Nathanson, 1990).

Reassortment can occur between the three segments of two different bunyaviruses. However, reassortment has not been detected between bunyaviruses of different serogroups and available evidence suggests that bunyavirus reassortment is limited to viruses belonging to the same serogroup, i.e., viruses more closely related genetically (Iroegbu and Pringle, 1981; discussed in more detail in Chapter 2). Naturally occurring reassortants of LAC virus have been detected by genotype analyses of field isolates (Klimas *et al.*, 1981; Ushijima *et al.*, 1981). There is inadequate information regarding the molecular mechanisms of persistent infections of arthropod vectors with LAC virus, other bunyaviruses, and other arboviruses. However, these infections are lifelong in the vectors and may exert little or no untoward effect on them. Indeed, that at least certain bunyaviruses are transovarially and venereally transmitted provides a mechanism with far-reaching evolutionary and overwintering implications (Watts *et al.*, 1973). Lifelong infection of the vector provides substantial opportunity for bunyaviruses to evolve by genetic drift and, under suitable circumstances (i.e., mixed infection), by segment reassortment (Beaty and Bishop, 1988).

Segment reassortants have been used to determine the biological role of each segment and have shown that the M RNA segment is a major determinant of peripheral virulence for mice and infectivity for mosquitoes. Distinct sites within the M RNA code for different genetic determinants of biological markers, including subcutaneous and intracranial mouse virulence and oral and intrathoracic infection of mosquitoes (Beaty *et al.*, 1982; Gonzalez-Scarano *et al.*, 1988). More recent evidence indicates an association of the polymerase with mouse neurovirulence and neuroinvasiveness (Endres *et al.*, 1991).

For many years there was no doubt that differences existed between and among the viruses of the family, but these differences were perplexing in that they did not provide an explanation for their occurrence, i.e., antigenic relationships between viruses of a serogroup might be detected by HI and neutralization and not by CF. The complexity of such antigenic relationships is exemplified by the Group C viruses (Shope and Causey, 1962). Karabatsos and Shope (1979) extended the latter studies and suggested that, because the CF antigen common to members of Group C antigenic pairs is not an antigen shared by all members of the serogroup, "pairing" might have resulted from natural genetic recombination (reassortment). Of course, it now is known that viruses of each genus within this family variously share G1, G2, N, and L structural proteins and few, some, or many nonstructural proteins and epitopes (Shope *et al.*, 1988).

There may be evolutionarily functional advantages of such reassortments. Among sympatric Group C bunyaviruses of Brazil, Itaqui virus is transmitted mainly by *Culex vomerifer*, a mosquito species apparently resistant to infection by Oriboca (Woodall, 1979). Vertebrate host specificity among the Group C viruses is illustrated by Apeu and Marituba viruses, which have been isolated from marsupials but not rodents, and Caraparu and Murutucu viruses, present in those same areas of Brazil but isolated frequently from rodents. If cross-protection between viruses and restricted vector–host relationships of viruses leads to mutual exclusion, the Group C viruses may be natural reassortants of each other or of now-extinct viruses. Further, restriction of a virus to a vertebrate–vector pairing with defined geographic distribution may lead to natural isolation and genetic stability. For example, African Bunyamwera serogroup viruses are found in distinct or overlapping geographic areas and ecosystems, but certain of the South American Bunyamwera serogroup viruses coexist in horizontally or vertically contiguous ecosystems. One mechanism for the separate maintenance of sympatric, closely related serotypes may be differences in vector or host susceptibility to them.

Among the Bunyamwera serogroup viruses occurring in North America are Cache Valley, Tensaw, Lokern, and Main Drain viruses. As with certain other Bunyamwera serogroup viruses, Cache Valley and Tensaw are essentially impossible to distinguish one from another by CF tests. Cache Valley, Lokern, and Tensaw viruses are more closely related to each other antigenically than they are to Main Drain virus (Hunt and Calisher, 1979).

Lokern and Main Drain viruses replicate in both hares and rabbits and are transmitted principally by *Culicoides* and mosquitoes, whereas Cache Valley and Tensaw viruses replicate in rabbits but not hares and in mosquitoes but not *Culicoides* (Calisher *et al.*, 1986). These biologic differences may account for the geographic distribution of these viruses. Lokern and Main Drain viruses, which are antigenically distant from each other, may coexist with each other because of their minimal cross-protectivity, whereas Cache Valley and Tensaw viruses may mutually exclude each other because of their significant cross-protectivity. Cache Valley virus occurs in areas where Lokern and Main Drain viruses are found, but its vector–host pairing is different and more restricted. Cache Valley virus is not found in the southeastern United States where Tensaw virus occurs, probably because of the range of the principal vectors (*Culiseta inornata, Anopheles quadrimaculatus*, and *Aedes sollicitans* for Cache Valley virus; *Anopheles crucians* for

Tensaw virus) and vertebrate hosts (rabbits for Cache Valley virus; rabbits and cotton rats, for Tensaw virus). Similar examples of vector–host pairing can be given for the California, Simbu, Guama, and Patois serogroup bunyaviruses (Beaty and Bishop, 1988; Calisher, 1988).

Thus, the dilemma posed by antigenic cross-reactions may have been a quandary for the serologist but the solution was key to understanding the biology of these viruses. Subsequent chapters in this volume provide details of the evolution, structure–function relationships, mechanisms of replication, pathogenetic mechanisms, and epidemiology of the fascinating viruses of this virus family.

At present, the International Catalogue of Arboviruses lists more than 525 viruses, (Karabatsos, 1985). Of these, 313 have been placed in the family *Bunyaviridae:* 172 in the genus *Bunyavirus*, 10 in the genus *Hantavirus*, 34 in the genus *Nairovirus*, 51 in the genus *Phlebovirus*, and 4 in the genus *Tospovirus*; an additional 19 viruses have characteristics of viruses of the family *Bunyaviridae* and have been placed in serogroups but are unassigned to a genus, and 23 others are both ungrouped and unassigned within the family (Table I). Among the members of this family are some significant pathogens: (genus *Bunyavirus*) La Crosse, Oropouche, and Akabane viruses; (genus *Nairovirus*) Congo–Crimean hemorrhagic fever and Nairobi sheep disease viruses; (genus *Phlebovirus*) Rift Valley fever and the sandfly fever viruses; (genus *Hantavirus*) Hantaan virus (Korean hemorrhagic fever, hemorrhagic fever with renal syndrome, epidemic hemorrhagic fever), Puumala virus (Nephropathia Epidemica), and Sin Nombre, Convict Creek, and other emerging viruses responsible for the recent outbreak of Hantavirus Pulmonary Syndrome in the western United States and isolated cases elsewhere in the United States; and (genus *Tospovirus*) tomato spotted wilt virus. Many other viruses of this family cause occasional cases of febrile illness or encephalitis in humans (Table I; Monath, 1988) or disease in plants (de Àvila *et al.*, 1993).

ACKNOWLEDGMENTS.   I thank Drs. Patricia A. Nuttall, NERC Institute of Virology, Oxford, England, Jean-Pierre Digoutte, Pasteur Institute, Dakar, Senegal, and Amelia P.A. Travassos da Rosa, Evandro Chagas Institute, Belém, Pará, Brazil, for sharing unpublished data on certain viruses and allowing me to cite them. This work was supported in part by USPHS Grant AI-34454 from the National Institutes of Health.

III. REFERENCES

Beaty, B. J., and Bishop, D. H., 1988, Bunyavirus–vector interactions, *Virus Res.* **10**:289.
Beaty, B. J., Miller, B. R., Shope, R. E., Rozhon, E. J., and Bishop, D. H. L., 1982, Molecular basis of bunyavirus *per os* infection of mosquitoes: Role of the middle-sized RNA segment, *Proc. Natl. Acad. Sci. USA* **79**:1295.

Bishop, D. H. L., 1990, Bunyaviridae and their replication, Part I. Bunyaviridae, in: *Virology*, 2nd ed. (B. N. Fields and D. M. Knipe, eds.), pp. 1155–1173, Raven Press, New York.

Bishop, D. H. L., Calisher, C. H., Casals, J., Chumakov, M. P., Gaidamovich, S. Y., Hannoun, C., Lvov, D. K., Marshall, I. D., Oker-Blom, N., Pettersson, R. F., Porterfield, J. S., Russell, P. K., Shope, R. E., and Westaway, E. G., 1980, *Bunyaviridae, Intervirology* **14**:125.

Calisher, C. H., 1988, Evolutionary significance of the taxonomic data regarding bunyaviruses of the family *Bunyaviridae, Intervirology* **29**:268.

Calisher, C. H., and Thompson, W. H., (eds.), 1983, *California Serogroup Viruses*, Liss, New York.

Calisher, C. H., Shope, R. E., Brandt, W., Casals, J., Karabatsos, N., Murphy, F. A., Tesh, R. B., and Wiebe, M. E., 1980, Proposed antigenic classification of registered arboviruses: I Togaviridae, alphavirus, *Intervirology* **14**:229.

Calisher, C. H., Francy, D. B., Smith, G. C., Muth, D. J., Lazuick, J. S., Karabatsos, N., Jakob, W. L., and McLean, R. G., 1986, Distribution of Bunyamwera serogroup viruses in North America, 1956–1984, *Am. J. Trop. Med. Hyg.* **35**:429.

Casals, J., 1963, New developments in the classification of arthropod-borne animal viruses, *An. Microbiol.* **11**:13.

Casals, J., and Tignor, G. H., 1980, The *Nairovirus* genus: Serological relationships, *Intervirology* **14**:144.

Casals, J., and Whitman, L. A., 1960, A new antigenic group of arthropod-borne viruses. The Bunyamwera group, *Am. J. Trop. Med. Hyg.* **9**:73.

Casals, J., and Whitman, L., 1961, Group C. A new serological group of hitherto undescribed arthropod-borne viruses. Immunological studies, *Am. J. Trop. Med. Hyg.* **10**:250.

de Àvila, A. C., de Haan, P., Smeets, M. L. L., Resende, R. de O., Kormelink, R., Kitajima, E. W., Goldbach, R. W., and Peters, D., 1993, Distinct levels of relationships between tospovirus isolates, *Arch. Virol.* **128**:211.

Elliott, R. M., 1990, Molecular biology of the Bunyaviridae, *J. Gen. Virol.* **71**:501.

Endres, M. J., Jacoby, D. R., Janssen, R. S., Gonzalez-Scarano, F., and Nathanson, N., 1989, The large viral RNA segment of California serogroup bunyaviruses encodes the large viral protein, *J. Gen. Virol.* **70**:223.

Endres, M. J., Griot, C., Gonzalez-Scarano, F., and Nathanson, N., 1991, Neuroattenuation of an avirulent bunyavirus variant maps to the L RNA segment, *J. Virol.* **65**:5465.

Francki, R. I. B., Fauquet, C. M., Knudson, D. L., and Brown, F., (eds.), 1991, Classification and nomenclature of viruses; Fifth report of the International Committee on Taxonomy of Viruses. *Arch. Virol.* (Suppl. 2), Springer-Verlag, Vienna.

Gentsch, J. R., Rozhon, E. J., Klimas, R. A., El Said, L. H., Shope, R. E., and Bishop, D. H. L., 1980, Evidence from recombinant bunyavirus studies that the M RNA gene products elicit neutralizing antibodies, *Virology* **102**:190.

Gonzalez-Scarano, F., and Nathanson, N., 1990, Bunyaviruses, in: *Virology*, 2nd ed. (B. N. Fields and D. M. Knipe, eds.), pp. 1195–1228, Raven Press, New York.

Gonzalez-Scarano, F., Shope, R. E., Calisher, C. H., and Nathanson, N., 1982, Characterization of monoclonal antibodies against the G1 and N proteins of La Crosse and Tahyna, two California serogroup bunyaviruses, *Virology* **120**:42.

Gonzalez-Scarano, F., Beaty, B., Sundin, D., Janssen, R., Endres, M. J., and Nathanson, N., 1988, Genetic determinants of the virulence and infectivity of La Crosse virus, *Microb. Pathogen.* **4**:1.

Grimstad, P. R., 1988, California group virus disease, in: *The Arboviruses: Epidemiology and Ecology*, Vol. II (T. P. Monath, ed.), pp. 99–136, CRC Press, Boca Raton, FL.

Hammond, W. M., and Reeves, W. C., 1952, California encephalitis virus—a newly described agent. I. Evidence of natural infection in man and other animals, *Calif. Med.* **77**:303.

Holmes, I. H., 1971, Morphological similarity of Bunyamwera Supergroup viruses, *Virology* **43**:708.

Hunt, A. R., and Calisher, C. H., 1979, Relationships of Bunyamwera group viruses by neutralization, *Am. J. Trop. Med. Hyg.* **28**:740.

Iroegbu, C. U., and Pringle, C. R., 1981, Genetic interactions among viruses of the Bunyamwera complex, *J. Virol.* **37**:383.

Karabatsos, N., (ed.), 1985, *International Catalogue of Arboviruses Including Certain Other Viruses of Vertebrates*, 3rd ed., Am. Soc. Trop. Med. Hyg., San Antonio.

Karabatsos, N., and Shope, R. E., 1979, Cross-reactive and type-specific complement-fixing structures of Oriboca virions, *J. Med. Virol.* **3**:167.

Klimas, R. A., Thompson, W. H., Calisher, C. H., Clark, G. G., Grimstad, P. R., and Bishop, D. H. L., 1981, Genotypic varieties of La Crosse virus isolated from different geographic regions of the continental United States and evidence for a naturally occurring intertypic recombinant La Crosse virus, *Am. J. Epidemiol.* **114**:112.

Kolakofsky, D., and Hacker, D., 1991, Bunyavirus RNA synthesis: Genome transcription and replication, in: *Bunyaviridae* (D. Kolakofsky, ed.), pp. 143–159, Springer-Verlag, Berlin.

Le Duc, J. W., 1979, The ecology of California group viruses, *J. Med. Entomol.* **16**:1.

Monath, T. P., (ed.), 1988, *The Arboviruses: Epidemiology and Ecology*, 5 vols., CRC Press, Boca Raton, FL.

Murphy, F. A., Harrison, A. K., and Whitfield, S. G., 1973, Bunyaviridae: Morphologic and morphogenetic similarities of Bunyamwera supergroup viruses and several other arthropod-borne viruses, *Intervirology* **1**:297.

Schmaljohn, C. S., and Patterson, J. L., 1990, Bunyaviridae and their replication, Part II. Replication of Bunyaviridae, in: *Virology*, 2nd ed. (B. N. Fields and D. M. Knipe, eds.), pp. 1175–1194, Raven Press, New York.

Shope, R. E., and Causey, O. R., 1962, Further studies on the serological relationships of group C arthropod-borne viruses and the application of these relationships to rapid identification of types, *Am. J. Trop. Med. Hyg.* **11**:283.

Shope, R. E., Woodall, J. P., and Travassos da Rosa, A., 1988, The epidemiology of diseases caused by viruses in groups C and Guama (Bunyaviridae), in: *The Arboviruses: Epidemiology and Ecology*, Vol. III (T. P. Monath, ed.), pp. 37–52, CRC Press, Boca Raton, FL.

Smithburn, K. C., Haddow, A. J., and Mahaffy, A. F., 1946, Neurotropic virus isolated from *Aedes* mosquitoes caught in Semliki Forest, *Am. J. Trop. Med. Hyg.* **26**:189.

Theiler, M., and Downs, W. G., (compilers and editors), 1973, *The arthropod-borne Viruses of Vertebrates*, Yale University Press, New Haven, CT.

Ushijima, H., Clerx-van Haaster, C. M., and Bishop, D. H. L., 1981, Analyses of the Patois group bunyaviruses: Evidence for naturally occurring recombinant bunyaviruses and existence of viral coded nonstructural proteins induced in bunyavirus infected cells, *Virology* **110**:318.

Watts, D. M., Pantuwatana, S., DeFoliart, G. R., Yuill, T. M., and Thompson, W. H., 1973, Transovarial transmission of La Crosse virus (California encephalitis group) in the mosquito *Aedes triseriatus*, *Science* **182**:1140.

Whitman, L., and Casals, J., 1961, The Guama group: A new serological group of hitherto undescribed viruses. Immunological studies, *Am. J. Trop. Med. Hyg.* **10**:259.

Whitman, L., and Shope, R. E., 1962, The California complex of arthropod-borne viruses and its relationship to the Bunyamwera group through Guaroa virus, *Am. J. Trop. Med. Hyg.* **11**:691.

Woodall, J. P., 1979, Transmission of group C arboviruses (Bunyaviridae) in: *Arctic and Tropical Arboviruses* (E. Kurstak, ed.), pp. 123–128, Academic Press, New York.

CHAPTER 2

Biology and Molecular Biology of Bunyaviruses

DAVID H. L. BISHOP

I. INTRODUCTION

The *Bunyavirus* genus is the largest genus in the *Bunyaviridae* family with more than 172 recognized virus serotypes and subtypes (Karabatsos, 1985; Chapter 1). As discussed in Chapter 1, they are organized into 18 serogroups with a small number of ungrouped viruses. The viruses have been placed into these serogroups on the basis of the results of a number of serological tests (Bishop and Shope, 1979). The data suggest that there are more significant antigenic differences between members of the different serogroups than between members of the same serogroup. Generally, there is no detectable cross-neutralization, or hemagglutinin inhibition, between members of different groups, but some degree of cross-neutralization, or hemagglutinin inhibition, between members of the same group (depending on the virus, see Hunt and Calisher, 1979). Some serogroups are more closely related to others than they are to the rest of the genus (Klimas *et al.*, 1981b). None of the viruses exhibit any antigenic relationship to members of other genera of the family (Bishop and Shope, 1979). Within a serogroup virus isolates may be considered as separate species, or relatives of such species. In antigenic terms, viruses are classified by serological tests into *serotypes* (the original virus of a grouping), *subtypes* (in which there are some antigenic differences to the founder virus as demonstrated using polyclonal sera), *variants* (where

DAVID H. L. BISHOP • NERC Institute of Virology and Environmental Microbiology, Oxford OX1 3SR, England.

The Bunyaviridae, edited by Richard M. Elliott, Plenum Press, New York, 1996.

such differences are only demonstrated using particular antibodies, such as a monoclonal antibody), or *varieties* (where the antigenic difference may be minor, or where sequence analyses indicate a small number of differences between virus isolates). These terms will be used in this chapter. However, what constitutes a bunyavirus species has yet to be defined (Bishop, 1985). It could be argued that each bunyavirus serogroup may include only a few virus species, or even only one if viruses that produce reassortants in dual infections are considered to constitute one species. This issue is, however, complicated by the fact that although certain viruses are genetically similar they may be ecologically distinct, thus reducing the opportunity for reassortment.

Morphologically, morphogenetically, and in terms of their coding and replication strategies, the viruses classified in the genus resemble each other (Martin *et al.*, 1985; Murphy *et al.*, 1973) and exhibit a number of common features of which their various antigenic relationships are a reflection. As a category, the viruses assigned to this genus are called bunyaviruses, although on occasion the same term is employed in the scientific literature to describe any member of the family. This is misleading. The preferred term is "member of the *Bunyaviridae*," and for each of the six recognized genera: "bunyavirus," "hantavirus," "nairovirus," "phlebovirus," "tospovirus," and "uukuvirus."

Despite their similarities, individual bunyaviruses differ to various extents in (1) the sequences of their genes and gene products (antigens, etc.), (2) their natural (hematophagous) arthropod vectors, (3) their usual vertebrate hosts, and (4) the places in the world where they occur. In general, the distribution of bunyaviruses reflects the distribution of their vectors and their wildlife hosts. Some bunyaviruses have only been isolated once or a limited number of times, while others have come from a number of different vectors or vertebrate hosts. Antibody surveys have also been used to determine which vertebrates may be infected. Although there are reports of the isolation of bunyaviruses from tabanids (e.g., horse- or deerflies; Wright *et al.*, 1970), the majority of bunyaviruses are from mosquitoes. Several bunyaviruses have been obtained from *Culicoides* species, others from ticks, and a few from phlebotomines, or bedbugs (Karabatsos, 1985).

In some cases the acquisition of a virus by a blood-feeding species may be incidental (i.e., the presence of virus in a recently acquired blood meal) and not relevant to virus transmission to another animal. Since virus isolation does not prove that the species is a vector, more rigorous criteria must be satisfied. One is whether virus acquired *per os* is subsequently transmitted to a vertebrate during the acquisition of a further blood meal (for those speciies that feed more than once). Relevant to this question is whether transmission also occurs between the different forms of a putative vector (eggs–juveniles–adults–eggs, etc.). For example, virus isolations from eggs, larvae, and male mosquitoes are taken as indicators of transovarial (vertical) transmission and the possible involvement of a particular mosquito species in transmission (likewise from the various forms of unfed juvenile or adult ticks, etc.).

Transovarial transmission has been established for a number of bunyaviruses and for some, virus isolations have been made from larvae and male mosquitoes. However, apart from when there is an epidemic, the recovery rate of viruses from arthropods is usually low (often < 1%, even in endemic areas). Rarely is enough information obtained to establish the competence of a particular arthropod species to act as a vector. As a result most bunyaviruses are described as "probable arboviruses." An exception is La Crosse (LAC) virus and its mosquito vector *Aedes triseriatus* in the midwestern United States (see later).

Some bunyaviruses occur in different countries and continents of the world [e.g., snowshoe hare (SSH) virus in Canada, the United States; and Russia, Calisher, 1983]. In different regions within or between countries, different vectors and hosts may be involved in the natural transmission cycles. Further, some bunyaviruses are sympatric in nature with other bunyaviruses, and infect the same arthropod and vertebrate hosts (Calisher, 1983).

In view of the wide distribution of arthropods, it is perhaps not surprising that bunyaviruses have been identified in every continent of the world (except Antarctica, and possibly there only because of a lack of investigation). Most bunyavirus isolates have little or no impact on humans or domestic animals: a few, however, are responsible for human disease (Chapters 1 and 10, e.g., LAC virus). Certain members of the genus cause disease in livestock (Chapters 1 and 10, e.g., Akabane virus).

Although important advances in our understanding and knowledge of the replication processes of bunyaviruses have been made over the last few years through genetic, molecular, and cellular studies, many gaps remain, providing subjects for future research.

The replication processes and coding information of selected members of the *Bunyavirus* genus will be discussed in this chapter after providing a brief review of the general properties of bunyaviruses as well as specific data concerning members that represent the different bunyavirus serogroups.

II. COMMON PROPERTIES AND FEATURES OF BUNYAVIRUSES

As discussed elsewhere, viruses have been placed in the *Bunyavirus* genus on the basis of a number of shared epitopes and common features. The presence of conserved sequences (Akashi and Bishop, 1983; Akashi *et al.*, 1984; Gerbaud *et al.*, 1987a; Dunn *et al.*, 1994) on the internal nucleocapsid (N) protein of certain bunyaviruses [e.g., Aino, Germiston (GER), and LAC viruses, members of the Simbu, Bunyamwera, and California serogroups, respectively] may correspond to some of the epitopes recognized in the cross-complement fixation analyses that have allowed the various serogroups to be placed together in the *Bunyavirus* genus.

As described above, bunyaviruses have been further categorized into serogroups, serotypes, and subtypes from the results of cross-neutralization and hemagglutination-inhibition studies (Bishop and Shope, 1979; Karabatsos, 1985). It should be borne in mind that these tests represent only the glycoprotein species; however, since the humoral responses to the viral glycoproteins play a part in the protection of vertebrates to subsequent infection, the serological relationships are important to consider. Parenthetically, cell-mediated responses to arbovirus infections are probably involved in resolving virus infections, although little has been reported in this connection with bunyaviruses, or other members of the *Bunyaviridae*.

The bunyavirus classification scheme has been confirmed by biochemical and genetic studies (Bishop *et al.*, 1980a,b, 1981, 1983a, 1984; Bishop and Shope, 1979; Elliott, 1990). As a result, a number of common features have been identified. These include the following:

1. The viruses are enveloped in a lipid membrane (Fig. 1). Their infectivity and structural integrity are sensitive to lipid solvents. Using negative staining, bunyaviruses are frequently visualized by electron microscopy as spherical, 90- to 100-nm, doughnut-shaped particles with a poorly defined arrangement of surface glycoproteins (Fig. 1; Murphy *et al.*, 1968; Obijeski *et al.*, 1976a; Martin *et al.*, 1985). However, by cryoelectron microscopy the particles exhibit a well-defined arrangement of surface spikes (Fig. 1; Talmon *et al.*, 1987).

2. There are two glycoproteins (G1, 108 to 125 kDa, and G2, 29 to 41 kDa) located on the surface of bunyavirus particles. They have covalently associated fatty acid residues (Madoff and Lenard, 1982) and relatively few carbohydrate side chains (Vorndam and Trent, 1979; Cash *et al.*, 1981; Eshita and Bishop, 1984; Lees *et al.*, 1986; Grady *et al.*, 1987; Pardigon *et al.*, 1988). The glycoproteins induce and interact with neutralizing antibodies (Gentsch *et al.*, 1980; Gonzalez-Scarano *et al.*, 1982; Grady *et al.*, 1983a,b). They may be removed from virus particles by protease treatment in the absence of detergent (Obijeski *et al.*, 1976a). Such treatment significantly reduces virus infectivity. The G1 and G2 proteins are transmembrane proteins that are present in virus particles in equimolar proportions (Obijeski *et al.*, 1976a, Gentsch *et al.*, 1977a; El Said *et al.*, 1979; Ushijima *et al.*, 1980, 1981; Klimas *et al.*, 1981b). When bunyaviruses are grown in mammalian cells, the glycans of these proteins are of the endoglycosidase-H (endo-H)-resistant, complex type (Madoff and Lenard, 1982). Since the form of glycosylation is determined by the host, for bunyaviruses grown in insect cells the glycans are of a simpler, principally mannose type. The glycoproteins are the major determinants of virulence (host cell/organ targeting) and influence the efficiency of virus transmission in arthropod vectors (Shope *et al.*, 1981a,b; Beaty *et al.*, 1981a,b, 1982, 1983b).

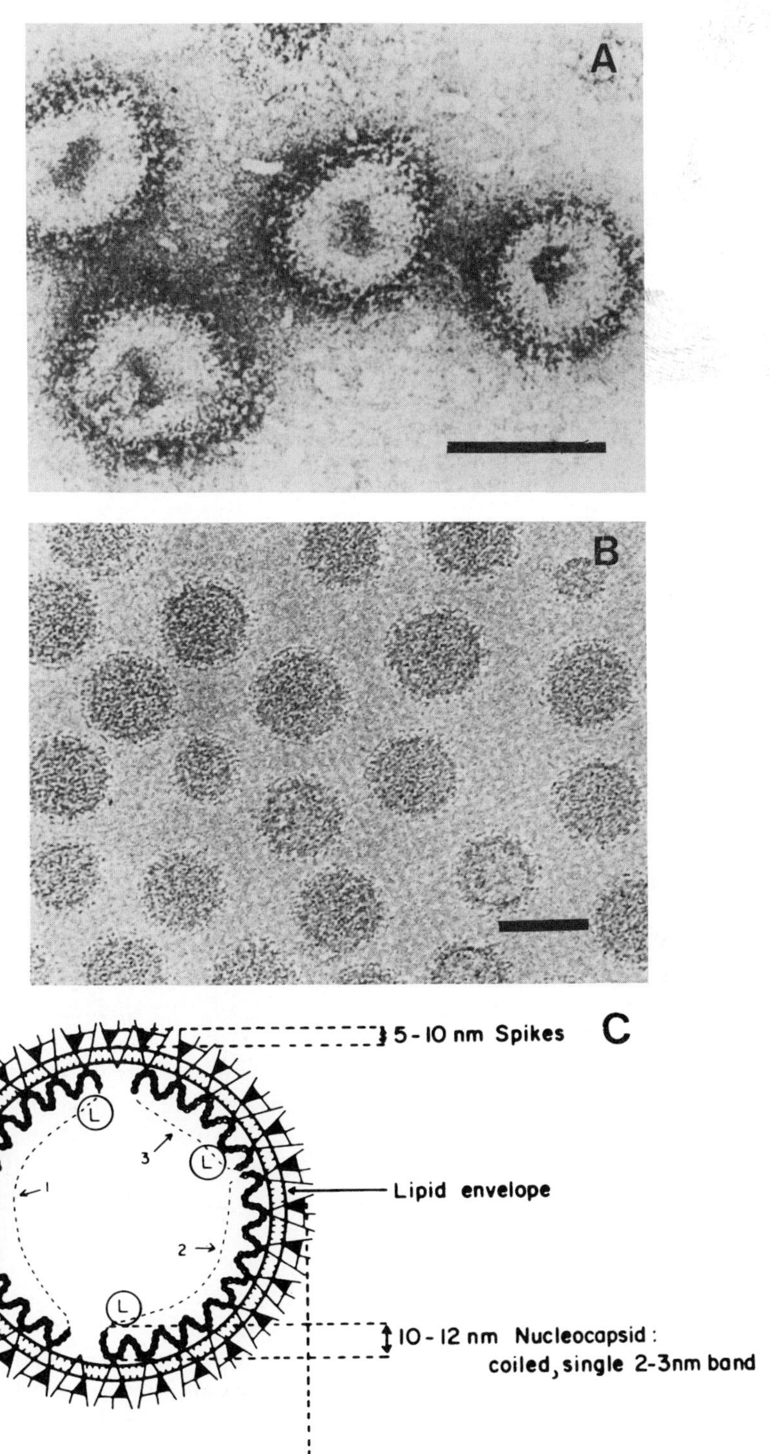

FIGURE 1. (A) Electron micrograph of glutaraldehyde-fixed, negatively stained LAC virus, magnification 153,360. (B) Cryoelectron micrograph of LAC virus. (C) Schematic of a bunyavirus.

3. The three internal viral nucleocapsids are circular and loosely helical in configuration (Fig. 1; Samso *et al.*, 1975; Bouloy and Hannoun,
   1976b; Obijeski *et al.*, 1976b; Raju and Kolakofsky, 1989). Each contains a unique species of negative-sense viral RNA species (Bouloy
   et al., 1973/4; Clewley *et al.*, 1977), many copies of a nucleocapsid
   protein (N, 19 to 26 kDa, depending on the virus), plus a few copies of
   a large protein (L protein, about 260 kDa) that is the virus transcriptase and probably serves as the virus replicase during the infection
   process (a virus-coded RNA-directed RNA polymerase; Bouloy and
   Hannoun, 1976a; Patterson *et al.*, 1984; Endres *et al.*, 1989). The
   three viral RNA species associated with the nucleocapsids are designated on the basis of their distinctive sizes as large (L), medium (M),
   and small (S). Rarely are the nucleocapsids, or viral RNA species,
   recovered from virus preparations in equimolar amounts. Usually
   the S species predominates (Bouloy *et al.*, 1973/4; Obijeski *et al.*,
   1976b; Gentsch *et al.*, 1977a). The significance of this observation is
   not known, but probably relates to the process of morphogenesis
   (Talmon *et al.*, 1987) and how nucleocapsids are packaged in virions.
   Recently, evidence for quasidiploid viruses has been obtained from
   analyses of the progeny of dual wild-type virus infections (Urquidi
   and Bishop, 1992).
4. The viral 0.28 to 0.36×10^6 Da (about 850–1000 nucleotides) S RNA
   codes for two proteins (Fig. 2) that are translated from overlapping
   reading frames of a single viral-complementary, subgenomic mRNA
   species (Bouloy and Hannoun, 1976b; Gentsch and Bishop, 1978;
   Clerx-van Haaster and Bishop, 1980; Bishop *et al.*, 1982a, 1984;
   Clerx-van Haaster *et al.*, 1982a; Fuller and Bishop, 1982; Akashi and
   Bishop, 1983; Fuller *et al.*, 1983; Cabradilla *et al.*, 1983; Akaski *et al.*,
   1984; Bouloy *et al.*, 1984; Elliott, 1985, 1989a; Gerbaud *et al.*, 1987b;
   Elliott and McGregor, 1989; Dunn *et al.*, 1994). These two proteins
   are N and a nonstructural protein designated NS_S (10–12 kDa) (Fuller
   and Bishop, 1982; Elliott, 1985). No other open reading frame in the S
   viral or viral-complementary sequences appears to be of significance.
5. The $1.4–1.6 \times 10^6$ Da (about 4300–4800 nucleotides) M RNA codes in
   its viral-complementary sequence for the precursor to both the G1
   and G2 proteins as well as a nonstructural protein designated NS_M
   (11–17 kDa) (Gentsch and Bishop, 1979; Fuller and Bishop, 1982;
   Eshita and Bishop, 1984; Lees *et al.*, 1986; Grady *et al.*, 1987; Pardigon *et al.*, 1988; Nakitare and Elliott, 1993) (Fig. 2). The gene order
   from amino to carboxy terminus is: $G2–NS_M–G1$ (Fazakerley *et al.*,
   1988). No gene product is coded in the M viral-sense sequence. The
   G1 protein has sequences recognized by neutralizing antibodies and
   mediates membrane fusion involved in the endosomal process of
   entry of bunyaviruses into cells (Gonzalez-Scarano *et al.*, 1984, 1985;
   Gonzalez-Scarano, 1985).

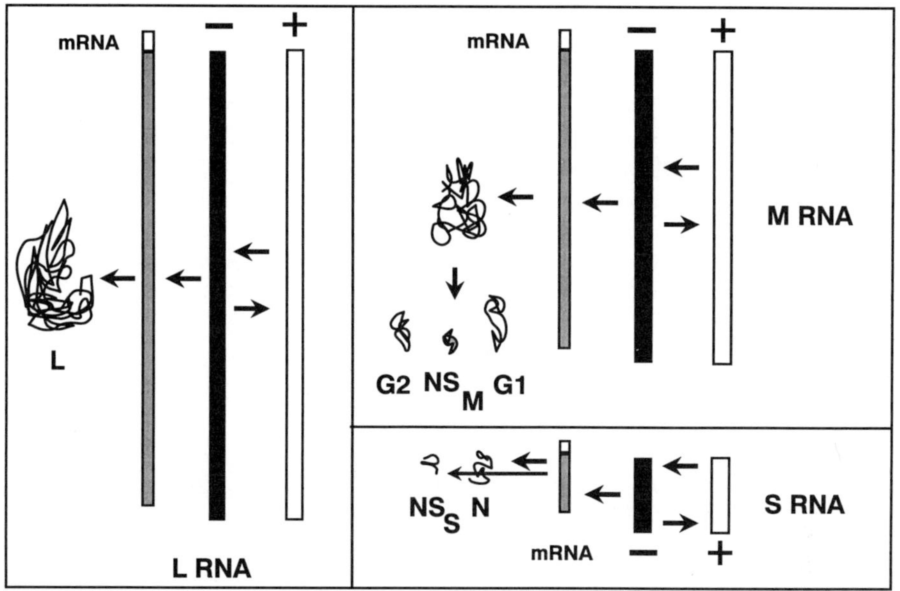

FIGURE 2. Bunyavirus RNA coding, transcription, and replication strategies.

6. The $2.0–2.5 \times 10^6$ Da (about 6000–7500 nucleotides) L RNA codes in its viral-complementary sequence for the virion L protein, about 260 kDa, that is the viral transcriptase-replicase (Clerx-van Haaster *et al.*, 1982a; Elliott, 1989b; Endres *et al.*, 1989; Jin and Elliott, 1991, 1992, 1993) (Fig. 2). A small, 129 amino acid, open reading frame in the Bunyamwera (BUN) L viral-sense sequence that overlaps the sequence encoding the carboxy terminus of the L protein has been identified by cDNA analyses (Elliott, 1989b). If it codes for a protein, then the BUN L RNA would have an ambisense coding strategy. However, to date there is no evidence that this protein is synthesized.

7. All three viral RNA species have a common consensus end sequence of (3′) UCAUCACAUG... and a complementary, uncapped 5′ sequence (Gentsch *et al.*, 1977a; Gentsch and Bishop, 1979; Clerx-van Haaster and Bishop, 1980; Obijeski *et al.*, 1980; Clerx-van Haaster *et al.*, 1982a,b; Patterson *et al.*, 1983; Akashi *et al.*, 1984; Eshita and Bishop, 1984; Lees *et al.*, 1986; Grady *et al.*, 1987; Pardigon *et al.*, 1988; Elliott, 1989a,b; Elliott and McGregor, 1989; Dunn *et al.*, 1994). These presumably reflect the requirements for protein recognition involved in the processes of RNA replication. It is presumed that for each RNA species hydrogen bonding of their 3′ and 5′ ends accounts for the observed noncovalently closed, circular forms of extracted

RNA and the corresponding nucleocapsids (Bouloy *et al.*, 1973/4; Samso *et al.*, 1976; Raju and Kolakofsky, 1989).

8. Viral mRNA species are made in the cytoplasm of infected cells (Lyons and Heyduk, 1973; Goldman *et al.*, 1977; Pennington and Pringle, 1978; Abraham and Pattnaik, 1983; Rossier *et al.*, 1986; Cunningham and Szilagyi, 1987). The mRNA species have heterogeneous, nonviral 5' end sequences (between 8 and 15 nucleotides in length) that are acquired from host mRNA sequences for the purposes of priming viral mRNA transcription (Bishop *et al.*, 1983c; Bouloy *et al.*, 1990; Jin and Elliott, 1993). Only one viral-complementary mRNA species has been identified for each virus RNA species (S mRNA, M mRNA, L mRNA). The mRNA species are shorter (at their 3' ends) than the corresponding virion RNAs (i.e., they are subgenomic) (Pattnaik and Abraham, 1983; Patterson and Kolakofsky, 1984; Eshita *et al.*, 1985; Bouloy *et al.*, 1990). The viral L protein exhibits a methylated cap-dependent endonuclease activity (Patterson *et al.*, 1984). There are conflicting data concerning whether protein synthesis antagonists inhibit mRNA synthesis (see later; Abraham and Pattnaik, 1983; Eshita *et al.*, 1985; Raju and Kolakofsky, 1986a,b; Bellocq and Kolakofsky, 1987). Transcription and viral RNA replication are not inhibited when either actinomycin D or α-amanitin is added to infected cell cultures (Bouloy and Hannoun, 1973; Vezza *et al.*, 1979).

9. Viral morphogenesis occurs in the Golgi apparatus of bunyavirus-infected cells (Murphy *et al.*, 1968; Nakitare and Elliott, 1993). The processes involved in subsequent virus release have not been delineated.

10. As discussed in Chapter 8, RNA segment reassortment has been documented between members of the California encephalitis serogroup (Gentsch and Bishop, 1976, 1980; Gentsch *et al.*, 1997b, 1978, 1979; Bishop and Gentsch, 1980; Bishop *et al.*, 1978, 1980a, 1981, 1987), or between members of the Bunyamwera serogroup (Iroegbu and Pringle, 1981a,b; Pringle *et al.*, 1984a,b), or between members of the Group C serogroup viruses (Bishop, 1982; Bishop *et al.*, 1983b), but not intergroup reassortment (i.e., not between members representing different serogroups). Nonrandom reassortment and the formation of quasidiploid viruses have been demonstrated (Pringle *et al.*, 1984a; Urquidi and Bishop, 1992). Apart from reassortment, no other form of recombination has been documented (i.e., recombination within a gene). Evidence for naturally occurring reassortants has been obtained from genotype analyses of field isolates of some viruses (Klimas *et al.*, 1981a; Ushijima *et al.*, 1981). However, reassortment in nature appears to be an exception rather than the rule. Genetic drift, involving the accumulation of point mutations and occasional sequence deletion or duplication, appears to be the principal factor in the evolution of bunyaviruses. The

extent to which this corresponds to the evolution of the host species is unknown.

11. Horizontal (arthropod vector to vertebrate host, male to female mosquito) and vertical (transovarial) transmission have been documented for some bunyaviruses involving particular arthropod species in particular localities (Pantuwatana *et al.*, 1972, 1974; Watts *et al.*, 1972, 1973a,b; Beaty and Thompson, 1976; Thompson and Beaty, 1977, 1978; Miller *et al.*, 1977, 1979; Thompson, 1979; Schopen *et al.*, 1991). The isolation of viruses from males, or during winter or spring months from mosquito ova and larvae has led to the conclusion that in temperate regions of the world at least some bunyaviruses over-winter *in ovo* (Berry *et al.*, 1974; Gauld *et al.*, 1974; Pantuwatana *et al.*, 1974; Watts *et al.*, 1974; Balfour *et al.*, 1975; Beaty and Thompson, 1975, 1976; Lisitza *et al.*, 1977).

A schematic representation and electron micrographs of bunyavirus particles, such as those of LAC virus, are provided in Fig. 1. Based on the virus structure and its components, the overall chemical composition is calculated to be 1–2% RNA, 58% protein, 33% lipid and 7% carbohydrate (Obijeski and Murphy, 1977). Like all other members of the family, but in contrast to members of four other negative-sense RNA virus families (*Filoviridae, Orthomyxoviridae, Paramyxoviridae,* and *Rhabdoviridae*), members of the *Bunyaviridae* lack an internal matrix protein. Members of the *Arenaviridae* also lack matrix proteins. While the absence of such proteins may correlate with the problems associated with the packaging of bunyavirus genomes (see above), it is likely that the processes of virus morphogenesis are quite distinct from those of other negative-sense RNA viruses. No enzymatic function (e.g., a neuraminidase) has been associated with the envelope glycoproteins of members of the *Bunyaviridae*. Although many of the viruses hemagglutinate certain species of red blood cells, they lack the magnitude of hemagglutinin activity characteristic of the orthomyxoviruses and certain paramyxoviruses (Beaty *et al.*, 1977).

In addition to size and coding differences, by comparison to members of other genera of the *Bunyaviridae* (Fig. 2) the RNA species of bunyaviruses have different consensus end sequences (Parker and Hewlett, 1981; Bishop *et al.*, 1982a; Clerx-van Haaster *et al.*, 1982b). There is, however, some similarity between the terminal sequences of the RNAs of hantaviruses (Schmaljohn and Dalrymple, 1983) and those of bunyaviruses (Clerx-van Haaster *et al.*, 1982b); nevertheless, the coding strategies of viruses assigned to the two groups are very different.

III. THE EPIDEMIOLOGY OF BUNYAVIRUSES

Readers are referred to the International Catalogue of Arboviruses (Karabatsos, 1985) and a previous review (Bishop and Shope, 1979) for further details and information concerning the occurrence, isolation, host range, and tropisms of bunyaviruses, and to Chapter 10 for descriptions of the disease capabilities of various bunyaviruses.

A. Anopheles A and B Serogroups

The Anopheles A and B groups include 15 and 2 viruses, respectively (Karabatsos, 1985). The viruses have been isolated from various mosquitoes recovered in parts of South America or, for Virgin River virus, North America. Of the listed Anopheles A (ANA) viruses (Chapter 1), Tacaiuma (TCM) virus has been obtained from human and primate infections as well as from sentinel animal studies. Data obtained from antibody surveys indicate that TCM and a number of other viruses of the serogroup cause natural infections in a range of species in Brazil. The sizes of the three RNA species of ANA virus have been reported to be 2.7×10^6 Da (L), 2.1×10^6 Da (M), and 0.43×10^6 Da (S) (Ushijima $et\ al.$, 1980). In view of the recent data from cDNA clones for other viruses in the genus (e.g., BUN virus; Lees $et\ al.$, 1986; Elliott, 1989b; Elliott and McGregor, 1989), these values are probably overestimates and the RNAs are more likely to be in the range: 2.3×10^6 Da (about 7000 nucleotides, L), 1.5×10^6 Da (about 4500 nucleotides, M), and 0.3×10^6 Da (about 1000 nucleotides, S). The ANA virus proteins are reported to be 118 kDa (G1), 39 kDa (G2), and 22 kDa (N) (Ushijima $et\ al.$, 1980). Other ANA serogroup viruses have not been analyzed.

Of the Anopheles B group, the proteins and three RNA species of Boraceia (BOR) virus have been analyzed (Klimas $et\ al.$, 1981b) and evidence obtained that their sizes are comparable to those of other members of the genus. For instance, the BOR G1 protein is estimated at 109 kDa, the G2 at 28 kDa, and N at 23 kDa. As in other analyses involving RNA electrophoretic comparisons, the reported BOR RNA sizes: 3.0×10^6 Da (L, about 9000 nucleotides), 2.1×10^6 Da (M, about 6300 nucleotides), and 0.40×10^6 Da (S, about 1200 nucleotides), may prove to be higher than the actual values when these are eventually determined by cDNA cloning. The 3' end sequences of the three BOR RNA species are UCAUCACAUG... (Klimas $et\ al.$, 1981b).

B. Bakau Serogroup

Of the five viruses assigned to the Bakau serogroup, one, Nola virus, has come from culicine mosquitoes collected in Africa, the others from mosquitoes or vertebrates in Malaysia (Karabatsos, 1985). Isolates of Bakau virus have also been reported from $Argas$ ticks collected in Pakistan. Apart from Nola virus, antibody surveys have indicated that the viruses infect humans, although to date no isolates have been made from humans. Bakau, Tanjong Rabok (TR), and Telok Forest viruses have been isolated from certain species of monkey. Serological surveys have established that TR virus infects a number of other animals, including birds, bats, flying squirrels, and other rodents. There are no reports on the RNA or protein species of these viruses.

C. Bunyamwera Serogroup

The prototype virus of the *Bunyaviridae* family is the mosquito-transmitted BUN virus that was isolated in 1943 from *Aedes* species collected in a tropical rain forest at Bunyamwera (Bwamba, Uganda) in Central Africa (Karabatsos, 1985). Other than providing a name for the family, the virus serves as the prototype for the BUN serogroup of some 32 described virus serotypes and subtypes (Karabatsos, 1985; Chapter 1). As discussed elsewhere, sequence data for the S RNA (Dunn *et al.*, 1994) indicate that Guaroa (GRO) virus should be classified with the BUN serogroup rather than the California serogroup (as listed in Chapter 1; Whitman and Shope, 1962). BUN group viruses have been isolated from a number of different mosquito species. Two viruses, Lokern and Main Drain, have been recovered from *Culicoides* as well as from mosquitoes. Eight of the viruses (BUN, Fort Sherman, GER, GRO, Ilesha, Shokwe, Wyeomyia, and Xingu) have been isolated from humans. Several have come from rodents and/or a number of other vertebrate species, including some from livestock (horses, cattle, sheep). Antibody surveys indicate that in addition to the above, Calovo (a subtype of Batai virus) and Tensaw viruses also infect humans. Certain BUN serogroup viruses have been obtained in South or North America, or Europe, or Asia, or Africa. Wyeomyia virus has been reported from South and North America while the Batai and Calovo viruses have come from Asia and Europe, respectively (Karabatsos, 1985).

The sequences and sizes of the complete genome and encoded products of BUN virus have been reported (Lees *et al.*, 1986; Elliott, 1989b; Elliott and McGregor, 1989). The L RNA is 6875 nucleotides long (2.3×10^6 Da), the M RNA is 4458 nucleotides (1.5×10^6 Da), and the S RNA is 961 nucleotides (0.3×10^6 Da), making a total size of 12,294 nucleotides (4.1×10^6 Da). As deduced from the coding data and/or protein analyses, the sizes of the BUN structural and nonstructural proteins are 259 kDa for the L RNA-coded L protein (2238 amino acids), 115 kDa (G1), 38 kDa (G2), and 16 kDa (NS_M) for the products of the M RNA-coded, 162-kDa glycoprotein precursor (1433 amino acids), and 19 kDa (N, 233 amino acids) and 11 kDa (NS_S, 101 amino acids) for the S RNA-coded proteins.

The sequences and sizes of the S RNAs of eight other BUN serogroup viruses have been reported (Batai, Cache Valley, GER, GRO, Kairi, Maguari, Main Drain, and Northway; Gerbaud *et al.*, 1987a; Elliott and McGregor, 1989; Dunn *et al.*, 1994). Their S RNAs have similar sizes (945–992 nucleotides, $0.30–0.35 \times 10^6$ Da) and code for similarly sized N proteins (233 amino acids), although their NS_S proteins are more variable in length (85–109 amino acids). In this connection, it is worth noting that the electrophoretic mobilities on SDS–PAGE of the virion proteins may differ significantly (and reproducibly) even for proteins of equivalent length (such as the N proteins of many bunyaviruses). Presumably this is related to the amount of SDS bound and hence the particular structure and charge of the protein under the electrophoretic conditions employed. Such phenotypes have aided in the determination of the coding assignments of the virus gene products, as discussed elsewhere.

The 4434-nucleotide M RNA sequence (1.49×10^6 Da) of GER virus has been determined (Pardigon *et al.*, 1988) and shown to encode a 1437-amino-acid precursor (162 kDa) to the viral glycoproteins. Protein analyses of other BUN serogroup viruses have indicated that the sizes of the G1 and G2 proteins of the member viruses are similar to those of GER and BUN, although minor differences in electrophoretic mobilities have been reported (see Ushijima *et al.*, 1980).

The L RNA of GER virus is reported to be 2.6×10^6 Da (about 7700 nucleotides), although its sequence has not been determined and its actual size might be lower. Generally for virus and infected cell analyses, the L protein is difficult to identify because of its low quantity and background problems, although where this has been investigated the sizes of the bunyavirus L proteins are similar to that reported for BUN virus (Elliott, 1989b).

The evolutionary relationships of the BUN serogroup viruses, as revealed by the sequence analyses of their genes and proteins, are discussed in Chapter 13.

D. Bwamba and C Group Serogroups

Bwamba virus was originally isolated in 1937 from an infected man working in Bwamba county, western Uganda. Subsequent isolations of Bwamba virus have originated from a number of human cases and from collections of anopheline mosquitoes in other parts of Central Africa, with serosurveys indicating a broad distribution of such viruses infecting humans, livestock, and other animals in Central, South, and West Africa (Karabatsos, 1985). No biochemical data have been reported for the two Bwamba serogroup viruses.

Of the 14 recognized C group viruses, most have been obtained from infected humans, and all from South or North America (Karabatsos, 1985). Many have been isolated from rodents and/or marsupials and a number of other vertebrates including sentinel animals. The viruses have also been isolated from culicine mosquitoes. Interestingly, closely related Group C virus isolates may induce quite distinct diseases in model animals. For example, some Caraparu virus isolates cause a rampant hepatitis in mice with death occurring by 2–4 days postinfection (depending on the dose), while other isolates which are genetically very similar (as indicated by their oligonucleotide fingerprints) cause encephalitis that is manifest after 1–2 weeks and no evidence of hepatitis, or other forms of liver infection (Bishop *et al.*, 1983b). Analyses of Caraparu intertypic reassortants have shown that these phenotypes segregate with the origins of their respective M RNAs (Bishop *et al.*, 1983b).

Only limited biochemical data are available for the viruses assigned to the C group. None of the genome segments has been cloned, so that the data reported for the Itaqui and Oriboca RNA sizes (L RNA: $2.7–3.0 \times 10^6$ Da, about 8000–9000 nucleotides; M RNA: $1.8–1.9 \times 10^6$ Da, about 5300–5700 nucleotides; S RNA: $0.43–0.50 \times 10^6$ Da, about 1280–1500 nucleotides) can only be considered to be approximate (Ushijima *et al.*, 1980). The proteins of these two viruses are in the range of 110–111 kDa (G1), 41 kDa (G2), and 22–23 kDa (N) (Ushijima *et al.*, 1980).

E. California Serogroup

The 15 recognized virus serotypes, subtypes, and variants of the California group (see Chapter 1; excluding GRO, but including Lumbo, a variant of Tahyña virus, as well as Jerry Slough, a variant of South River) have been isolated from various culicine and, for some, anopheline mosquitoes as well as from a number of natural vertebrate infections including humans [LAC and Tahyña (TAH) viruses] (Karabatsos, 1985). Several viruses have been obtained from rodents and other animals, including sentinels. Antibody surveys have shown that a range of wildlife species may be involved in the natural transmission and amplification of these viruses and that several of the viruses infect humans [in addition to the above, California encephalitis (CAL), Inkoo, Jamestown Canyon, and SSH viruses]. Some of the member viruses are broadly distributed (SSH virus in North America and Russia), while others, such as the prototype CLA virus, appear to be more limited in distribution (e.g., for CAL virus, regions of the United States west of the Rocky Mountains) (Calisher, 1983).

The sequences and sizes of the M and S RNAs and encoded products of LAC and SSH viruses have been reported, but only partial data for their L RNAs (Clerx-van Haaster and Bishop, 1980; Bishop *et al.*, 1982a; Clerx-van Haaster *et al.*, 1982a,b; Akashi and Bishop, 1983; Cabradilla *et al.*, 1983; Eshita and Bishop, 1984; Grady *et al.*, 1987). The LAC and SSH M RNAs (Eshita and Bishop, 1984; Grady *et al.*, 1987) are 4526 and 4527 nucleotides in length (1.5×10^6 Da) and their S RNAs are 984 and 982 nucleotides long, respectively (0.3×10^6 Da; Bishop *et al.*, 1982a; Akashi and Bishop, 1983; Cabradilla *et al.*, 1983). The L RNA, which codes for the L protein (Endres *et al.*, 1989), is estimated to be 2.8×10^6 Da (about 8300 nucleotides), although this is probably an overestimate. The M-coded proteins, G1 (115 kDa), G2 (38 kDa), and NS_M (15 kDa), derive from a 1441-amino-acid precursor (162 kDa, gene order G2, NS_M, G1) (Fazakerley *et al.*, 1988). The S-coded N proteins are 235 amino acids in length (27 kDa), while the overlapping NS_S protein is 92 amino acids long (11 kDa) (Bishop *et al.*, 1982a). The sizes of the structural components (RNAs and proteins) of other California group viruses are similar, albeit there are minor differences as evident by analyses of viral and infected cell proteins and RNAs resolved by appropriate gels (El Said *et al.*, 1979; Klimas *et al.*, 1981b). Such differences have allowed the coding assignments of the viruses to be established using reassortants (as discussed in Chapter 8).

Interestingly, considerable divergence has been reported between the genomes of LAC virus isolates obtained from various regions of the continental United States (Midwest, East, South, and Southwest) (Klimas *et al.*, 1981a). Although many of the more than 20 LAC isolates analyzed could be geographically grouped (topotypes), no two isolates were actually identical when analyzed by oligonucleotide fingerprinting. Some evidence for natural reassortment between certain topotypes was obtained in areas of the United States where two groups were sympatric (Klimas *et al.*, 1981a). Surprisingly, in view of the extent of divergence, two isolates obtained from human cases were quite similar despite a separation of some 18 years in their isolation (Klimas *et al.*, 1981a).

F. Capim, Gamboa, Guama, Koongol, Minatitlan, Nyando, and Olifantsvlei Serogroups

The ten viruses and subtypes assigned to the Capim (CAP) serogroup (and found primarily in South but also North America; Karabatsos, 1985) have been isolated from culicine mosquitoes, as well as from a number of sentinel and other animals, including rodents. None of the viruses has been obtained from a human infection. Further, antibody surveys have not indicated that human infections are common. The reported sizes of the three CAP viral RNA species are 3.1×10^6 Da (L, about 9200 nucleotides), 2.3×10^6 Da (M, about 6800 nucleotides), and 0.43×10^6 Da (S, about 1280 nucleotides), although as noted above these sizes are probably overestimates (Ushijima *et al.*, 1980). CAP protein analyses have provided values of 114 kDa (G1), 36 kDa (G2), and 21 kDa (N). The eight Gamboa group viruses and subtypes have been isolated from culicine mosquitoes collected in South or North America (Karabatsos, 1985). Their natural vertebrate hosts are not known. No data are available for the RNA or proteins of the Gamboa group viruses.

The Guama group of bunyaviruses includes 12 viruses that have been obtained from culicine and in one case (Catu virus) anopheline mosquitoes (Karabatsos, 1985). Guama (GMA) virus has been isolated from *Culex* and other mosquitoes, as well as from *Lutzomys* phlebotomines. Two of the viruses have been recovered from natural human infections (GMA, Catu). Depending on the virus, others have come from rodents, bats, birds, and marsupials, as well as from a number of sentinel animals mostly in South America, but also North America for GMA and Mahogany Hammock viruses. The sizes of the RNA species of GMA virus are reported to be 2.9×10^6 Da (L, about 8600 nucleotides), 1.9×10^6 Da (M, about 5600 nucleotides), and 0.49×10^6 Da (S, about 1400 nucleotides) (Ushijima *et al.*, 1980) and are also probably overestimated (see above). The G1, G2, and N proteins (114, 41, and 21 kDa, respectively) of GMA are comparable to those of other bunyaviruses (Ushijima *et al.*, 1980).

The two Koongol group viruses were isolated from mosquitoes collected in Australia and although not associated with human infection, serological surveys indicate that a range of mammals, marsupials, and possibly birds may be infected (Karabatsos, 1985). There are no reported data on the RNAs or structural proteins of these viruses. Of the two Minatitlan serogroup viruses, Minatitlan came from a sentinel hamster in Veracruz, Mexico (vector unknown), while Palestina came from mosquitoes collected in Ecuador, South America as well as from sentinel hamsters (Karabatsos, 1985). There are no data on the RNAs or structural proteins of these viruses.

The two Nyando serogroup viruses have come from mosquito collections obtained in Eastern (Ethiopia), Central (e.g., Kenya), or Western Africa (Senegal). Nyando has also been recovered from an infected human in the Central African Republic and antibodies to the virus have been detected in surveys of human sera both in Kenya and in Uganda (Karabatsos, 1985). There are no reported data on the RNAs or structural proteins of these viruses.

The five Olifantsvlei group viruses have all been recovered from culicine mosquitoes collected in the Republic of South Africa and other regions of Africa (Karabatsos, 1985). Their vertebrate hosts are unknown. There are no reported data on their RNAs or structural proteins.

G. Patois Serogroup

The seven viruses assigned to the Patois serogroup have been obtained either from mosquitoes collected in North or South America, or from ticks (Estero Real virus), or from sentinel animals and in some cases wildlife, e.g., cotton rats for Patois, Shark River (SR), and Zelga viruses (Karabatsos, 1985). For Zelga virus, the potential arthropod vectors are unknown. Serological surveys indicate that Patois and Zelga viruses infect humans. The RNAs of SR (L: 3.1×10^6 Da, about 9200 nucleotides; M: 2.3×10^6 Da, about 6800 nucleotides; S: 0.48×10^6 Da, about 1400 nucleotides), Pahayokee (PAH), and Zelga viruses have been analyzed. Their RNA sizes are similar to those of other bunyaviruses although, as discussed above, probably overestimated (Ushijima *et al.*, 1980, 1981). The proteins of SR virus are reported to be 113 kDa (G1), 35 kDa (G2), and 22 kDa (N). For PAH the sizes are 118 kDa (G1), 35 kDa (G2), and 22 kDa (N). A number of nonstructural proteins have been identified for both viruses. The S RNA of PAH virus has a 3′ terminal sequence of UCAUCAAAUGA..., i.e., similar to those of other bunyaviruses. Interestingly, analyses of the RNAs of SR and PAH viruses by oligonucleotide fingerprinting indicated that although their L and S RNAs are almost identical, their M RNAs are quite distinct suggesting that one or the other virus represents a naturally occurring reassortant (Ushijima *et al.*, 1981).

H. Simbu Serogroup

The 25 viruses classified to the Simbu serogroup make this group the second largest in the genus. Half of the viruses have been recovered from *Culicoides* species (Karabatsos, 1985). Five of these have also been recovered from culicine mosquitoes. Four other viruses have come from culicine mosquitoes; for the rest the arthropod vectors are unknown (Karabatsos, 1985). Oropouche and Shuni viruses have been isolated from humans while Manzanilla virus has been obtained from other primates. Other viruses have been recovered from birds and a number of vertebrate species, including livestock (cattle, pigs). The geographic distribution of the Simbu group viruses is one of the widest of the genus, with viruses isolated from Asia, Australia, Africa, North and South America (Karabatsos, 1985). From a medical view, Oropouche virus is the most important virus of the group. This virus has caused epidemics involving hundreds of human cases in Brazil (Karabatsos, 1985). It has also been isolated from a human case in Trinidad. Serological studies indicate that humans, monkeys, birds (domestic and wild), and, on occasion, rodents may be infected by Oropouche virus. The virus has been recovered from a range of mosquito species as well as *Culicoides* species. By contrast, Shuni virus has only been recovered once from a human although it has been obtained several times from livestock and this is confirmed by the results of serological surveys which indicate that in certain parts of West Africa the virus frequently infects livestock. Akabane and Aino viruses are important pathogens of livestock in Australia and Japan. They also infect domestic animals (including pigs, horses) in other parts of Southeast Asia. Related viruses have been detected in Kenya. The viruses are transmitted by *Culicoides* and certain mosquito species, and in nonendemic regions can cause fetal abnormalities (arthrogryposis-hydranencephaly) following infection of nonimmune pregnant cattle or sheep (Karabatsos, 1985).

Molecular data have been reported for Aino, Simbu, and Mermet viruses (Ushijima *et al.*, 1980). Their virion proteins include G1 proteins of 108–120 kDa, G2 proteins of 29–34 kDa, and N proteins of 19–21 kDa. The RNA species of Aino and Mermet viruses are similar in size to those of other bunyaviruses. The Aino S RNA has been sequenced (Akashi *et al.*, 1984). It has the conserved ends typical of other bunyaviruses. The Aino S RNA is 850 nucleotides in length (0.28×10^6 Da, the smallest so far described) and codes for a 233-amino-acid N protein and 91-amino-acid NS_S protein that are similar in size to those of other bunyaviruses analyzed. The Aino S RNA has a shorter downstream noncoding sequence in the viral-complementary sequence than the other bunyaviruses, and the second shortest upstream noncoding sequence of those examined (Dunn *et al.*, 1994). The Aino L and M RNA species are estimated to be 2.8×10^6 Da (about 8300 nucleotides) and 1.9×10^6 Da (about 5600 nucleotides), respectively, although these are probably too high.

I. Tete and Turlock Serogroups and the Ungrouped Viruses

The Tete serogroup viruses include six viruses that have either been isolated from ixodid ticks (two viruses) or *Culicoides* species (one virus), or for which an arthropod vector has yet to be identified (Karabatsos, 1985). All of the viruses have been isolated from birds. Four have been obtained in Africa, two of these also from Europe. One virus has come from Asia, another from North America. None is associated with human disease or infection. The five Turlock (TUR) serogroup viruses have been obtained from culicine mosquitoes in either Europe, Africa, Asia, or North and South America (TUR virus) (Karabatsos, 1985). The structural components of TUR virus have been analyzed (Klimas *et al.*, 1981b); the data indicate that the viral components are comparable in size to those of other members of the genus. For example, the TUR G1 protein is estimated at 127 kDa, the G2 protein at 34 kDa, and the N protein at 24 kDa (Klimas *et al.*, 1981b). As in other analyses involving RNA electrophoretic comparisons, the reported TUR RNA sizes of 2.9×10^6 Da (L, about 8600 nucleotides), 2.2×10^6 Da (M, about 6500 nucleotides), and 0.45×10^6 Da (S, about 1300 nucleotides) may prove to be higher than the actual values when these are eventually determined by cDNA cloning. The 3' end sequences of the three TUR RNAs (3' UCAUCACAUG...) are comparable to those of other members (Klimas *et al.*, 1981b). No data have been reported for other members of the Turlock group or any member of the Tete serogroup.

Of the four ungrouped viruses (Chapter 1) assigned to the genus, Kaeng Khoi virus has been isolated a number of times in Thailand from the bedbugs *Stricticimex parvus* and *Cimex insectus*. Other isolates have been obtained from bats and sentinel rodents (Karabatsos, 1985). Serological studies indicate that the virus infects humans and wild rodents.

Apart from the unassigned bunyaviruses, as discussed in Chapter 1 there are a number of grouped but unassigned viruses in the family as well as ungrouped and unassigned viruses (Karabatsos, 1985). When further information is obtained, many of these viruses will no doubt turn out to be bunyaviruses.

IV. VIRION STRUCTURE AND ORGANIZATION

As described above, members of the *Bunyaviridae* are lipid-enveloped viruses (Fig. 1, about 100-nm diameter) with an outer surface covered with protein (von Bonsdorff *et al.*, 1969) that includes two glycoprotein species. The lipid is derived from the cellular site of virus maturation (usually the Golgi saccules and vesicles). On average, within the virus particle there are three, apparently circular, nucleocapsids composed of RNA (L, M, S), nucleoprotein (N), and a few molecules of L protein. A transcriptase-replicase enzyme activity is located within the virions. How the nucleocapsids are packaged and how the virions are otherwise organized are unknown.

V. STAGES OF REPLICATION

Not much is understood about the infection processes of members of the *Bunyaviridae*; what little is known stems principally from work undertaken on members of the *Bunyavirus* genus and in relation to the infection course in vertebrate cells. In invertebrate cells the infection course may differ in detail.

The principal stages of the replication process for viruses in the *Bunyaviridae* are illustrated in Fig. 3 and can be summarized as follows:

1. Attachment is mediated via an interaction of the viral glycoproteins and host receptors of unknown function.
2. Entry and uncoating occurs by endocytosis of virions and fusion of the viral membrane with the endosomal membrane to release the three nucleocapsids into the cell cytoplasm.
3. Primary transcription, i.e., the synthesis of three subgenomic, viral-complementary mRNA species from the three genome templates, and involving host cell mRNA-derived primers and the virion-associated polymerase, occurs in the cytoplasm.
4. Translation of primary L and S segment mRNAs by free ribosomes and translation of M segment mRNAs by membrane-bound ribosomes yield the viral structural and nonstructural proteins which are required for the subsequent replication steps. The function of the NS_S protein is unknown. Primary glycosylation and cotranslational processing of nascent envelope proteins occur in the Golgi saccules and vesicles.
5. It is assumed that this is followed by synthesis and the encapsidation by N protein of full-length, exact-copy viral-complementary RNA (replicative RNA, i.e., lacking primers on their 5' ends) to serve as templates for the subsequent synthesis of genomic RNA. Whether NS_S is involved at this stage is unknown. Presumably the viral template may be used simultaneously for both mRNA and replicative RNA synthesis, with the initiation events of the process involving either primers, or N protein encapsidation of an exact-copy product, determining the outcome.
6. Genome RNA replication using the exact-copy replicative inter-

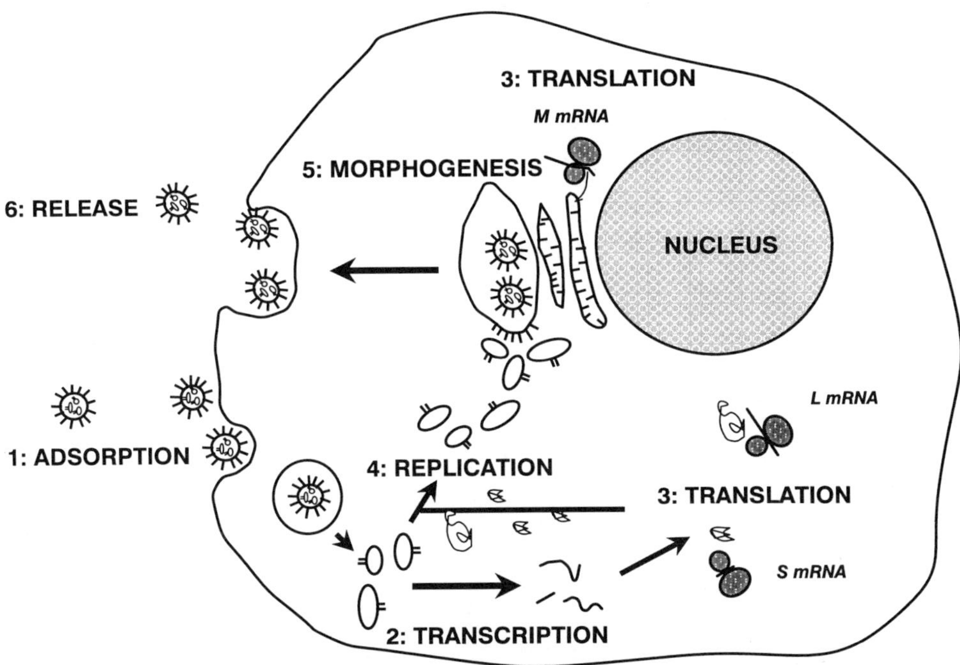

FIGURE 3. Schematic of the infection course of a bunyavirus.

mediate subsequently provides full-length copies of each of the viral RNA species. These RNA species are also in the form of nucleo-capsids.

7. Secondary transcription involves the amplified synthesis of the three mRNA species when the newly synthesized viral RNA species are available. There is at least an order of magnitude (in molar terms) more S mRNA species than M mRNA species, which in turn is more abundant than the L mRNA species (Vezza *et al.*, 1979; Rossier *et al.*, 1988). The reasons for these differences are unknown.

8. Further translation, RNA replication and transcription occur in the final stages of infection and during the time when progeny viruses are released.

9. Morphogenesis, including the accumulation of viral glycoproteins in the Golgi apparatus, terminal glycosylation, acquisition of modified host membranes, and budding into the Golgi cisternae leads to the accumulation of virions in cytoplasmic vesicles. Possibly NS_M functions at this stage to regulate the form of the glycoproteins as in other viruses (e.g., influenza virus).

10. Finally, fusion of cytoplasmic vesicles containing viruses with the surface membrane of the cell results in the release of mature virions.

A. Attachment and Entry

The early events in the infection process of members of the *Bunyaviridae* are not well defined. Like other enveloped viruses, one or both of the integral viral envelope proteins mediate attachment to host cell receptors. This was first demonstrated by proteolytic enzyme treatment of purified LAC virus, which resulted in "spikeless" virus particles, and a five log reduction in infectivity of mammalian cells as compared to nondigested virions (Obijeski *et al.*, 1976a). The nature of cell receptors involved in attachment has not been identified for any member of the family. The viral proteins involved in attachment to vertebrate cells have been examined indirectly by using polyclonal and monoclonal antibodies to block infection. Neutralization of LAC virus infectivity has been reported using antibodies specific for the LAC virus G1 protein, indicating a relationship of that protein to infectivity in vertebrate cells (Gonzalez-Scarano *et al.*, 1982; Grady *et al.*, 1983a,b; Kingsford and Hill, 1983; Najjar *et al.*, 1985). Neutralizing monoclonal antibodies to LAC virus G2 have not been reported. It has been suggested that the G1 protein is more actively involved in binding to vertebrate host cells than is G2, based on the finding that treatment of virus with bromelain (or pronase), which degrades portions of G1, but leaves G2 uncleaved, rendered the virus noninfectious for mammalian cells (Kingsford and Hill, 1983). The situation with invertebrates and invertebrate cells may be different. From studies with LAC virus, Ludwig *et al.*, (1989) proposed that after G1 cleavage the exposed G2 might facilitate binding to specific receptors on mosquito cells since they observed an enhanced infectivity posttreatment.

The infection process of bunyaviruses has been shown to require an acidic pH (Gonzalez-Scarano *et al.*, 1984, 1985). pH-dependent fusion is generally believed to relate to the early events in the infection process, including a change in the structure of the viral glycoprotein(s) (Gonzalez-Scarano, 1985) and fusion of the virus envelope with that of the endosomal vesicle to allow the translocation of viral nucleocapsids into the cell cytoplasm. Direct evidence for this process for viruses of the *Bunyaviridae* has not yet been obtained. Analysis of a LAC G1 mutant virus with a defective fusion function has also indicated that this protein mediates fusion (Gonzalez-Scarano *et al.*, 1985). Whether G2 is involved is unknown.

B. Transcription of mRNA Species

Following uncoating of viral genomes, transcription of the negative-sense viral RNA to complementary mRNA involves an interaction of the virion-associated polymerase with the RNA templates in the individual nucleocapsids (Bouloy *et al.*, 1975; Bouloy and Hannoun, 1976a). The mechanism by which viruses in the *Bunyaviridae* family initiate transcription of their mRNAs is not well defined, but at least for bunyaviruses and phleboviruses (Ihara *et al.*, 1985), some similarities exist with the mRNA transcription processes of the influenza viruses. Influenza viruses cleave capped and methylated oligonucleotides from host cell mRNAs and use them to prime transcription of the viral genome. However, unlike influenza virus, the transcription of bunyavirus mRNA species occurs in the cell cytoplasm (Rossier *et al.*, 1986). Nevertheless, like influenza virus the mRNAs of both bunyaviruses and phleboviruses possess 5' terminal extensions of approximately 8–15 nucleotides in length that are heterogeneous in sequence and not templated by the viral RNA (Bishop *et al.*, 1983c; Patterson and Kolakofsky, 1984; Bouloy *et al.*, 1990; Jin and Elliott, 1993). The mRNA 5' extensions are analogous to the primers acquired by influenza virus in the cell nucleus (where influenza viral transcription occurs).

Direct evidence for capped, methylated structures on the 5' termini of the intracellular mRNA species of *Bunyaviridae* members has yet to be demonstrated. Indirect *in vitro* evidence has been obtained with LAC virus indicating that the primers are acquired from mature mRNA species. The *in vitro* RNA polymerase activity associated with LAC virions may be stimulated by oligonucleotides such as A_nG, cap analogues such as $m^7GpppAm$, and alfalfa mosaic virus (ALMV) 4 RNA (Patterson *et al.*, 1984). As with influenza virus, the A_nG stimulation of LAC virus transcription results in the incorporation of these compounds into nascent transcripts. The stimulation by cap analogues appears to involve a different, as yet unknown mechanism. ALMV 4 RNA apparently stimulates LAC virus transcription by providing transcriptional primers which are specifically cleaved 10 to 14 nucleotides downstream from the ALMV cap group by the LAC virion-associated endonuclease. For this endonuclease activity to be functional, there must be at least one methyl group on the cap (Patterson *et al.*, 1984). Goldman *et al.*, (1977) reported that the California encephalitis virus could produce progeny in enucleated cells. Although this has not been confirmed with LAC virus (Pennington *et al.*, 1977), other evidence also suggests that bunyavirus replication occurs exclusively in the cytoplasm. Pulse-labeling of LAC virus-infected cells at various times after infection, and examination of both cytoplasmic and nuclear fractions for labeled mRNA, has demonstrated viral S genome transcription only in the cytoplasm (Rossier *et al.*, 1986). Moreover, unlike influenza virus, SSH (Vezza *et al.*, 1979) and TAH (Bouloy and Hannoun, 1973) viruses are resistant to the effects of actinomycin D, a drug that inhibits DNA-dependent RNA polymerases. It is believed, therefore, that bunyaviruses acquire the primers needed for transcription from a cytoplasmic pool of host cell messages rather than from newly synthesized nuclear transcripts and, consequently, have no requirement for ongoing host RNA synthesis as is the case for the influenza viruses. Also unlike the influenza viruses, none of the bunyavirus mRNA species are spliced, which could be another reason why bunyavirus infections do not involve the cell nucleus.

Inhibitors of host cell protein synthesis, such as cycloheximide and puromycin, have no effect on primary transcription in other negative-strand RNA virus families such as the *Orthomyxoviridae*, *Rhabdoviridae*, and *Paramyxoviridae*. Conflicting data have been reported concerning the requirement for ongoing host protein synthesis and primary transcription for viruses in the *Bunyavirus* genus. Abraham and Pattnaik (1983) first reported cycloheximide sensitivity for both BUN and Akabane viruses. Similarly, either no or greatly reduced amounts of S segment transcripts were detected in LAC virus-infected cell cultures treated with cycloheximide or puromycin, as compared to untreated cultures, even using sensitive hybridization procedures (Patterson and Kolakofsky, 1984; Raju and Kolakofsky, 1986a). With this virus, the virion-associated polymerase was found to produce only aberrant, incomplete transcripts *in vitro* unless rabbit reticulocyte lysates were added to provide a coupled transcription–translation system (Raju and Kolakofsky, 1986b). In the coupled system, drugs that inhibited protein synthesis also inhibited full-length mRNA synthesis and resulted in the reappearance of the incomplete transcripts (Raju and Kolakofsky, 1986b; Bellocq and Kolakofsky, 1987). This led these investigators to postulate that translation of the nascent bunyavirus S segment mRNA is required to prevent premature termination of primary transcription product, a surprising conclusion. In contrast to these data, although S segment mRNA synthesis of GER virus was inhibited in cell culture by either anisomycin or cycloheximide, full-length S transcripts were obtained in an *in vitro* transcription system without added translational capabilities (Gerbaud *et al.*, 1987b). Further, in infections with SSH bunyavirus in the presence of puromycin or cycloheximide, strand-specific cDNA probes readily detected full-length S segment mRNA (but not, as expected, viral RNA transcripts; Vezza *et al.*, 1979; Eshita *et al.*, 1985). Similar results on the lack of inhibition of primary transcription by protein synthesis inhibitors have been reported for Punta Toro phlebovirus (Ihara *et al.*, 1985). In summary, it is not clear whether there really is a dependence for bunyavirus mRNA synthesis on translation or whether host factors are required. If the latter is the case, then they may be limiting in some cell types but not others (see Raju *et al.*, 1989).

Differences between bunyavirus viral RNA and mRNA are found not only at their 5' termini but also at their 3' termini. The S segment mRNAs of the bunyaviruses LAC (Raju and Kolakofsky, 1986b), SSH (Cash *et al.*, 1979), Akabane (Pattnaik and Abraham, 1983), and GER (Bouloy *et al.*, 1984), as well as the M segment mRNA of SSH virus (Eshita *et al.*, 1985) are all truncated at their 3' termini by about 100 nucleotides as compared to the viral RNA sequence. Potential transcription termination sites have been proposed for the S segments of LAC (Patterson and Kolakofsky, 1984) and SSH (Eshita *et al.*, 1985) bunyaviruses at or near the genomic sequence 3'-G/CUUUUU, which is similar to other negative strand RNA virus transcription termination–polyadenylation signals. Whether this site templates polyadenylation is not known; however, the low affinity of bunyavirus mRNAs for oligo(dT) columns suggests that the mRNAs are not extensively polyadenylated (Cash *et al.*, 1979; Pattnaik and Abraham, 1983; Pettersson *et al.*, 1985). Also, the S segment of Aino bunyavirus lacks such U-rich tracks (Akashi *et al.*, 1984). The M segment mRNA of SSH virus, while shorter than that of corresponding viral RNA, has no analogous 3'-CUUUUU template sequence and, therefore, a homopolymeric U tract is not involved in the termination of transcription (Bishop and Shope, 1979; Bishop *et al.*, 1984). Although the transcription termination signals are not known for the bunyavirus S or M segment mRNA species, there is a conserved purine-rich sequence located immediately after the transcription termination sites of the S segment RNAs of SSH and LAC bunyaviruses (Bishop *et al.*, 1982a; Akashi and Bishop, 1983; S nucleotides 903–914, M nucleotides 4454–4467: 5'-GGUGGGGGGUGGGG). It has been suggested that this sequence plays a role in transcription termination (Bishop *et al.*, 1982a). If correct, then the mechanism by which the sequence is recognized and how it is involved will require further study., Similar sequences are present in the Aino, Batai, Cache Valley, GRO, Kairi, Lumbo, Maguari, Main Drain, and Northway S RNAs (Akashi *et al.*, 1984; Dunn *et al.*, 1994).

C. Genome RNA Replication

In negative-strand viruses the change from primary transcription to replication requires a switch from mRNA synthesis to the synthesis of full-length viral-complementary RNA templates and then viral RNA synthesis. The processes involved in these switches have not been defined for any member of the *Bunyaviridae* family. Most likely viral factors are required (1) to initiate RNA replication, (2) to prevent the use of primers, and (3) to suppress the mRNA transcription termination signals. Genome RNA replication and subsequent secondary transcription is prevented by translational inhibitors such as cycloheximide (Vezza *et al.*, 1979; Eshita *et al.*, 1985). These results indicate that viral protein synthesis is required for replication of the genome. While it is not known which protein(s) are used, they are likely to include N and L, and possibly NS_S.

For the rhabdovirus vesicular stomatitis virus (VSV), the switch from RNA transcription to antigenome synthesis appears to be controlled by the availability of VSV N protein. Encapsidation by the VSV N also seems to serve as an antitermination signal, thus allowing full-length genome synthesis. It has been suggested that the VSV NS proteins are also involved and act to control the availability and delivery of N to the RNA species. A similar mechanism is plausible for the *Bunyaviridae*, whereby the bunyavirus N protein functions to regulate transcription and replication, possibly involving the NS_S protein. The factors dictating how replicative RNAs react with N to form nucleocapsid structures, while mRNAs do not, are unknown. It has been suggested that the added (presumably capped) host cell sequences on the 5′ ends of viral messages may somehow prevent encapsidation (Raju and Kolakofsky, 1986b). These are subjects for future research.

VI. MORPHOGENESIS

A. Synthesis and Processing of Viral Proteins

Viral polypeptides are synthesized shortly after infection suggesting that mRNAs are transcribed and translated rapidly. Time-course studies of the synthesis of viral polypeptides have revealed that at a high multiplicity of infection, radiolabeled N and NS_S proteins can both be detected as early as 2 hr after infection and the envelope glycoproteins shortly thereafter with a maximum rate reached between 3 and 5 hr postinfection (Pennington *et al.*, 1977; Madoff and Lenard, 1982). For bunyaviruses no posttranslational modifications of the S-coded proteins have been demonstrated (e.g., phosphorylation), apart from evidence that the amino terminus of N is blocked with regard to Edman degradation, suggesting that it is processed like many other eukaryotic proteins and probably acetylated (D. H. L. Bishop, unpublished data).

In contrast, for all bunyaviruses and all members of the *Bunyaviridae* studied to date, the M segment gene products are both processed and modified. G2, NS_M, and G1 are translated from a single M segment mRNA species as a polyprotein precursor that is cotranslationally cleaved. No polyprotein precursor to these proteins has been demonstrated in virus-infected cells. Examination of the predicted amino acid sequences of the polyprotein precursors of SSH, LAC, GER, and BUN viruses indicates that distinct hydrophobic regions precede both the encoded G1 and G2 proteins and appear to represent signal sequences (Eshita and Bishop, 1984; Lees *et al.*, 1986; Grady *et al.*, 1987; Pardigon *et al.*, 1988). This suggests that in addition to the first glycoprotein in the polyprotein, the second glycoprotein has its own signal sequence analogous to the internal signal sequence described for the E1 protein of Semliki Forest and Sindbis alphaviruses. Between the bunyavirus G2 and G1 is NS_M (Fazakerley *et al.*, 1988). This region of the polyprotein has a number of hydrophobic domains suggesting that NS_M may also be an integral membrane protein. The protein has been localized to the Golgi apparatus of virus-infected cells (Nakitare and Elliott, 1993). However, as noted elsewhere the function of NS_M is unknown. The proteases (most probably cellular) that are involved in the M-coded polyprotein processing have not been defined.

Also inferred from the predicted amino acid sequences are hydrophobic, carboxy-terminal anchor regions on the glycoproteins of SSH, LAC, GER, and BUN viruses. A common property of all M segment gene products predicted from cDNA sequences studied so far is their high cysteine content

(5–7%); and, in related viruses, the conservation of the positions of the cysteine residues (Eshita and Bishop, 1984; Grady *et al.*, 1987; Elliott, 1990). These findings suggest that disulfide bridge formation may occur and that the positions of these bridges may be crucial for providing the correct polypeptide folds.

B. Glycosylation

All of the bunyavirus envelope proteins examined to date possess N-linked oligosaccharides. The gene sequences which define a potential N-linked glycosylation site are those which encode Asn-X-Ser/Thr (where X is not proline). For SSH, LAC, GER, and BUN viruses two conserved glycan addition sites are present in all of the G2 sequences, none in the NS_M sequences, and, depending on the virus and the comparisons, there are one or two conserved glycan addition sites in their G1 proteins (Eshita and Bishop, 1984; Lees *et al.*, 1986; Grady *et al.*, 1987; Pardigon *et al.*, 1988). The number of sites actually utilized in mature virion proteins has not been defined for all viruses; however, analyses of the G2 protein of SSH virus (Cash *et al.*, 1981; Eshita and Bishop, 1984; Fazakerley *et al.*, 1988) suggest that all of the potential glycosylation sites on G2 are used and at least one of the potential sites on G1.

In general, two classes of asparagine-linked oligosaccharides, complex or predominantly mannose in composition (large or trimmed), are normally found on mature glycoproteins. Often both types are attached to the same polypeptide chain and, for the population of protein molecules, they may be of one or another type. As described for the hemagglutinin protein of influenza virus, as well as other glycoproteins, in order for oligosaccharides to evolve from the high-mannose type to the complex type, they are normally transported through the Golgi, where mannose residues are trimmed and terminal residues added. Examination of the oligosaccharides attached to the G1 and G2 proteins of LAC and Inkoo viruses grown in mammalian cells have revealed that G2 possessed mostly high-mannose glycans, whereas G1 contained both complex as well as an intermediate-type oligosaccharide (Madoff and Lenard, 1982). The type and amount of oligosaccharides attached to viral proteins correlate to some extent with the cell type and site of viral maturation. In general, shortly after primary glycosylation of nascent proteins at the rough endoplasmic reticulum (ER), oligosaccharides are susceptible to cleavage by endo-H, an enzyme that cleaves only high-mannose residues. Later, after removal of glucose residues at the rough ER, migration of the glycoproteins to the smooth ER and Golgi, trimming of residues, and attachment of the peripheral sugars, the oligosaccharides are no longer susceptible to endo-H cleavage. This acquired resistance to endo-H, therefore, generally indicates that the proteins have been processed through the Golgi. The endo-H susceptibility of the high-mannose and intermediate-type glycans found on the envelope proteins of viruses in the *Bunyaviridae* suggests that they are incompletely processed through the Golgi, a finding that may relate to their mode of morphogenesis. More information is required to determine the exact nature of the Golgi-associated processing events.

C. Transport

One of the earliest features described that distinguished members of the *Bunyaviridae* from all other negative-strand RNA viruses is that the viral particles are formed intracellularly by a budding process at smooth-surface vesicles in the Golgi area (Lyons and Heyduk, 1973; Murphy *et al.*, 1973; Bishop and Shope, 1979). In certain other negative-sense RNA viruses, matrix or membrane (M) proteins are believed to be responsible for bridging the gap between the viral integral envelope proteins and their nucleocapsids and possibly also act as nucleation sites to signal assembly of virions at the cell surface. Alternatively, M may act as a spacer within the nucleocapsid and thereby facilitate nucleocapsid interaction with the envelope. For members of the *Bunyaviridae* there is no M protein. How nucleocapsids recognize the viral glycoproteins in the progenitor envelope is unknown. What viral structures dictate the site of maturation are also unknown. LAC virus development is inhibited by the ionophore monensin (Cash, 1982). Possibly some or all of the viral glycoproteins possess sequences that target the Golgi, like those found in cellular glycosyltransferases. Alternatively, the bunyavirus glycoproteins may lack the signals required for transport from the Golgi to the plasma membrane. Recent studies using vaccinia expression vectors have shown that the BUN G2 and NS_M proteins can individually locate to the Golgi whereas G1 does not unless coexpressed with G2 (Nakitare and Elliott, 1993; Lappin *et al.*, 1994; see Chapter 7).

D. Assembly and Release

Detailed electron microscopic studies of bunyavirus maturation similar to those reported for the maturation of phleboviruses (Smith and Pifat, 1982), and uukuviruses (Kuismanen *et al.*, 1982) have not been reported. These other viruses mature mostly, but not exclusively, in the perinuclear regions of infected cells in association with smooth membranes, presumed to be Golgi membranes. In this process, electron-dense, ribonucleoprotein (nucleocapsid) structures have been observed immediately beneath the membrane where virus budding occurs. The viral nucleocapsid and spike structures were only seen on the portion of the Golgi vesicle membrane directly involved in the budding process and not on adjacent areas of the same membrane (Smith and Pifat, 1982). Nucleocapsids were not found under membranes that did not have spikes, suggesting that an interaction of transmembrane regions of the viral glycoproteins and the nucleocapsids is prerequisite to budding. Candidate transmembrane regions have been predicted from hydropathic analyses of derived amino acid sequences representing the envelope proteins of all members of the *Bunyaviridae* sequenced to date. Direct examination of the phlebovirus Karimabad by enzymatic digestion of exposed proteins embedded in intracellular membranes, demonstrated that approximately 12% of G1 and/or G2 was exposed on the cytoplasmic face of membranes in infected cells and accessible to digestion (Smith and Pifat, 1982). A large protease-resistant fragment was also identified, which was presumably sequestered in the membrane in a manner that rendered it safe from the enzymatic digestion (Smith and Pifat, 1982). These enzyme-resistant fragments may represent transmembrane regions of proteins which could provide the interaction between nucleocapsids and the cellular membranes required for envelopment.

After the particles bud into the Golgi cisternae, it is believed that they are released in individual small vesicles in a manner analogous to secretory granules of other cell types (Smith and Pifat, 1982). The release of virus from infected cells presumably occurs when the cytoplasmic, virus-containing vesicles fuse with the cellular plasma membrane, i.e., via normal exocytosis.

VII. EFFECTS OF VIRUS REPLICATION ON HOST CELLS

The cytopathic effects observed in cultured cells infected with members of the *Bunyaviridae* vary widely depending on both the virus and the type of host cell studied. Bunyaviruses are capable of alternately replicating in vertebrates and arthropods (Nuttall *et al.*, 1991). Generally the viruses are cytolytic for their vertebrate host cells in tissue culture, but cause little or no cytopathology in their invertebrate host cells (Lyons and Heyduk, 1973; James and Millican, 1986).

Some viruses display a very narrow host range, especially for their arthropod vectors. Although the reason for this has not been completely defined, studies on LAC variant and revertant viruses have suggested that the specificity is related to the glycoproteins, probably at the level of viral attachment to susceptible cells (Sundin *et al.*, 1987). In natural virulent or avirulent infections of mammals, bunyaviruses are often targeted to cells of particular organs, or to neurons and the brain. For example, LAC virus is neurotropic in mice, although it will replicate in various other cell types (Bishop and Shope, 1979). The importance of the observed neurotropism, if any, to LAC virus is unknown since it is often a late phenotype and apparent only after the major viremic phase of infection. It will be interesting to determine whether such targeting is related solely to the avidity of the respective viruses for host cell receptors on cells of particular organs, or whether there are other factors involved such as differences in virus replication efficiency in different cell types.

In vertebrates, including *Xenopus* cells, bunyaviruses have been shown to cause a reduction in host-cell protein synthesis (Pennington *et al.*, 1977;

Watret *et al.*, 1985; Rossier *et al.*, 1988). A decline in host-cell protein synthesis with BUN virus-infected BSC-1 cells was observed at 5 hr postinfection and by 7 hr host protein synthesis was almost completely abolished (Pennington *et al.*, 1977). Similar results were obtained in LAC virus-infected BHK cells (Madoff and Lenard, 1982). In contrast, mosquito cells infected with Marituba bunyavirus displayed no reduction in host macromolecular synthesis and the cells readily became persistently infected (Carvalho *et al.*, 1986). Similar data have been reported for BUN and LAC virus infections of insect cells (Elliott and Wilkie, 1986; Rossier *et al.*, 1988). Mosquito cells infected with BUN virus have been reported to generate defective virus RNAs (Scallan and Elliott, 1992). Other information on defective bunyaviruses is provided in Chapter 8.

VIII. VIRUS REPLICATION IN THE ARTHROPOD VECTOR

The abilities of a bunyavirus to infect and to be transmitted by a particular arthropod species depend on the virus, the arthropod and its opportunities for virus acquisition. As an example, for LAC virus *Ae. triseriatus* is the usual vector in the midwestern United States (Thompson and Evans, 1965; Thompson *et al.*, 1967, 1972; Watts *et al.*, 1972, 1973a,b, 1974; Berry *et al.*, 1974; Gauld *et al.*, 1974; Pantuwatana *et al.*, 1974; Balfour *et al.*, 1975; Beaty and Thompson, 1975; Grimstad *et al.*, 1977; Miller *et al.*, 1977; Thompson and Beaty, 1977, 1978; Beaty and Bishop, 1988). In the same and other areas, other vectors may be involved (e.g., *Ae. trivittatus*, *Ae. communis*, *Culex pipiens*). In some cases these alternate vectors may transmit LAC virus less efficiently than *Ae. triseriatus* mosquitoes (either transovarially, venereally, or via blood meals). The reasons for the different tropisms and efficiencies of these vectors are unknown. The roles of arthropod vectors in maintaining arboviruses, including bunyaviruses, have been recently reviewed (Nuttall *et al.*, 1991).

A. The Infection Course in Arthropods

Following ingestion of a contaminated blood meal, virus replication can be identified in cells and tissues of a permissive vector using suitable probes (e.g., monoclonal antibodies, tagged DNA or RNA probes). In some cases, virus infection may be observed in certain but not all types of midgut cells of an arthropod early during the infection course. Subsequently, virus RNA and viral antigens may be found in various tissues of the infected vector (depending on the virus and the arthropod), such as ovary-associated cells (thereby facilitating transovarial transmission), male gonad and associated tissues (to facilitate venereal transmission), or salivary gland and associated cells (to allow horizontal transmission). The reasons for the various cellular tropisms of bunyaviruses are unknown.

Since certain blood-feeding (female) dipterans only take a single blood meal, transovarial transmission may be an important feature in bunyavirus maintenance for such species. There is no evidence that juvenile stages of dipteran vectors are involved in the horizontal transmission of bunyaviruses, nor males, other than transmission to females during insemination (i.e., for those males that acquire virus transovarially). For dipterans that feed more than once on a vertebrate, horizontal transmission from recently fed females may play an important role in the infection cycle between arthropods and vertebrates. Where this has been examined, transmission occurs after an intrinsic period of virus replication in the vector (e.g., after a week postexposure), indicating that regurgitation of virus is probably not involved and that transmission normally involves virus delivered in saliva. Although no work has been reported on the tick-transmitted bunyaviruses, horizontal transmission may be particularly relevant in the case of ticks, since these arthropods feed at various stages of their development (larvae, nymphs, adults), and probably pass the viruses transstadially (see Nuttall *et al.*, 1994).

The efficiencies of transovarial transmission vary between vectors. For those that have been investigated (e.g., LAC virus), virus replication has been shown to occur during the juvenile phases of the arthropods eventually resulting in infected adults that may transmit the virus to their offspring, or to a vertebrate host depending on the vector's feeding preferences. The abilities of bunyaviruses to be transmitted vertically from parent to offspring and to establish long-lived, persistent infections in particular arthropods (involving many tissues) are probably the most important factors in their survival in nature (Watts *et al.*, 1972, 1973a,b, 1974; Berry *et al.*, 1974; Beaty and Thompson, 1976, 1977; Miller *et al.*, 1977, 1979; Thompson and Beaty, 1977; Thompson, 1979). Although it has been reported that viral infection of mosquitoes induces changes in feeding behavior (Grimstad *et al.*, 1980), neither the molecular basis of this, nor the biological significance is known.

Ae. triseriatus mosquitoes are not the usual vectors of SSH virus, although as noted above they are vectors of LAC virus in the midwestern United States. Using SSH, LAC, and all forms of SSH–LAC reassortant, it has been shown that the viral M RNA gene products are the principal determinants of efficient virus infection in the colonized vector and the transmission of viruses to an animal (Beaty *et al.*, 1981a,b, 1982, 1983b, 1985). Thus, following oral infection, both SSH and LAC viruses can equally well establish a midgut infection in *Ae. triseriatus* mosquitoes (Beaty *et al.*, 1982). However, LAC virus and SSH–LAC reassortants with a LAC M RNA species (e.g., L/M/S genotypes such as SSH/LAC/SSH) are more efficient at establishing disseminated infections in these mosquitoes than SSH virus, or LAC–SSH reassortants with an SSH M RNA species (Beaty *et al.*, 1982). Even if the viruses are delivered by inoculation into the thorax of *Ae. triseriatus* mosquitoes, LAC virus and SSH–LAC reassortants with a LAC M RNA are more successfully transmitted to an animal host than SSH virus, or reassortants with an SSH M RNA species (Beaty *et al.*, 1981a). Since the viral glycoproteins are coded by the M RNA species, these LAC–SSH tropisms may reflect receptor–glycoprotein interactions in the various mosquito cell types. LAC viruses with a mutated G1 protein have been obtained that exhibit an altered ability to infect mosquitoes (Sundin *et al.*, 1987). It is not known if there are bunyaviruses that do not involve vertebrate hosts in their natural cycles, i.e., viruses for which transmission occurs only from arthropod generation to generation (as for Sigma rhabdovirus). Probably there are, although if they cannot infect vertebrates then they would not have been identified by the procedures usually employed to isolate arboviruses (i.e., assays in suckling mice or other vertebrate species).

B. Coinfection of Vector Species

Coinfections of *Ae. triseriatus* mosquitoes with temperature-sensitive (*ts*) or wild-type viruses have produced evidence for the formation of reassortant viruses, as well as their transmission to vertebrates and succeeding mosquito generations following transovarial transmission (Beaty *et al.*, 1981a,b,c, 1982, 1983b, 1985; Chandler *et al.*, 1990, 1991). As mentioned above, analyses of natural virus isolates have also shown that reassortment occurs in nature (Klimas *et al.*, 1981a).

Studies with mosquitoes have shown that when a bunyavirus establishes an infection, by 48 hr the opportunity for a successful superinfection by a genetically compatible bunyavirus is reduced, although unrelated viruses can still establish superinfections (Beaty *et al.*, 1983a; Bishop and Beaty, 1986, 1988). This restriction reduces the opportunity for reassortment. The molecular basis of the observed inhibition of superinfection is unknown; it may involve defective interfering particles (see Chapter 8), or the removal of virus receptors from previously infected target cells.

A factor that may also reduce the opportunities for reassortment is that even for closely related bunyaviruses they can vary in their ability to infect particular mosquito species. For example, for LAC virus, mosquitoes of certain species (e.g., *Ae. triseriatus* from the East Coast of the United States) may be infected at a lower efficiency than another variety of the mosquito (e.g., *Ae. triseriatus* from the midwestern United States), even with the same virus topotype. With another LAC topotype, a different result may be obtained (B. J. Beaty, personal communication). Also, using related bunyaviruses (e.g., LAC and SSH viruses) and different vectors (e.g., *Ae. triseriatus* and *Culiseta inornata*), different efficiencies of infection are obtained with the two bunyaviruses (Schopen *et al.*, 1991).

IX. VIRUS REPLICATION IN THE VERTEBRATE SPECIES

Viruses are delivered to a vertebrate host during the course of the vector acquiring a blood meal. The infection of a vertebrate depends on the feeding preferences of the arthropod and the ability of the vertebrate to sustain that virus infection (hence transmit the virus to another vector). For LAC virus, small mammals such as chipmunks and tree squirrels are the usual amplifying hosts of LAC virus in the midwestern United States (Thompson *et al.*, 1967; Pantuwatana *et al.*, 1972; Gauld *et al.*, 1974, 1975). Because both these hosts and the mosquitoes that feed on them are parochial in their life-styles, the ability of LAC virus varieties to be disseminated over large areas of country may be restricted. Nevertheless, serological surveys indicate that LAC virus also infects other, more wide-ranging species, such as foxes (Amundson and Yuill, 1981). Thus even where the preferred vectors and hosts may not facilitate virus dissemination, the involvement of other hosts (and vectors) with different habits may allow the virus to be disseminated.

It is not known if some LAC virus varieties are more virulent for humans than others. By fingerprint analyses, LAC virus isolates obtained from two fatal human infections (Thompson *et al.*, 1965; Klimas *et al.*, 1981a) are apparently more closely related to each other than to LAC viruses obtained from mosquitoes over a similar time span (18 years) and in the same region of the United States (Wisconsin/Minnesota). It is possible, but not proven, that LAC virus varieties virulent for humans are present in certain localities in the United States. Although LAC virus infections of humans are self-limiting and rarely fatal, some evidence has been obtained for recurrent seizures in patients who have had prior LAC virus infections (Gundersen and Brown, 1983).

As discussed in Chapters 8 and 10, bunyavirus virulence has been investigated in model and natural animal hosts (Shope *et al.*, 1981a,b, 1982; Seymour *et al.*, 1983). In young mice, virus virulence can be measured by dose-related death, or survival following intraperitoneal inoculation (e.g., for LAC and Tahyña viruses, respectively) (Shope *et al.*, 1981a; Bishop *et al.*, 1987). Following intracranial inoculation, virus virulence can also be measured by the length of survival of the infected animals before death (Shope *et al.*, 1981a). By both criteria, the data have shown that for the California group viruses, the bunyavirus M RNA gene products are the principal determinants of virulence (Shope *et al.*, 1981a). Similar data have been obtained for parental and reassortant group C bunyaviruses (Bishop *et al.*, 1983b).

Evidence has been obtained that the sites of infection within tissues (and organs) of a vertebrate host can vary for the different California group viruses (Tignor *et al.*, 1983). For the group C viruses, organ targets can be radically different, even for two closely related strains of a virus serotype (Bishop *et al.*, 1983b). By parent and reassortant virus analyses, it has been shown that the virulence properties of viscerotropic and neurotropic varieties of Caraparu viruses are determined by their M RNA gene products (Bishop, 1982; Bishop *et al.*, 1983b).

Despite the evidence that the viral glycoproteins are the principal determinants of virulence, other gene products can mitigate this property. This has been shown by the demonstration in mice, chipmunks (*Tamias striatus*), and snowshoe hares (*Lepus americanus*) that LAC–SSH reassortants with a non-*ts* mutation in the LAC L RNA species are avirulent by comparison with the parental viruses and without respect to the origin of their M RNA species (Rozhon *et al.*, 1981; Shope *et al.*, 1981a; Seymour *et al.*, 1983). It is not known to what extent, if any, viral RNA replication and/or transcription are impaired by the L mutation, or how the defective L RNA gene products cause attenuation of virulence. Reassortant LAC–SSH viruses with this particular mutant L RNA are also inefficient at establishing productive *Ae. triseriatus* infections (Beaty *et al.*, 1981a, 1982). Thus, although the M RNA gene products are the principal determinants of virulence, changes in proteins coded by other RNA species can attenuate virulence (Janssen *et al.*, 1986). Whether there are circumstances in which bunyavirus L or S RNA gene products can enhance virulence is unknown.

X. GENETIC AND BIOLOGICAL ATTRIBUTES

The genetic attributes of bunyaviruses are comprehensively reviewed in Chapter 8. The data obtained to date support the view that viruses assigned to a serogroup constitute one or more gene pools as reflected by their abilities to produce reassortants in dual virus infections. Further, evidence for quasi-diploid viruses (with more than one copy of an RNA size class), representing up to 10% of the progeny of dual infections, has been obtained (Urquidi and Bishop, 1992).

Genetic studies provided the first data on the coding assignments of the bunyavirus RNAs and proteins (Gentsch and Bishop, 1976, 1978, 1979, 1980; Gentsch *et al.*, 1977a,b, 1978, 1979, 1980; Bishop, 1982; Bishop and Gentsch, 1980; Bishop *et al.*, 1978, 1980a, 1981, 1983a,b). Using appropriate reassortants, the principal determinants of virus pathogenicity and vector competence have been identified (Beaty *et al.*, 1981a,b,c, 1982, 1983b, 1985; Shope *et al.*, 1981a,b, 1982; Rozhon *et al.*, 1981; Tignor *et al.*, 1983; Janssen *et al.*, 1986; Gonzalez-Scarano *et al.*, 1988; Endres *et al.*, 1991; Griot *et al.*, 1993).

A. Common Ancestors and Evolution of Bunyaviruses

Virus isolation studies have shown that certain California group viruses are sympatric in nature (e.g., SSH and LAC viruses in the United States east of the Rockies) (Calisher, 1983). Other California group viruses are located in separate regions of the world (e.g., TAH in central Europe, Inkoo in Scandinavia, Lumbo in South Africa, CAL in the western United States) (Calisher, 1983; Karabatsos, 1985). Each of these viruses is serologically related and genetically compatible with all of the other members of the serogroup (Bishop *et al.*, 1987). Such attributes imply a common ancestor for the California group viruses. How can this be the case? One possibility is if, at some time, ancestral isolates were dispersed to other localities in the world with the result that different routes of evolution occurred leading to the various virus serotypes and subtypes that are currently recognized. From what is known about the biology and virus–vector–host relationships of bunyaviruses, it can be speculated that, among other means, the transfer of materials (e.g., wood containing infected mosquito eggs), infected animals, or arthropods may have been instrumental in establishing new foci of enzootic viruses. Bunyaviruses whose hosts are birds, or other animals capable of extensive dispersal, or whose vectors are ticks, may be more easily disseminated.

While there is no indication of the time it has taken the California serogroup of bunyaviruses to evolve into the species, subtypes, variants, and varieties currently recognized, and to occupy the various ecological niches in which the known viruses are found, it is most likely that the viruses have been around for a very long time. Further, when one considers the diversity between the different bunyavirus serogroups, and the distinctive features between the different *Bunyaviridae* genera, it is probable that the family of viruses has experienced a long evolutionary trail.

B. Evolution Rates of Bunyaviruses

The genetic diversity of individual bunyavirus species, as indicated in their gene sequences (Klimas *et al.*, 1981a; Hewlett *et al.*, 1992; see above for a discussion on LAC virus varieties), reflects the mutations that occur during RNA replication and the opportunities afforded in the particular virus–vector–host relationships involved in the normal infection cycles (Bishop and Beaty, 1988). The basis of the mutations is the mistake rate of the virus polymerase when copying its template. Without an editing ability, mutations that occur will accumulate. Apart from the occasional involvement of sequence deletions and duplications and other forms of change, the 1–2% rate of recovery of spontaneous *ts* mutants among the progeny of a wild-type virus infection indicates that *by this criterion* the mutation rate is about 1 in 10^6 nucleotides (1% of progeny with a genome of 1.2×10^4 nucleotides). However, this figure does not take into account point mutations that are lethal, or those that cause protein sequence changes that are not manifest in relation to the virus infection in culture. The latter types of mutant probably account for most of the changes that are found in viable progeny (i.e., not all mutants involving amino acid changes provide a *ts* phenotype). Nor does the figure take into account other mutations of the genome, e.g., those that do not affect an encoded amino acid (e.g., third position changes of certain codons). In other systems (e.g., influenza virus isolates taken over a period of time), the latter types of mutant have been estimated to be at least ten times more abundant than those mutations that change amino acids (Bishop *et al.*, 1982b). Thus, it is not unreasonable to argue that a large proportion, if not all, of the copies of a bunyavirus genome that are found among the progeny viruses, are genetic variants of their parent virus and that the derived population represents a genetic "swarm."

As a result of the mutation rates, progeny viruses will have mutations distributed over their genomes. Where structural, or gene product constraints work against the acceptability of a particular mutation, then the result may be fatal for that particular virus and contribute toward the proportion of the progeny that are noninfectious (often very high for RNA viruses). In some cases, mutations will have no effect, in others there may be a change in the virus phenotype which is manifested in a number of ways (e.g., *ts*, slow growth as evidenced by a small plaque size, pathogenicity changes, etc.). In evolutionary terms, the high mutation rates and the effects of accumulated changes are probably the reasons why there are no RNA viruses with large genomes comparable in size to the DNA genomes of baculoviruses, herpesviruses, or poxviruses.

A final point concerning bunyavirus evolution is the issue of virus cloning *in natura*. In the natural course of events, bunyaviruses can infect arthropods during the process of the arthropod acquiring a blood meal from an infected vertebrate host, e.g., during the host's viremic phase. The amount of blood obtained is in the microliter range for mosquitoes and certain other blood-feeding insects. For ticks, it may involve much larger quantities of blood (and infected host cells) and obtained over a number of days. The amount of virus acquired depends on both the titer of virus in the meal and the meal volume. Particularly for gnats, mosquitoes, and phlebotomines, the small amounts of blood taken must lead to many occasions where a single virus establishes an infection in the insect, i.e., de facto cloning of the virus. Since a bunyavirus population may consist of a swarm of genotypes, such cloning will facilitate the establishment of mutant genotypes in the absence of competing viruses. Other forms of natural selection probably occur (e.g., antibody selection in a preinfected vertebrate that is exposed again to a bunyavirus); however, natural cloning may be one reason why so many virus serotypes, subtypes, variants, and varieties exist in the *Bunyavirus* genus.

XI. CONCLUSIONS

The study of bunyaviruses offers a fertile area of research that extends from the molecular to the cellular, from the vector to the host, and to the environment. While there are many different bunyaviruses that have already been identified, these probably represent just the tip of a mountain of viruses that have yet to be obtained from wildlife. The involvement of all forms of blood-feeding arthropods, the many animals on which they feed, and the various regions of the world where viruses have been sought, indicate the dimension of the subject. Many bunyaviruses appear to be innocuous in their vertebrate hosts, while others are highly virulent. In the invertebrate, few effects have been recorded, other than certain behavioral changes of which the significance is not known.

Despite the ability of certain bunyaviruses to be transmitted transovarially and with considerable efficiency in some species, the frequency of virus isolation from these insects in enzootic areas is often very low (< 1%). Why? Is it because there are costs to the vectors that have yet to be identified? Or are there insect defense mechanisms that are as yet unknown? Thus, in addition to the questions that have yet to be answered concerning the molecular biology of the viruses, there are many outstanding and important questions that remain to be addressed concerning the biology of bunyavirus–vector–host relationships.

XII. REFERENCES

Abraham, G., and Pattnaik, G., 1983, Early RNA synthesis in Bunyamwera virus-infected cells, *J. Gen. Virol.* **64:**1277.

Akashi, H., and Bishop, D. H. L., 1983, Comparison of the sequences and coding of La Crosse and snowshoe hare bunyavirus S RNA species, *J. Virol.* **45:**1155.

Akashi, H., Gay, M., Ihara, T., and Bishop, D. H. L., 1984, Localized conserved regions of the S RNA gene products of bunyaviruses are revealed by sequence analyses of the Simbu serogroup Aino viruses, *Virus Res.* **1:**51.

Amundson, T. E., and Yuill, T. M., 1981, Natural La Crosse virus infection in the red fox (*Vulpes fulva*), gray fox (*Urocyon cinereoargenteus*), raccoon (*Procyon lotor*) and opossum (*Didelphis virginiana*), *Am. J. Trop. Med. Hyg.* **30:**706.

Balfour, H. H., Edelman, C. K., Cook, F. E., Barton, W. I., Buzicky, A. W., Siem, R. A., and Bauer, H., 1975, Isolates of California encephalitis (La Crosse) virus from field-collected eggs and larvae of *Aedes triseriatus*: Identification of the overwintering site of California encephalitis, *J. Infect. Dis.* **131:**712.

Beaty, B. J., and Bishop, D. H. L., 1988, Bunyavirus–vector interactions, *Virus Res.* **10:**289.

Beaty, B. J., and Thompson, W. H., 1975, Emergence of La Crosse virus from endemic foci. Fluorescent antibody studies of overwintered *Aedes triseriatus*, *Am. J. Trop. Med. Hyg.* **24:**685.

Beaty, B. J., and Thompson, W. H., 1976, Delineation of La Crosse virus in developmental stages of transovarially infected *Aedes triseriatus*, *Am. J. Trop. Med. Hyg.* **25:**505.

Beaty, B. J., and Thompson, W. H., 1977, Tropisms of La Crosse virus in *Aedes triseriatus* following infective blood meals, *J. Med. Entomol.* **14:**499.

Beaty, B. J., Shope, R. E., and Clarke, D. H., 1977, Salt-dependent hemagglutination with *Bunyaviridae* antigens, *J. Clin. Microbiol.* **5:**548.

Beaty, B. J., Holterman, M., Tabachnick, W., Shope, R. E., Rozhon, E. J., and Bishop, D. H. L., 1981a, Molecular basis of bunyavirus transmission by mosquitoes: The role of the M RNA segment, *Science* **211:**1433.

Beaty, B. J., Miller, B. R., Holterman, M., Tabachnick, W. J., Shope, R. E., Rozhon, E. J., and Bishop, D. H. L., 1981b, Pathogenesis of bunyavirus reassortants in *Aedes triseriatus* mosquitoes, in: *Replication of Negative Strand Viruses* (D. H. L. Bishop and R. W. Compans, eds.), pp. 153–158, Elsevier, Amsterdam.

Beaty, B. J., Rozhon, E. J., Gensemer, P., and Bishop, D. H. L., 1981c, Formation of reassortant bunyaviruses in dually infected mosquitoes, *Virology* **111:**662.

Beaty, B. J., Miller, B. R., Shope, R. E., Rozhon, E. J., and Bishop, D. H. L., 1982, Molecular basis of bunyavirus *per os* infection of mosquitoes: The role of the M RNA segment, *Proc. Natl. Acad. Sci. USA* **79:**1295.

Beaty, B. J., Bishop, D. H. L., Gay, M., and Fuller, F., 1983a, Interference between bunyaviruses in *Aedes triseriatus* mosquitoes, *Virology* **127**:83.

Beaty, B. J., Fuller, F., and Bishop, D. H. L., 1983b, Bunyavirus gene structure–function relationships and potential for RNA segment reassortment in the vector: La Crosse and snowshoe hare reassortant viruses in mosquitoes, in: *California Serogroup Viruses* (C. H. Calisher and W. H. Thompson, eds.), pp. 119–128, Liss, New York.

Beaty, B. J., Sundin, D. R., Chandler, L., and Bishop, D. H. L., 1985, Evolution of bunyaviruses via genome segment reassortment in dually-infected (per os) *Aedes triseriatus* mosquitoes, *Science* **230**:548.

Bellocq, C., and Kolakofsky, D., 1987, Translational requirement for La Crosse virus S-mRNA synthesis: A possible mechanism, *J. Virol.* **61**:3960.

Berry, R. L., Lalonde, B. J., Stegmiller, H. W., Parsons, M. A., and Bear, G. T., 1974, Isolation of La Crosse virus (California encephalitis group) from field collection *Aedes triseriatus* (Say) larvae in Ohio (Diptera: Culicidae), *Mosq. News* **34**:454.

Bishop, D. H. L., 1982, The molecular properties of bunyaviruses including group C isolates, and the genetic capabilities of bunyaviruses to form recombinants, in: *International Symposium on Tropical Arboviruses and Hemorrhagic Fevers* (F. de P. Pinheiro, ed.), pp. 117–133, Impresso Nat. Fund. Sci. Dev. Tech., Belem, Brazil.

Bishop, D. H. L., 1985, The genetic basis for describing viruses as species, *Intervirology* **24**:79.

Bishop, D. H. L., and Beaty, B. J., 1986, Interference—immunity of mosquitoes to bunyavirus superinfection, *Symp. Zool. Soc. London* **56**:95.

Bishop, D. H. L., and Beaty, B. J., 1988, Molecular and biochemical studies of the evolution, infection and transmission of insect bunyaviruses, *Philos. Trans. R. Soc. London Ser. B* **321**:463.

Bishop, D. H. L., and Gentsch, J., 1980, The genetic recombination potential of bunyaviruses, in: *Arboviruses in the Mediterranean Countries* (J. Vesenjak-Hirjan, ed.), pp. 81–96, Gustav Fischer Verlag, Stuttgart.

Bishop, D. H. L., and Shope, R. E., 1979, Bunyaviridae, in: *Comprehensive Virology*, Vol. 14 (H. Fraenkel-Conrat and R. R. Wagner, eds.), pp. 1–156, Plenum Press, New York.

Bishop, D. H. L., Gentsch, J. R., and El-Said, L. H., 1978, Genetic capabilities of bunyaviruses, *J. Egypt. Public Health Assoc.* **53**:217.

Bishop, D. H. L., Beaty, B. J., and Shope, R. E., 1980a, Recombination and gene coding assignments of bunyaviruses and arenaviruses, *Ann. N.Y. Acad. Sci.* **354**:84.

Bishop, D. H. L., Calisher, C., Casals, J., Chumakov, M. P., Gaidamovich, S. Y., Hannoun, C., Lvov, D. K., Marshall, I. D., Oker-Blom, N., Pettersson, R. F., Porterfield, J. S., Russell, P. K., Shope, R. E., and Westaway, E. G., 1980b, Bunyaviridae, *Intervirology* **14**:125.

Bishop, D. H. L., Clerx, J. P. M., Clerx-van Haaster, C. M., Robeson, G., Rozhon, E. J., Ushijima, H., and Veerisetty, V., 1981, Molecular and genetic properties of members of the Bunyaviridae, in: *Replication of Negative Strand Viruses* (D. H. L. Bishop and R. W. Compans, eds.), pp. 135–145, Elsevier, Amsterdam.

Bishop, D. H. L., Gould, K. G., Akashi, H., and Clerx-van Haaster, C. M., 1982a, The complete sequence and coding content of snowshoe hare bunyavirus small (S) viral RNA species, *Nucleic Acids Res.* **10**:3703.

Bishop, D. H. L., Huddleston, J. A., and Brownlee, G. G., 1982b, The complete sequence of RNA segment 2 of influenza A/NT/60/68 and its encoded P1 protein, *Nucleic Acids Res.* **10**:1335.

Bishop, D. H. L., Fuller, F. J., and Akashi, H., 1983a, Coding assignments of the RNA genome segments of California serogroup viruses, in: *California Serogroup Viruses* (C. H. Calisher and W. H. Thompson, eds.), pp. 107–117, Liss, New York.

Bishop, D. H. L., Fuller, F. J., Akashi, H., Beaty, B., and Shope, R. E., 1983b, The use of reassortant bunyaviruses to deduce their coding and pathogenic potentials, in: *Mechanisms of Viral Pathogenesis: From Gene to Pathogen* (A. Kohn and P. Fuchs, eds.), pp. 49–60, Nijhoff, The Hague.

Bishop, D. H. L., Gay, M. E., and Matsuoko, Y., 1983c, Nonviral heterogeneous sequences are present at the 5' ends of one species of snowshoe hare bunyavirus S complementary RNA, *Nucleic Acids Res.* **11**:6409.

Bishop, D. H. L., Rud, E., Belloncik, S., Akashi, H., Fuller, G., Ihara, T., Matsuoko, Y., and Eshita, Y., 1984, Coding analyses of bunyavirus RNA species, in: *Segmented Negative Strand Viruses, Arenaviruses, Bunyaviruses and Orthomyxoviruses* (R. W. Compans and D. H. L. Bishop, eds.), pp. 3–11, Academic Press, New York.

Bishop, D. H. L., Shope, R. E., and Beaty, B. J., 1987, Molecular mechanisms in arbovirus disease, in: *Molecular Basis of Virus Disease* (W. C. Russell and J. W. Almond, eds.), pp. 135–166, Cambridge University Press, London.

Bouloy, M., and Hannoun, C., 1973, Effet de l'actinomycine D sur la multiplication du virus Tahyna, *Ann. Microbiol. Inst. Pasteur* **124b:**547.

Bouloy, M., and Hannoun, C., 1976a, Studies on Lumbo virus replication. I. RNA-dependent RNA polymerase associated with virions, *Virology* **69:**258.

Bouloy, M., and Hannoun, C., 1976b, Studies on Lumbo virus replication. II. Properties of viral ribonucleoproteins and characterization of messenger RNAs, *Virology* **71:**363.

Bouloy, M., Krams-Ozden, S., Horodniceanu, F., and Hannoun, C., 1973/4, Three-segment RNA genome of Lumbo virus (bunyavirus), *Intervirology* **2:**173.

Bouloy, M., Colbère, F., Krams-Ozden, S., Vialat, P., Garapin, A.-C., and Hannoun, C., 1975, Activité RNA polymérasique associée à un Bunyavirus (Lumbo), *C. R. Acad. Sci. Ser. D* **280:**213.

Bouloy, M., Vialat, P., Giriard, M., and Pardigon, N., 1984, A transcript from the S segment of the Germiston bunyavirus is uncapped and codes for the nucleoprotein and a nonstructural protein, *J. Virol.* **49:**717.

Bouloy, M., Pardigon, N., Vialat, P., Gerbaud, S., and Girard, M., 1990, Characterization of the 5' and 3' ends of viral messenger RNAs isolated from BHK21 cells infected with Germiston virus (Bunyavirus), *Virology* **75:**50.

Cabradilla, C. D., Holloway, B. P., and Obijeski, J. F., 1983, Molecular cloning and sequencing of the La Crosse virus S RNA, *Virology* **128:**463.

Calisher, C. H., 1983, Taxonomy, classification, and geographic distribution of California serogroup bunyaviruses, in: *California Serogroup Viruses* (C. H. Calisher and W. H. Thompson, eds.), pp. 1–16, Liss, New York.

Carvalho, M. G. C., Frugulhetti, I. C., and Rebello, M. A., 1986, Marituba (Bunyaviridae) virus replication in cultured *Aedes albopictus* cells and in L-A9 cells, *Arch. Virol.* **90:**325.

Cash, P., 1982, Inhibition of La Crosse virus replication by monensin, a monovalent ionophore, *J. Gen. Virol.* **59:**193.

Cash, P., Vezza, A. C., Gentsch, J. R., and Bishop, D. H. L., 1979, Genome complexities of the three mRNA species of snowshoe hare bunyavirus and *in vitro* translation of S mRNA to viral N polypeptide, *J. Virol.* **31:**685.

Cash, P., Hendershot, L., and Bishop, D. H. L., 1981, The effects of glycosylation inhibitors on the maturation and intracellular polypeptide synthesis induced by snowshoe hare bunyavirus, *Virology* **103:**235.

Chandler, L. J., Beaty, B. J., Baldridge, G. D., Bishop, D. H. L., and Hewlett, M. J., 1990, Heterologous reassortment of bunyaviruses in *Aedes triseriatus* mosquitoes and transovarial and oral transmission of newly evolved phenotypes, *J. Gen. Virol.* **71:**1045.

Chandler, L. J., Hodge, G., Endres, M., Jacoby, D. R., Nathanson, N., and Beaty, B. J., 1991, Reassortment of La Crosse and Tahyna bunyaviruses in *Aedes triseriatus* mosquitoes, *Virus Res.* **20:**181.

Clerx-van Haaster, C., and Bishop, D. H. L., 1980, Analyses of the 3'-terminal sequences of snowshoe hare and La Crosse bunyaviruses, *Virology* **105:**564.

Clerx-van Haaster, C. M., Akashi, H., Auperin, D. D., and Bishop, D. H. L., 1982a, Nucleotide sequence analyses and predicted coding of bunyavirus genome RNA species, *J. Virol.* **41:**119.

Clerx-van Haaster, C. M., Clerx, J. P. M., Ushijima, H., Akashi, H., Fuller, F., and Bishop, D. H. L., 1982b, The 3' terminal RNA sequences of bunyaviruses and nairoviruses (Bunyaviridae). Evidence of end sequence generic differences within the virus family, *J. Gen. Virol.* **61:**289.

Clewley, J., Gentsch, J., and Bishop, D. H. L., 1977, Three unique viral RNA species of snowshoe hare and La Crosse bunyaviruses, *J. Virol.* **22:**459.

Cunningham, C., and Szilagyi, J. F., 1987, Viral RNAs synthesized in cells infected with Germinston virus, *Virology* **157**:431.

Dunn, E. D., Pritlove, D. C., and Elliott, R. M., 1994, The S RNA genome segments of Batai, Cache Valley, Guaroa, Kairi, Lumbo, Main Drain and Northway bunyaviruses: Sequence determination and analysis, *J. Gen. Virol.* **75**:597.

Elliott, R. M., 1985, Identification of nonstructural proteins encoded by viruses of the Bunyamwera serogroup (family Bunyaviridae), *Virology* **143**:119.

Elliott, R. M., 1989a, Nucleotide sequence analysis of the small (S) RNA segment of Bunyamwera virus, the prototype of the family Bunyaviridae, *J. Gen. Virol.* **70**:1281.

Elliott, R. M., 1989b, Nucleotide sequence analysis of the large (L) genomic RNA segment of Bunyamwera virus, the prototype of the family Bunyaviridae, *Virology* **173**:426.

Elliott, R. M., 1990, Molecular biology of the Bunyaviridae, *J. Gen. Virol.* **71**:501.

Elliott, R. M., and McGregor, A., 1989, Nucleotide sequence and expression of the small (S) RNA segment of Maguari bunyavirus, *Virology* **171**:516.

Elliott, R. M., and Wilkie, M. L., 1986, Persistent infection of *Aedes albopictus* C6/36 cells by Bunyamwera virus, *Virology* **150**:21.

El Said, L. H., Vorndam, V., Gentsch, J. R., Clewley, J. P., Calisher, C. H., Klimas, R. A., Thompson, W. H., Grayson, M., Trent, D. W., and Bishop, D. H. L., 1979, A comparison of La Crosse virus isolates obtained from different ecological niches and an analysis of the structural components of California encephalitis serogroup viruses and other bunyaviruses, *Am. J. Trop. Med. Hyg.* **28**:364.

Endres, M. J., Jacoby, D. R., Janssen, R. S., Gonzalez-Scarano, F., and Nathanson, N., 1989, The large viral RNA segment of California serogroup bunyaviruses encodes the large viral protein, *J. Gen. Virol.* **70**:223.

Endres, M. J., Griot, C., Gonzalez-Scarano, F., and Nathanson, N., 1991, Neuroattenuation of an avirulent bunyavirus maps to the L RNA segment, *J. Virol.* **65**:5465.

Eshita, Y., and Bishop, D. H. L., 1984, The complete sequence of the M RNA of snowshoe hare bunyavirus reveals the presence of internal hydrophobic domains in the viral glycoproteins, *Virology* **137**:227.

Eshita, Y., Ericson, B., Romanowski, V., and Bishop, D. H. L., 1985, Analyses of the mRNA transcription processes of snowshoe hare bunyavirus S and M RNA species, *J. Virol.* **55**:681.

Fazakerley, J. K., Gonzalez-Scarano, F., Strickler, J., Dietzschold, B., Karush, F., and Nathanson, N., 1988, Organization of the middle RNA segment of snowshoe hare bunyavirus, *Virology* **167**:422.

Fuller, F., and Bishop, D. H. L., 1982, Identification of viral coded non-structural polypeptides in bunyavirus infected cells, *J. Virol.* **41**:643.

Fuller, F., Bhown, A. S., and Bishop, D. H. L., 1983, Bunyavirus nucleoprotein, N, and a nonstructural protein, NS_S, are coded by overlapping reading frames in the S RNA, *J. Gen. Virol.* **64**:1705.

Gauld, L. W., Hanson, R. P., Thompson, W. H., and Sinha, S. K., 1974, Observations on a natural cycle of La Crosse virus (California group) in southwestern Wisconsin, *Am. J. Trop. Med. Hyg.* **23**:983.

Gauld, L. W., Yuill, T. M., Hanson, R. P., and Sinha, S. K., 1975, Isolation of La Crosse virus (California encephalitis group) from the chipmunk (*Tamias striatus*), an amplifier host, *Am. J. Trop. Med. Hyg.* **24**:999.

Gentsch, J., and Bishop, D. H. L., 1976, Recombination and complementation between temperature-sensitive mutants of the bunyavirus snowshoe hare virus, *J. Virol.* **20**:351.

Gentsch, J., and Bishop, D. H. L., 1978, Small viral RNA segments of bunyaviruses codes for viral nucleocapsid protein, *J. Virol.* **28**:417.

Gentsch, J., and Bishop, D. H. L., 1979, M viral RNA segment of bunyaviruses codes for two unique glycoproteins: G1 and G2, *J. Virol.* **30**:767.

Gentsch, J., and Bishop, D. H. L., 1980, The genetic recombination potential of bunyaviruses, in: *Arboviruses in the Mediterranean Countries* (J. Vesenjak-Hirjan, ed.), pp. 81–96, Gustav Fischer Verlag, Stuttgart.

Gentsch, J., Bishop, D. H. L., and Obijeski, J. F., 1977a, The virus particle nucleic acids and proteins of four bunyaviruses, *J. Gen. Virol.* **34:**257.

Gentsch, J., Wynne, L. R., Clewley, J. P., Shope, R. E., and Bishop, D. H. L., 1977b, Formation of recombinants between snowshoe hare and La Crosse bunyaviruses, *J. Virol.* **24:**893.

Gentsch, J., Clewley, J. P., Wynne, L. R., and Bishop, D. H. L., 1978, Genetic and molecular studies of bunyaviruses, in: *Negative Strand Viruses and the Host Cell* (B. W. J. Mahy and R. D. Barry, eds.), pp. 61–71, Academic Press, New York.

Gentsch, J. R., Robeson, G., and Bishop, D. H. L., 1979, Recombination between snowshoe hare and La Crosse bunyaviruses, *J. Virol.* **31:**707.

Gentsch, J., Rozhon, E. J., Klimas, R. A., El Said, L. H., Shope, R. E., and Bishop, D. H. L., 1980, Evidence from recombinant bunyavirus studies that the M RNA gene products elicit neutralizing antibodies, *Virology* **102:**190.

Gerbaud, S., Pardigon, N., Vialat, P., and Bouloy, M., 1987a, The S segment of Germiston bunyavirus genome: Coding strategy and transcription, in: *The Biology of Negative Strand Viruses* (D. Kolakofsky and B. Mahy, eds.), pp. 191–198, Elsevier, Amsterdam.

Gerbaud, S., Vialat, P., Pardigon, N., Wychowski, C., Girard, M., and Bouloy, M., 1987b, The S segment of Germiston virus RNA genome can code for three proteins, *Virus Res.* **8:**1.

Goldman, N., Presser, I., and Sreenvalson, T., 1977, California encephalitis virus: Some biological and biochemical properties, *Virology* **76:**352.

Gonzalez-Scarano, F., 1985, La Crosse virus G1 glycoprotein undergoes a conformational change at the pH of fusion, *Virology* **140:**209.

Gonzalez-Scarano, F., Shope, R. E., Calisher, C. E., and Nathanson, N., 1982, Characterization of monoclonal antibodies against the G1 and N proteins of La Crosse and Tahyna, two California serogroup bunyaviruses, *Virology* **120:**42.

Gonzalez-Scarano, F., Pobjecky, N., and Nathanson, N., 1984, La Crosse bunyavirus can mediate pH-dependent fusion from without, *Virology* **132:**222.

Gonzalez-Scarano, F., Janssen, R. S., Najjar, J. A., Pobjecky, N., and Nathanson, N., 1985, An avirulent G1 glycoprotein variant of La Crosse bunyavirus with defective fusion function, *J. Virol.* **54:**757.

Gonzalez-Scarano, F., Beaty, B., Sundin, D., Janssen, R., Endres, M. J., and Nathanson, N., 1988, Genetic determinants of the virulence and infectivity of La Crosse virus, *Microbiol. Pathogen* **4:**1.

Grady, L. J., Sanders, M. L., and Campbell, W. P., 1983a, Evidence for three separate antigenic sites on the G_1 protein of La Crosse virus, *Virology* **126:**395.

Grady, L. J., Srihongse, S., Grayson, M. A., and Deibel, R., 1983b, Monoclonal antibodies against La Crosse virus, *J. Gen. Virol.* **64:**1699.

Grady, L. J., Sanders, M. L., and Campbell, W. P., 1987, The sequence of the M RNA of an isolate of La Crosse virus, *J. Gen. Virol.* **68:**3057.

Grimstad, P. R., Craig, G. B., Ross, Q. E., and Yuill, T. M., 1977, *Aedes triseriatus* and La Crosse virus: Geographic variation in vector susceptibility and ability to transmit, *Am. J. Trop. Med. Hyg.* **26:**990.

Grimstad, P. R., Ross, Q. E., and Craig, G. B., 1980, *Aedes triseriatus* (Diptera: Culicidae) and La Crosse virus. II. Modification of mosquito feeding behavior by virus infection, *J. Med. Entomol.* **17:**1.

Griot, C., Pekosz, A., Lukac, D., Scherer, S. S., Stillmock, K., Schmeidler, D., Endres, M. J., Gonzalez-Scarano, F., and Nathanson, N., 1993, Polygenic control of neuroinvasiveness in California serogroup bunyaviruses, *J. Virol.* **67:**3861.

Gundersen, C. B., and Brown, K. L., 1983, Clinical aspects of La Crosse encephalitis: Preliminary report, in: *California Serogroup Viruses* (C. H. Calisher and W. H. Thompson, eds.), pp. 169–177, Liss, New York.

Hewlett, M. J., Clerx, J. P. M., Clerx-van Haaster, C. M., Chandler, L. J., McLean, D. M., and Beaty, B. J., 1992, Genomic and biologic analyses of snowshoe hare virus field and laboratory strains, *Am. J. Trop. Med. Hyg.* **46:**524.

Hunt, A. R., and Calisher, C. H., 1979, Relationships of Bunyamwera serogroup viruses by neutralisation, *Am. J. Trop Med. Hyg.* **28**:740.

Ihara, T., Matsuura, Y., and Bishop, D. H. L., 1985, Analyses of the mRNA transcription processes of Punta Toro phlebovirus (Bunyaviridae), *Virology* **147**:317.

Iroegbu, C. U., and Pringle, C. R., 1981a, Genetic interactions among viruses of the Bunyamwera complex, *J. Virol.* **37**:383.

Iroegbu, C. U., and Pringle, C. R., 1981b, Genetics of the Bunyamwera complex, in: *The Replication of Negative Strand Viruses* (D. H. L. Bishop and R. W. Compans, eds.), pp. 159–165, Elsevier, Amsterdam.

James, W. S., and Millican, D., 1986, Host-adaptive antigenic variation in bunyaviruses, *J. Gen. Virol.* **67**:2803.

Janssen, R. S., Nathanson, N., Endres, M. J., and Gonzalez-Scarano, F., 1986, Virulence of La Crosse virus is under polygenic control, *J. Virol.* **59**:1.

Jin, H., and Elliott, R. M., 1991, Expression of functional Bunyamwera virus L protein by recombinant vaccinia viruses, *J. Virol.* **65**:4182.

Jin, H., and Elliott, R. M., 1992, Mutagenesis of the L protein encoded by Bunyamwera virus and production of monoclonal antibodies, *J. Gen. Virol.* **73**:2235.

Jin, H., and Elliott, R. M., 1993, Characterization of Bunyamwera virus S RNA that is transcribed and replicated by the L protein expressed from recombinant vaccinia virus, *J. Virol.* **67**:1396.

Karabatsos, N., (ed.), 1985, *International Catalogue of Arboviruses including certain other viruses of vertebrates*, The American Society of Tropical Medicine and Hygiene, San Antonio.

Kingsford, L., and Hill, D. W., 1983, The effect of proteolytic cleavage of La Crosse virus G1 glycoprotein on antibody neutralization, *J. Gen. Virol.* **64**:2147.

Klimas, R. A., Thompson, W. H., Calisher, C. H., Clark, G. G., Grimstad, P. R., and Bishop, D. H. L., 1981a, Genotypic varieties of La Crosse virus isolated from different geographic regions of the continental United States and evidence for a naturally occurring intertypic recombinant La Cross virus, *Am. J. Epidemiol.* **114**:112.

Klimas, R. A., Ushijima, H., Clerx-van Haaster, C. M., and Bishop, D. H. L., 1981b, Radioimmune assays and molecular studies that place Anopheles B and Turlock serogroup viruses in the *Bunyavirus* genus (Bunyaviridae), *Am. J. Trop. Med. Hyg.* **30**:876.

Kuismanen, E., Hedman, K., Saraste, J., and Pettersson, R. F., 1982, Uukuniemi virus maturation, accumulation of virus particles and viral antigens in the Golgi complex, *Mol. Cell. Biol.* **2**:1444.

Lappin, D. F., Nakitare, G. W., Palfreyman, J. W., and Elliott, R. M., 1994, Localisation of Bunyamwera bunyavirus G1 glycoprotein to the Golgi requires association with G2 but not NSm, *J. Gen. Virol.* **75**:3441–3451.

Lees, J. F., Pringle, C. R., and Elliott, R. M., 1986, Nucleotide sequence of the Bunyamwera virus M RNA segment: Conservation of structural features in the bunyavirus glycoprotein gene product, *Virology* **148**:1.

Lenard, J., and Miller, D. K., 1982, Uncoating of enveloped viruses, *Cell* **28**:5.

Lisitza, M. A., DeFoliart, G. R., Yuill, T. M., and Karandinos, M. G., 1977, Prevalence rates of La Crosse virus (California encephalitis group) in larvae from overwintered eggs of *Aedes triseriatus*, *Mosq. News* **37**:745.

Ludwig, G. V., Christensen, B. M., Yuill, T. M., and Schultz, K. T., 1989, Enzyme processing of La Crosse virus glycoprotein G1: A bunyavirus–vector infection model, *Virology* **171**:108.

Lyons, M. J., and Heyduk, J., 1973, Aspects of the developmental morphology of California encephalitis virus in cultured vertebrate and arthropod cells and in mouse brain, *Virology* **54**:37.

Madoff, D. H., and Lenard, J., 1982, A membrane glycoprotein that accumulates intracellularly: Cellular processing of the large glycoprotein of La Crosse virus, *Cell* **28**:821.

Martin, M. L., Lindsey-Regnery, H., Sasso, D. R., McCormick, J. B., and Palmer, E., 1985, Distinction between Bunyaviridae genera by surface structure and comparison with Hantaan virus using negative stain electron microscopy, *Arch. Virol.* **86**:17.

Miller, B. R., DeFoliart, G. R., and Yuill, T. M., 1977, Vertical transmission of La Crosse virus (California encephalitis group). Transovarial and filial infection rates in *Aedes triseriatus* (Diptera: Culicidae), *J. Med. Entomol.* **14**:437.

Miller, B. R., DeFoliart, G. R., and Yuill, T. M., 1979, *Aedes triseriatus* and La Crosse virus: Lack of infection in eggs of the first ovarian cycle following oral infection of females, *Am. J. Trop. Med. Hyg.* **28**:897.

Murphy, F. A., Whitfield, S. G., Coleman, P. H., Calisher, C. H., Rabin, E. R., Jenson, A. B., Melnick, J. L., Edwards, M. R., and Whitney, E., 1968, California group arboviruses: Electron microscopic studies, *Exp. Mol. Pathol.* **9**:44.

Murphy, F. A., Harrison, A. K., and Whitfield, S. G., 1973, Morphologic and morphogenetic similarities of Bunyamwera serological supergroup viruses and several other arthropod-borne viruses, *Intervirology* **1**:297.

Najjar, J. A., Gentsch, J. R., Nathanson, N., and Gonzalez-Scarano, F., 1985, Epitopes of the G1 glycoprotein of La Crosse virus form overlapping clusters within a single antigenic site, *Virology* **144**:426.

Nakitare, G. W., and Elliott, R. M., 1993, Expression of the Bunyamwera virus M genome segment and intracellular localization of NS_M, *Virology* **195**:511.

Nuttall, P. A., Jones, L. D., and Davies, C. R., 1991, The role of arthropod vectors in arbovirus evolution, in: *Advances in Disease Vector Research*, Vol. 8 (K. F. Harris, ed.), pp. 15–45, Springer-Verlag, Berlin.

Nuttall, P. A., Jones, L. D., Labuda, M., and Kaufman, W. R., 1994, Adaptations of arboviruses to ticks, *J. Med. Entomol.* **31**:1.

Obijeski, J. F., and Murphy, F. A., 1977, Bunyaviridae: Recent biochemical developments, *J. Gen. Virol.* **37**:1.

Obijeski, J. F., Bishop, D. H. L., Murphy, F. A., and Palmer, E. L., 1976a, Structural proteins of La Crosse virus, *J. Virol.* **19**:985.

Obijeski, J. F., Bishop, D. H. L., Palmer, E. L., and Murphy, F. A., 1976b, Segmented genome and nucleocapsid of La Crosse virus, *J. Virol.* **20**:664.

Obijeski, J. F., McCauley, J., and Skehel, J. J., 1980, Nucleotide sequences at the termini of La Crosse virus RNAs, *Nucleic Acids Res.* **8**:2431.

Pantuwatana, S., Thompson, W. H., Watts, D. M., and Hanson, R. P., 1972, Experimental infection of chipmunks and squirrels with La Crosse and trivittatus viruses and biological transmission of La Crosse virus by *Aedes triseriatus*, *Am. J. Trop. Med. Hyg.* **21**:476.

Pantuwatana, S., Thompson, W. H., Watts, D. M., Yuill, T. M., and Hanson, R. P., 1974, Isolation of La Crosse virus from field collected *Aedes triseriatus* larvae, *Am. J. Trop. Med. Hyg.* **23**:246.

Pardigon, N., Vialat, P., Gerbaud, S., Girard, M., and Bouloy, M., 1988, Nucleotide sequence of the M segment of Germiston virus: Comparison of the M gene product of several bunyaviruses, *Virus Res.* **11**:73.

Parker, M. D., and Hewlett, M. J., 1981, The 3'-terminal sequences of Uukuniemi and Inkoo virus RNA genome segments, in: *The Replication of Negative Strand Viruses* (D. H. L. Bishop and R. W. Compans, eds.), pp. 125–145, Elsevier, Amsterdam.

Patterson, J. L., and Kolakofsy, D., 1984, Characterization of La Crosse virus small-genome segment transcripts, *J. Virol.* **49**:680.

Patterson, J. L., Kolakofsy, D., Holloway, B. P., and Obijeski, J. F., 1983, Isolation of the ends of La Crosse virus small RNA as a double-stranded structure, *J. Virol.* **45**:882.

Patterson, J. L., Holloway, B., and Kolakofsy, D., 1984, La Crosse virions contain a primer-stimulated RNA polymerase and a methylated cap-dependent endonuclease, *J. Virol.* **52**:215.

Pattnaik, A. K., and Abraham, G., 1983, Identification of four complementary RNA species in Akabane virus-infected cells, *J. Virol.* **47**:452.

Pennington, T. H., and Pringle, C. R., 1978, Negative strand viruses in enucleate cells, in: *Negative Strand Viruses and the Host Cell* (W. J. Mahy and R. D. Barry, eds.), pp. 457–464, Academic Press, New York.

Pennington, T. H., Pringle, C. R., and McCrae, M. A., 1977, Bunyamwera virus-induced polypeptide synthesis, *J. Virol.* **24:**397.

Pettersson, R. F., Kuismanen, E., Rönnholm, R., and Ulmanen, I., 1985, In: *Viral Messenger RNA: Transcription, Processing, Splicing and Molecular Structure* (Y. Becker, ed.), pp. 283–300, Nijhoff, The Hague.

Pringle, C. R., Clark, W., Lees, J. F., and Elliott, R. M., 1984a, Restriction of sub-unit reassortment in the *Bunyaviridae,* in: *Segmented Negative Strand Viruses* (R. W. Compans and D. H. L. Bishop, eds.), pp. 45–50, Academic Press, New York.

Pringle, C. R., Lees, J. F., Clark, W., and Elliott, R. M., 1984b, Genome subunit reassortment among bunyaviruses analysed by dot hybridization using molecularly cloned complementary DNA probes, *Virology* **135:**244.

Raju, R., and Kolakofsky, D., 1986a, Inhibitors of protein synthesis inhibit both La Crosse virus S-mRNA and S genome syntheses *in vivo, Virus Res.* **5:**1.

Raju, R., and Kolakofsky, D., 1986b, Translational requirement of La Crosse virus S-mRNA synthesis: *In vivo* studies, *J. Virol.* **61:**96.

Raju, R., and Kolakofsky, D., 1989, The ends of La Crosse virus genome and antigenome RNAs within nucleocapsids are base paired, *J. Virol.* **63:**122.

Raju, R., Raju, L., and Kolakofsky, D., 1989, The translation requirement for complete La Crosse virus mRNA synthesis is cell-type dependent, *J. Virol.* **63:**5159.

Rossier, C., Patterson, J., and Kolakofsky, D., 1986, La Crosse virus small genome mRNA is made in the cytoplasm, *J. Virol.* **58:**647.

Rossier, C., Raju, R., and Kolakofsky, D., 1988, La Crosse virus gene expression in mammalian and mosquito cells, *Virology* **165:**539.

Rozhon, E. J., Gensemer, P., Shope, R. E., and Bishop, D. H. L., 1981, Attenuation of virulence of a bunyavirus involving an L RNA defect and isolation of LAC/SSH/LAC and LAC/SSH/SSH reassortants, *Virology* **111:**125.

Samso, A., Bouloy, M., and Hannoun, C., 1975, Présence de ribonucléoprotéines circulaires dans le virus Lumbo (Bunyavirus), *C. R. Acad. Sci. Ser. D* **280:**779.

Samso, A., Bouloy, M., and Hannoun, C., 1976, Mise en évidence de molécules d'acide ribonucléique circulaire dans le virus Lumbo (Bunyavirus), *C. R. Acad. Sci. Ser. D* **282:**1653.

Scallan, M. F., and Elliott, R. M., 1992, Defective RNAs in mosquito cells persistently infected with Bunyamwera virus, *J. Gen. Virol.* **73:**53.

Schmaljohn, C. S., and Dalrymple, J. M., 1983, Analysis of Hantaan virus RNA: Evidence for a new genus of Bunyaviridae, *Virology* **131:**482.

Schopen, S., Labuda, M., and Beaty, B. J., 1991, Vertical and venereal transmission of California group viruses by *Aedes triseriatus* and *Culiseta inornata* mosquitoes, *Acta Virol.* **35:**373.

Seymour, C., Amundson, T. E., Yuill, T. M., and Bishop, D. H. L., 1983, Experimental infection of chipmunks and snowshoe hares with La Crosse and snowshoe hare viruses and four of their reassortants, *Am. J. Trop. Med. Hyg.* **32:**1147.

Shope, R. E., Rozhon, E. J., and Bishop, D. H. L., 1981a, Role of the middle-sized bunyavirus RNA segment in mouse virulence, *Virology* **114:**273.

Shope, R. E., Tignor, G. H., Rozhon, E. J., and Bishop, D. H. L., 1981b, The association of the bunyavirus middle-sized RNA segment with mouse pathogenicity, in: *Replication of Negative Strand Viruses* (D. H. L. Bishop and R. W. Compans, eds.), pp. 146–152, Elsevier, Amsterdam.

Shope, R. E., Tignor, G. H., Jacoby, R. O., Watson, H., Rozhon, E. J., and Bishop, D. H. L., 1982, Pathogenicity analyses of reassortant bunyaviruses: Coding assignments, in: *International Symposium on Tropical Arboviruses and Hemorrhagic Fever* (F. de P. Pinheiro, ed.), pp. 135–146, Impresso Nat. Fund. Sci. Dev. Tech., Belem, Brazil.

Smith, J. F., and Pifat, D. Y., 1982, Morphogenesis of sandfly viruses (Bunyaviridae family), *Virology* **121:**61.

Sundin, D. R., Beaty, B. J., Nathanson, N., and Gonzalez-Scarano, F., 1987, A G1 glycoprotein epitope of La Crosse virus: A determinant of infection of *Aedes triseriatus, Science* **235:**591.

Talmon, Y., Prasad, B. V. V., Clerx, J. P. M., Wang, G.-J., Wah, C., and Hewlett, M., 1987, Electron microscopy of vitrified-hydrated La Crosse virus, *J. Virol.* **61:**2319.

Thompson, W. H., 1979, Higher venereal infection and transmission rates with La Crosse virus in *Aedes triseriatus* engorged before mating, *Am. J. Trop. Med. Hyg.* **28:**890.

Thompson, W. H., and Beaty, B. J., 1977, Venereal transmission of La Crosse (California encephalitis) arbovirus in *Aedes triseriatus* mosquitoes, *Science* **196:**530.

Thompson, W. H., and Beaty, B. J., 1978, Venereal transmission of La Crosse virus from male to female *Aedes triseriatus*, *Am. J. Trop. Med. Hyg.* **27:**187.

Thompson, W. H., and Evans, A. S., 1965, California encephalitis virus studies in Wisconsin, *Am. J. Epidemiol.* **81:**230.

Thompson, W. H., Kalfayan, B., and Anslow, R. O., 1965, Isolation of California encephalitis group virus from a fatal human illness, *Am. J. Epidemiol.* **81:**245.

Thompson, W. H., Engeseth, D. J., Jackson, J. O., DeFoliart, G. R., and Hanson, R. P., 1967, California encephalitis-group virus infections in small mammals and insects in a deciduous forest area of southwestern Wisconsin, *Bull. Wildl. Dis. Assoc.* **3:**92.

Thompson, W. H., Anslow, R. O., Hanson, R. P., and DeFoliart, G. R., 1972, La Crosse virus isolations from mosquitoes in Wisconsin, 1964–68, *Am. J. Trop. Med. Hyg.* **21:**90.

Tignor, G. H., Burrage, T. C., Smith, A. L., and Shope, R. E., 1983, California serogroup gene structure–function relationships: Virulence and tissue tropisms, in: *California Serogroup Viruses* (C. H. Calisher and W. H. Thompson, eds.), pp. 129–138, Liss, New York.

Urquidi, V., and Bishop, D. H. L., 1992, Non-random reassortment between the tripartite RNA genomes of La Crosse and snowshoe hare viruses, *J. Gen. Virol.* **73:**2255.

Ushijima, H., Klimas, R., Kim, S., Cash, P., and Bishop, D. H. L., 1980, Characterization of the viral ribonucleic acids and structural polypeptides of Anopheles A, Bunyamwera, Group C, California, Capim, Guama, Patois and Simbu bunyaviruses, *Am. J. Trop. Med. Hyg.* **29:**1441.

Ushijima, H., Clerx-van Haaster, C. M., and Bishop, D. H. L., 1981, Analyses of Patois group bunyaviruses: Evidence for naturally occurring recombinant bunyaviruses and existence of immune precipitable and non-precipitable non-virion proteins induced in Bunyavirus-infected cells, *Virology* **110:**318.

Vezza, A. C., Repik, P. M., Cash, P., and Bishop, D. H. L., 1979, *In vivo* transcription and protein synthesis capabilities of bunyaviruses: Wild-type snowshoe hare virus and its temperature-sensitive group I, group II and group I/II mutants, *J. Virol.* **31:**426.

von Bonsdorff, C.-H., Saikku, P., and Oker-Blom, N., 1969, The inner structure of Uukuniemi and two Bunyamwera supergroup arboviruses, *Virology* **39:**342.

Vorndam, A. V., and Trent, D. W., 1979, Oligosaccharides of the California encephalitis viruses, *Virology* **95:**1.

Watret, G. E., Pringle, C. R., and Elliott, R. M., 1985, Synthesis of bunyavirus-specific proteins in a continuous cell line (XTC-2) derived from *Xenopus laevis*, *J. Gen. Virol.* **66:**473.

Watts, D. M., Morris, C. D., Wright, R. E., DeFoliart, G. R., and Hanson, R. P., 1972, Transmission of La Crosse virus (California encephalitis group) by the mosquito *Aedes triseriatus*, *J. Med. Entomol.* **9:**125.

Watts, D. M., Grimstad, P. R., DeFoliart, G. R., Yuill, T. M., and Hanson, R. P., 1973a, Laboratory transmission of La Crosse encephalitis virus by several species of mosquitoes, *J. Med. Entomol.* **10:**583.

Watts, D. M., Pantuwatana, S., DeFoliart, G. R., Yuill, T. M., and Thompson, W. H., 1973b, Transovarial transmission of La Crosse virus (California encephalitis group) in the mosquito, *Aedes triseriatus*, *Science* **182:**1140.

Watts, D. M., Thompson, W. H., Yuill, T. M., DeFoliart, G. R., and Hanson, R. P., 1974, Overwintering of La Crosse virus in *Aedes triseriatus*, *Am. J. Trop. Med. Hyg.* **23:**694.

Whitman, L., and Shope, R. E., 1962, The California complex of arthropod-borne viruses and its relationship to the Bunyamwera serogroup through Guaroa virus, *Am. J. Trop. Med. Hyg.* **11:**691.

Wright, R. E., Anslow, R. W., Thompson, W. H., DeFoliart, G. R., Seawright, G., and Hanson, R. P., 1970, Isolations of La Crosse virus of the California group from Tabanidae in Wisconsin, *Mosq. News* **30:**600.

CHAPTER 3

Molecular Biology of Hantaviruses

CONNIE S. SCHMALJOHN

I. INTRODUCTION

The genus *Hantavirus* includes several of the most significant human pathogens in the family *Bunyaviridae*. Certain hantaviruses are known to cause such deadly illnesses as hemorrhagic fever with renal syndrome (HFRS) and hantavirus pulmonary syndrome (HPS). Unlike viruses in other genera, hantaviruses are not transmitted by arthropods, but instead, persistently infect rodents and are transmitted in infectious aerosols of the animals' excreta. A wide array of antigenically and genetically distinct hantaviruses have been detected in numerous rodent species throughout the world (Lee and van der Groen, 1989; Chu *et al.*, 1994; Liang *et al.*, 1994; Xiao *et al.*, 1994). Among the hantaviruses described to date at least nine antigenically and genetically distinct viruses have been propogated in cell culture. HTN, Seoul (SEO), Puumala (PUU), and Dobrava (DOB) viruses have all been incriminated in HFRS, and Sin Nombre (SN)* and Black Creek Canal (BCC) viruses are linked to HPS. Prospect Hill (PH), Thailand (THAI), and Thottapalayam

---

*Provisional nomenclature. Convict Creek 107 (CC107), was obtained by direct inoculation of cell cultures with infectious materials from *Peromyscus maniculatus* trapped in California (Schmaljohn, *et al.*, 1994) #2926]. Shortly thereafter, Four Corners virus, briefly called Muerto Canyon virus, and finally renamed Sin Nombre virus, was obtained from *P. maniculatus* trapped in New Mexico by passage through laboratory kept *Peromyscus* prior to cell culture isolation (Elliott, *et al.*, 1994).

CONNIE S. SCHMALJOHN • Virology Division, United States Army Medical Research Institute of Infectious Diseases, Fort Detrick, Frederick, Maryland 21702.

The Bunyaviridae, edited by Richard M. Elliott, Plenum Press, New York, 1996.

(TPM) viruses are not known to cause human disease. Interestingly, serologically distinct viruses are usually isolated from different rodent species. Thus, most viruses closely related to HTN virus have been isolated from *Apodemus agrarius* (striped field mice) and those most similar to SEO, PUU, DOB, and PH viruses from *Rattus norvegicus* (domestic rats), *Clethrionomys glareolus* (bank voles), *Apodemus flavicollis* (yellow-necked mouse), and *Microtus pennsylvanicus* (meadow voles), respectively. Two hantavirus isolates from Thailand, from *Bandicota* (bandicoot) or *Rattus* (Elwell *et al.*, 1985), can be distinguished from SEO virus by PRNT (Chu *et al.*, 1994). An Indian isolate from *Suncus* (tree shrew), TPM virus (Carey *et al.*, 1971), the most unique hantavirus isolated to date, does not cross-react with other hantaviruses by PRNT (Chu *et al.*, 1994), and has a clearly different genetic lineage (Xiao *et al.*, 1994). The primary rodent host of SN virus is *Peromyscus maniculatus* (deer mouse) and that of BCC virus, *Sigmodon hispidus* (cotton rat) (Chapman and Khabbaz, 1994; Childs *et al.*, 1994; Nerurkar *et al.*, 1994; Ravkov *et al.*, 1995; Turell *et al.*, 1995). A number of other hantaviruses that appear genetically distinct, but have not yet been isolated in cell culture include El Moro Canyon (ELMC), Rio Segundo (RIOS), and New York-1 (NY-1)[*] viruses in the U.S. (Hjelle *et al.*, 1994; Song *et al.*, 1994; Hjelle *et al.*, 1995), and Tula (TUL) virus in Europe (Plyusnin *et al.*, 1994).

Hantaviruses are distributed throughout the world and have been detected in various rodent species in Asia, Europe, Africa, North and South America, Scandinavia, and Australia (Lee *et al.*, 1990). Although the factors contributing to the distinctive ecological and pathogenic characteristics of hantaviruses have not yet been defined at the molecular level, much has been learned about the structure and function of viral genes and gene products. The purpose of this chapter is to review information that contributes toward our understanding of the molecular nature of hantaviruses, and which may provide clues for preventing hantaviral diseases.

II. VIRION STRUCTURE

A. Morphology

Hantaviruses generally appear, by electron microscopy, to be spherical, with an average diameter of approximately 100 nm; however, pleomorphic forms ranging from 78 to 210 nm have been reported (McCormick *et al.*, 1982; White *et al.*, 1982; Hung *et al.*, 1985; Martin *et al.*, 1985). Negatively stained particles display a surface with a uniquely square, gridlike structure (Fig. 1A). Virions have a distinct bilayered membrane surrounding a granulofilamentous interior consisting of ribonucleocapsids. Surface projections, approximately 6 nm long, are composed of two virus-specified glycoproteins, designated G1 and G2 (Fig. 1B). A large (L) protein, the viral transcriptase, is associ-

---

[*]This virus was originally called Shelter Island RI-1 [Song, 1994 #3011] but was later renamed.

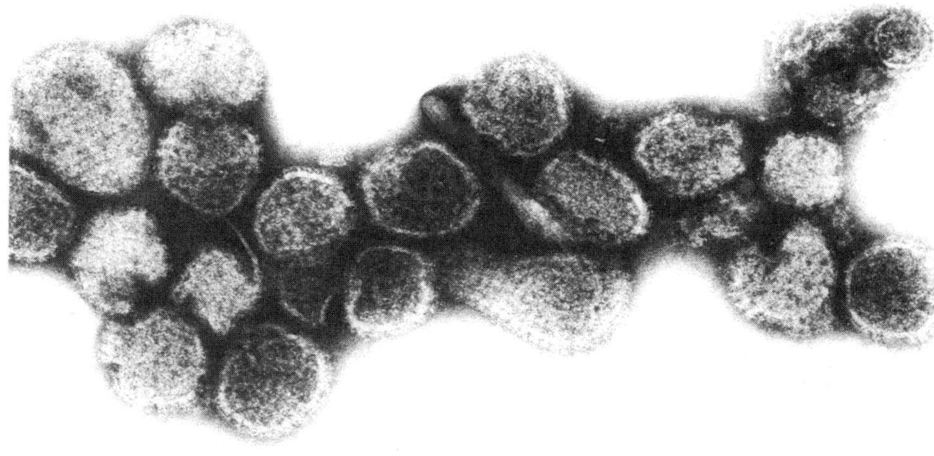

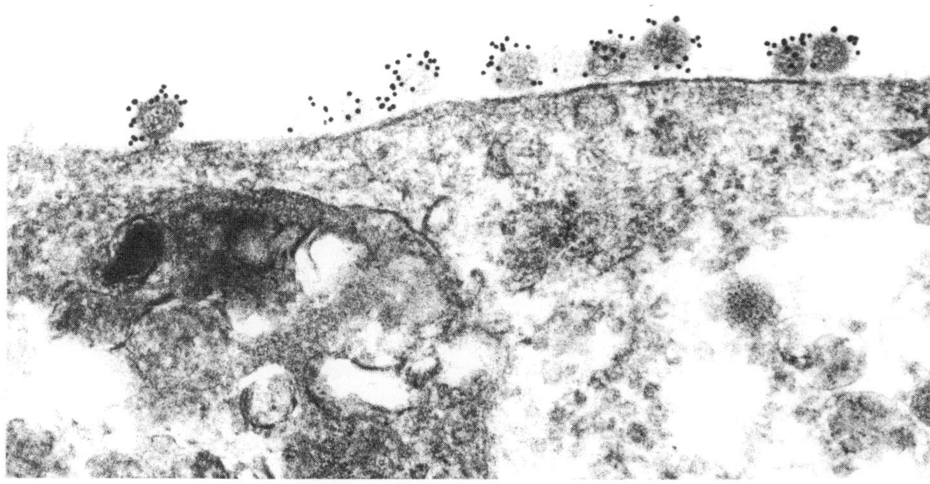

FIGURE 1. Morphology of Hantaan virus. (A) Negatively stained virion particles. (B) Virion particles on the surface of Vero E6 cells decorated with gold-labeled G2-specific monoclonal antibodies to Hantaan virus. (Photographs courtesy of Kathy Kuehl and John White.)

ated with virions (Schmaljohn and Dalrymple, 1983), but its precise location is unknown.

B. Physical Characteristics of Virions

The physical properties of hantaviruses are similar to those of other viruses in the family (Murphy *et al.*, 1995). Prototype HTN virus (strain 76-118) sediments to a density of 1.16 to 1.17 in sucrose or 1.20 to 1.21 in cesium chloride. Treating virions with non-ionic detergents releases three ribonucleocapsids, which can then be separated by rate-zonal centrifugation, sedimenting to a density of 1.18 in sucrose, or 1.25 in cesium chloride (Schmaljohn *et al.*, 1983).

For transmission, hantaviruses must remain stable in rodent urine, feces, and saliva, and in dust aerosols. Laboratory studies revealed HTN virus to remain viable for at least 30 min in buffers from pH 6.6 to 8.8, and, in the presence of 10% fetal bovine serum, from pH 5.8 to 9.0 (J. Huggins, unpublished information). Urine pH in rodents can vary from 5.5 to 8.0 and may contain up to 30 mg protein (Kaplan *et al.*, 1982), conditions well within the tolerance limits of HTN virus. Hantaan virus is also infectious after a 30 min incubation, either with or without added serum, at temperatures from 4°C to 42°C. Dried HTN virus is infectious for 1 to 3 days when 10% serum is present. These studies indicate that HTN virus could remain viable as an infectious aerosol for several days after excretion by rodents into the environment.

III. CODING STRATEGIES AND GENETIC PROPERTIES

Among the coding strategies described for the L, M, and S segments of viruses in the five genera of the family *Bunyaviridae*, those of the hantaviruses apparently are the simplest. All three genome segments encode only structural proteins in the virus complementary-sense RNA (cRNA). As described for all other viruses in the family, each of the three hantaviral genome segments has a consensus 3'-terminal nucleotide sequence, which is complementary to 5'-terminal sequence and is distinct from those of viruses in other genera (Fig. 2) (Schmaljohn *et al.*, 1985; Schmaljohn *et al.*, 1986b; Schmaljohn *et al.*, 1987; Schmaljohn, 1990). These complementary sequences may allow the viral ribonucleocapsids to form noncovalently closed circular structures, and although not proven, may serve a role in transcription or other aspects of replication (Raju and Kolakofsky, 1989).

A. S Segment Coding Strategy

Comparison of the nucleotide sequences of cDNA representing the S genome segments of a number of hantaviruses revealed a single long open

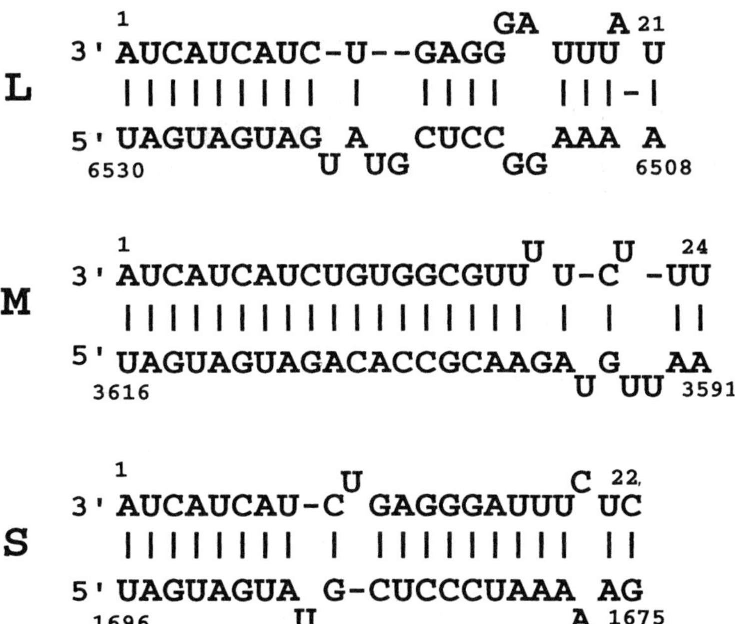

FIGURE 2. Terminal nucleotide sequences of the L, M, and S genome segments of Hantaan virus. Numbers indicate nucleotide sequence with respect to the 3′ terminus of virus-sense RNA.

reading frame (ORF) in cRNA (Table I, Fig. 3). The HTN virus S segment ORF was first demonstrated to code for the nucleocapsid protein (N) by immune precipitation of cell-free translation products from rabbit reticulocyte lysates programmed with RNA transcribed from S segment cDNA (Schmaljohn et al., 1986b). A small ORF, which could code for a 6 kD polypeptide, is present in the same reading frame, immediately following the termination codon of HTN N (Fig. 3); but a protein of this size was not detected in HTN virus-infected cells (Schmaljohn et al., 1986b). Moreover, this ORF is not conserved in the S genome segments of other hantaviruses such as SEO (Arikawa et al., 1990), PUU (Stohwasser et al., 1990; Xiao et al., 1993), PH (Parrington and Kang, 1990), or SN (Spiropoulou et al., 1994) viruses. PUU, PH, and SN viruses all have a small ORF, overlapping that encoding N, similar to the ORF that generates the nonstructural protein (NBs) of bunyaviruses (Fig. 3). These small hantaviral S segment ORFs potentially encode from 63 (SN virus) to 90 (PUU and PH viruses) amino acids. Although analysis of third-position base changes in these predicted ORFs suggested a high degree of conservation (Spiropoulou et al., 1994), no polypeptides corresponding to the predicted sizes have yet been detected in infected cell cultures (Parrington and Kang, 1990; Stohwasser et al., 1990; Spiropoulou et al., 1994). It remains to be determined if these ORFs have significance; nevertheless, it is clear that synthesis of an NSs protein is not a generic property of

TABLE I. Coding Capacity of Hantavirus Genome Segments[a]

| Segment virus | Total | Noncoding[b] | | Number of amino acids | M_r | GB/EMBL Accession number |
		5′	3′			
L						
HTN (76-118)	6533	37	40	2151	246,500	X55901
SEO (80–39)	6530	36	38	2151	246,660	X56492
PUU (Sotkamo)	6550	36	43	2156	246,000	M63194
M						
HTN (76-118)	3616	40	168	1136	126,300	M14627
DOB (3970/87)	3644	40	199	1134	125,500	L33685
SEO (80-39)	3651	46	203	1133	125,500	X56493
THAI (749)	3613	46	168	1132	125,700	L08756
PUU (Sotkamo)	3682	40	195	1148	126,800	X61034
PH (PHV-1)	3707	49	229	1142	125,800	X55129
SN (CC107)	3696	51	222	1140	125,700	L08804
BCC	3668	51	191	1141	125,500	L39950
S						
HTN (76-118)	1696	36	370	429	48,100	M14626
SEO (SR-11)	1769	42	437	429	48,100	M34881
PUU (Sotkamo)	1830	42	486	433	49,400	X61035
PH (MP-20)	1675	42	331	433	49,000	X55128
SN (CC107)	2083	42	730	428	48,200	L33683
BCC	1989	42	660	429	47,800	L39949

[a]Data listed are for representative rodent virus isolates that have been molecularly cloned and sequenced.
[b]Noncoding sequences at the 5′ and 3′ termini are listed with respect to virus-complementary sense RNA.

hantaviruses. As a result, as much as one-third of the 3′-terminal sequences in the cRNA are noncoding (Table I). Because the function of NSs has not yet been determined for any virus in the family, it is not known if hantaviruses replicate in the absence of a function that may be provided by NSs, or perhaps may perform that function with their much larger N protein or a different polypeptide.

B. M Segment Coding Strategy

Like the S segments of hantaviruses, the M genome segments do not appear to encode nonstructural proteins. Only one significant ORF, with a coding capacity of approximately 126 kDa, was detected in the cRNA of the M segments of HTN, SEO, PUU, PH, THAI, and SN viruses (Schmaljohn et al., 1987; Yoo and Kang, 1987; Giebel et al., 1989; Arikawa et al., 1990; Antic et al., 1991a; Parrington et al., 1991; Spiropoulou et al., 1994; Xiao et al., 1994) (Table I). Amino-terminal sequence analysis of the G1 and G2 envelope glycoproteins of HTN (76-118) and SEO (SR-11) viruses conclusively demonstrated that the hantaviral M segment codes for the envelope glycoproteins

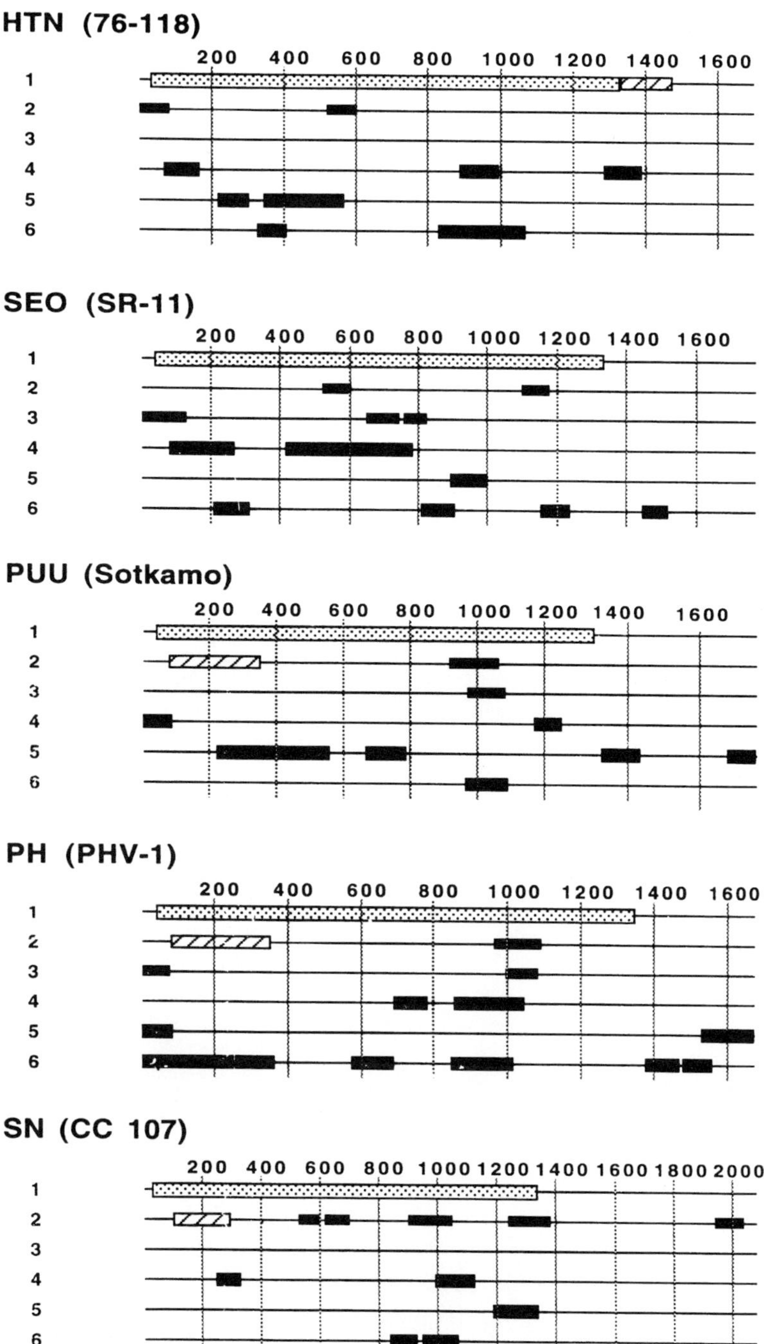

FIGURE 3. Potential open reading frames in the S genome segments of hantaviruses. Numbers listed to the left indicate the six potential reading frames. Reading frames one to three are virus complementary-sense RNA and frames four to six are virus-sense RNA. Stippled bars indicate the nucleocapsid protein ORF. Striped bars represent reading frames suggested to encode non-structural proteins (see text). Solid bars indicate other potential reading frames.

and has a gene order of 5′ G1–G2 3′ with respect to virus-complementary sense RNA (Schmaljohn *et al.*, 1987; Arikawa *et al.*, 1990). Synthetic peptides representing amino acids encoded by the exact 3′ terminus of the HTN M segment ORF elicited antibodies in rabbits that recognized HTN G2. Therefore, the carboxy terminus of G2 extends to the end of the coding region of the M segment. Similarly, synthetic peptides representing the last predicted hydrophilic region before the amino terminus of G2 (amino acids 588–614 of the ORF) induced antibodies in rabbits that recognized HTN virus G1, indicating that the carboxy terminus of G1 extends at least to amino acid residue 588. Consequently, if an intergenic region is present, it could not code for more than 6 kDa of protein (Schmaljohn *et al.*, 1987). Only a small amount of noncoding information is present at either the 3′ or 5′ termini of the hantavirus M genome segments (Table I).

Although not proven, the expression strategy of the hantaviral M segment is believed to be similar to that of other viruses in the family, in that a polyprotein precursor is cleaved by cotranslational cleavage. This is supported by the inability to observe a polyprotein of the size expected for a G1–G2 precursor, suggesting that posttranslational cleavage either occurs very rapidly or not at all. Because the sequence encoding both the G1 and G2 proteins of HTN virus are preceded by typical signal sequences and translation initiation codons, it is possible that an alternative means of expression is the independent initiations of translation for the first (G1) and second (G2) genes. Gene expression studies with vaccinia virus recombinants (Schmaljohn *et al.*, 1990; Pensiero and Hay, 1992), or baculovirus recombinants (Schmaljohn *et al.*, 1990), indicated that both proteins could be independently expressed from their natural translation initiation codon. However, more detailed studies revealed that G2 is expressed efficiently even if the ATG preceding its signal sequence is removed; therefore, independent gene expression is not required to generate the HTN virus envelope glycoproteins (Kamrud and Schmaljohn, 1994).

C. L Segment Coding Strategy

The complete nucleotide sequences of three hantaviral L segments are known: HTN, SEO, and PUU viruses. A single ORF, in the cRNA of L, encodes a polypeptide of approximately 246 kDa (Schmaljohn, 1990; Antic *et al.*, 1991b; Stohwasser, *et al.*, 1991). Almost all of the cRNA of the hantaviral L segment encodes the L protein (Table I), and there is no evidence that other, minor ORFs encode additional polypeptides.

D. Genetic Evolution

Genetic diversity among hantaviruses was examined by comparing 330 M segment nucleotides encoding the amino terminus of G2 (Xiao *et al.*,

1994). This region could be amplified with a single oligonucleotide primer pair, by reverse transcriptase-polymerase chain reaction (RT-PCR), for most hantaviruses tested (Xiao *et al.*, 1992). A phylogenetic tree derived for 13 hantaviruses by pairwise comparison of this region is analogous to a tree obtained by comparing the entire G1 and G2 amino acid sequence of the same viruses; therefore, this small gene region can serve as a practical surrogate for the M segment in assessing the genetic relationships among hantaviruses. A phylogenetic tree of 32 hantaviruses, obtained by comparison of this gene region, has seven distinct genetic clusters represented by HTN, DOB, SEO, THAI, PUU, PH, and SN viruses (Fig. 4). Viruses in the HTN, DOB, SEO, and THAI clusters share one genetic lineage, while PUU, PH, SN, and other new world hantaviruses share a different lineage. These results correlate with the antigenic groupings of these same viruses (Chu *et al.*, 1994). Because the primers would not amplify TPM RNA in RT-PCR, TPM virus could not be compared to the other viruses with this method. However, comparing a small portion of TPM virus S segment to those of other hantaviruses demonstrated that TPM virus is genetically unique among hantaviruses (Xiao *et al.*, 1994).

Nucleotide sequences of the M and/or S segments of a number of hantaviruses have been determined. Comparison of the M segments of any two viruses and the S segments of the same viruses results in approximately the same percentage homology, suggesting that the M and S segments have evolved at about the same rate (Table II). This observation differs from findings with viruses such as influenza virus, which have higher rates of mutation in genes encoding surface proteins than in genes encoding internal proteins (Smith and Palese, 1989). Although far fewer data are available, the L segments of hantaviruses appear to have slightly higher nucleotide sequence homologies than the M or S segments, perhaps indicating a greater need for conservation of structure and/or function of the polymerase protein than for the nucleocapsid (N) or envelope (G1 and G2) proteins.

The accumulation of point mutations can eventually lead to significant genetic diversity among viruses (genetic drift), and genome segment reassortment can rapidly alter their genetic properties (genetic shift). The most striking example of such changes are seen with the influenza viruses, which routinely undergo both drift and shift changes that alter their genetic and antigenic structure. Although reassortment has been demonstrated experimentally for certain closely related viruses in the family *Bunyaviridae* (Bishop *et al.*, 1980; Pringle *et al.*, 1984a; Pringle *et al.*, 1984b; Beaty *et al.*, 1985; Saluzzo and Smith, 1990; Chandler *et al.*, 1991; Urquidi and Bishop, 1992; Chapter 8, this volume), such studies with hantaviruses have not been reported. Evidence for reassortment among SN hantaviruses in nature was obtained by analyzing two isolates from *Peromyscus* mice trapped within 5 miles of each other in eastern California. These viruses, CC107 and CC74, had M and S segment homologies of 99% and 87%, respectively, suggesting that reassortment of gene segments had occurred in the mice (Table II). Support for this finding was obtained by comparing the CC107 and CC74

　　　　　　　　　　　　　　　　　　　　CONNIE S. SCHMALJOHN

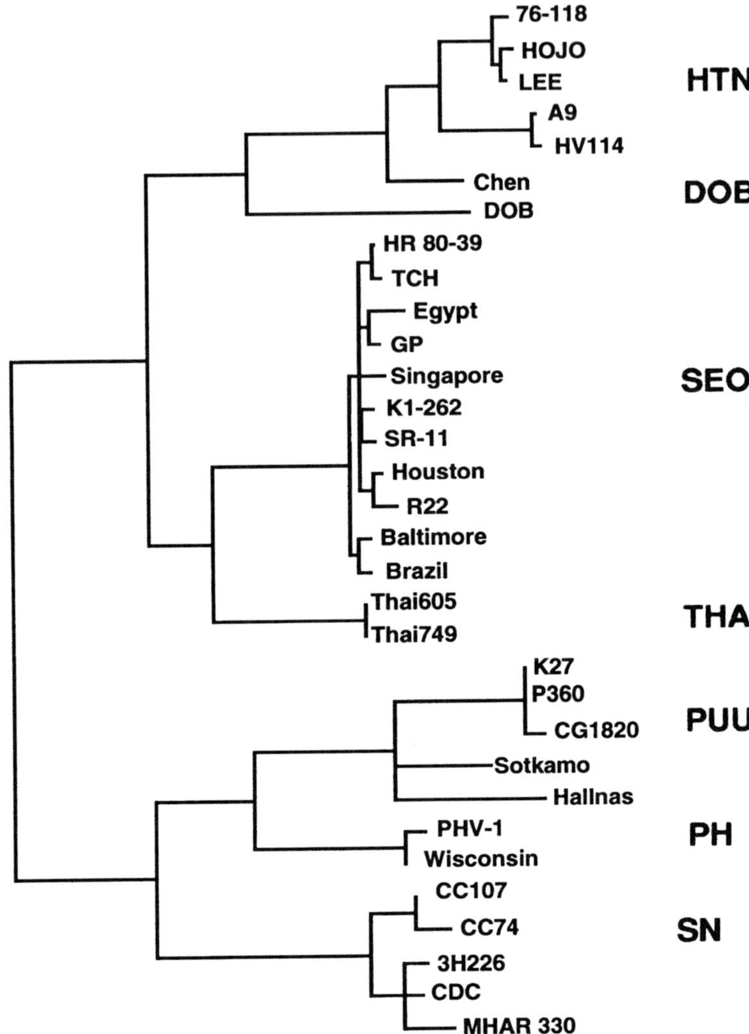

FIGURE 4. Phylogenetic relationships of a portion of the M segment nucleotide sequence of 32 hantaviruses. The tree was prepared by the maximum parsimony method using PAUP 3.0 software. Sequences compared are 330 nucleotides encoding the amino terminus of G2 (Xiao *et al.*, 1994). Bootstrap confidence limits were all in excess of 50%. The horizontal lines are proportional to the number of nucleotide substitutions for each virus.

sequences to partial sequences of the M and S segments of another virus that was detected in rodents trapped seven years earlier in a nearby region. This virus, Sweetwater Canyon (SWC) virus, has an S segment that is 99% homologous to that of CC107 and an M segment only 89% homologous. Both the M and S segments of SWC virus are 89% homologous with those of CC74, a typical finding when reassortment is not observed (Nerurkar *et al.*, 1994).

TABLE II. M and S Segment Nucleotide and Amino Acid Sequence Homologies (%identity)[a]

Virus strain	HTN 76–118		SEO 80–39		SEO SR-11		PUU CG18–20		PUU P360		PUU[b] K27		PUU Sotkamo		PH PHV-1		SN CC107		SN CC74		SN Case H	
Segment	M	S	M	S	M	S	M	S	M	S	M	S	M	S	M	S	M	S	M	S	M	S
HTN (76–118)	—	—	71	71	72	71	58	56	48	55	58	60	58	56	58	59	58	56	57	56	57	57
SEO (80–39)	77	80	—	—	96	94	58	55	58	55	58	60	58	53	57	56	57	54	57	54	56	55
SEO (SR–11)	77	82	99	95	—	.—	58	57	58	57	58	62	58	55	57	58	57	55	57	55	57	56
PUU (CG18–20)	53	61	63	59	53	62	—	—	99	99	99	99	83	82	70	67	66	61	66	61	66	61
PUU (P360)	53	61	63	59	53	62	99	99	—	—	99	99	83	83	70	67	66	61	66	61	66	61
PUU (K27)	53	60	63	58	53	61	99	98	99	99	—	—	83	85	71	73	66	68	66	68	66	68
PUU (Sotkamo)	53	60	63	59	53	62	94	96	94	96	96	96	—	—	70	67	65	60	65	60	65	60
PH (PHV-1)	54	62	63	60	54	63	75	80	76	80	76	79	75	79	—	—	65	61	65	61	65	62
SN (CC107)	55	63	63	59	53	61	67	72	67	71	67	70	66	70	67	73	—	—	99	87	88	86
SN (CC74)	55	63	53	59	53	61	67	72	67	71	67	70	66	70	67	73	99	99	—	—	89	89
SN (Case H)	55	63	53	59	53	61	67	71	67	71	67	70	66	70	67	73	99	99	99	99	—	—

[a]Values above dashes are nucleotide sequence homologies and those below dashes are amino acid sequence homologies. All values were calculated by using PAUP 3.1.1 program and are based on the shorter of the two sequences in the pairwise comparison.
[b]The noncoding regions of PUU K27 virus have not been determined and were not included in the analysis.

The overall significance of such genome segment reassortment among hantaviruses is not known. Interestingly, despite the nucleotide differences, there are few resultant amino acid differences among any of the SN viruses reported to date, perhaps indicating a strong evolutionary pressure to maintain the sequence integrity of N, G1, and G2.

IV. PROPERTIES OF VIRAL PROTEINS

A. Nucleocapsid Protein

The N proteins of HTN, SEO, PUU, PH, and SN viruses have an overall amino acid sequence identity of 50%. Certain regions of the proteins, however, display much higher or lower homologies (Fig. 5). For example, the carboxy termini of the proteins (amino acids 340–433 with respect to the amino terminus of HTN N) are 85% homologous, while amino acids 240–310, are only 11% homologous. A computer prediction of antigenicity, based hydropholicity, surface probability, backbone flexibility, and secondary structure, indicates a number of potentially antigenic regions on hantaviral N proteins, one of which is in the highly conserved carboxy terminal region, and is present in all five of these viruses (Fig. 6). Thus, despite the low homology of the complete N proteins, the presence of these highly conserved regions and resultant conserved antigenic sites evidently is sufficient to generate immune responses that are cross-reactive among numerous hantaviruses. This is especially true for closely related viruses, such as HTN and SEO viruses, which have N amino acids that are 82% identical (Arikawa *et al.*, 1990), and PUU and SN viruses, which have N amino acids that are 83% identical (Spiropoulou *et al.*, 1994). Studies with panels of monoclonal antibodies (MAbs) generated to N of HTN, SEO, and PUU viruses revealed certain antibodies that could cross-react with all other viral N proteins tested in immunofluorescent antibody tests (IFAT), and others that were specific for particular hantaviruses (Dantas *et al.*, 1987; Sugiyama *et al.*, 1987; Van der Groen *et al.*, 1989; Lundkvist *et al.*, 1991; Ruo *et al.*, 1991). One of the cross-reactive antibodies—to the nucleocapsid protein of PUU virus (Ruo *et al.*, 1991—also reacted with N of SN virus (Zaki *et al.*, 1994; Schmaljohn *et al.*, 1995; Zaki *et al.*, 1995). Thus, diagnostic tests based on recognition of N are generally quite good at detecting cross-reactive hantaviral antibodies. For example, diagnostic antigen prepared by expressing the S segment of HTN virus with baculovirus recombinants was nearly as sensitive as authentic viral antigens for detecting HTN and SEO virus antibodies in ELISA, but was not useful for differentiating between them (Rossi *et al.*, 1990). *E. coli* expression of the S segments of HTN, PUU, or SEO viruses also provided diagnostic ELISA antigen that was cross-reactive between HTN and SEO viruses, but was less reactive with antibodies to PUU virus (Gött *et al.*, 1991; Wang *et al.*, 1993b). Likewise, *E. coli* expression of the S segment of SN virus resulted in an antigen that reacted with sera from patients infected with PUU or SN

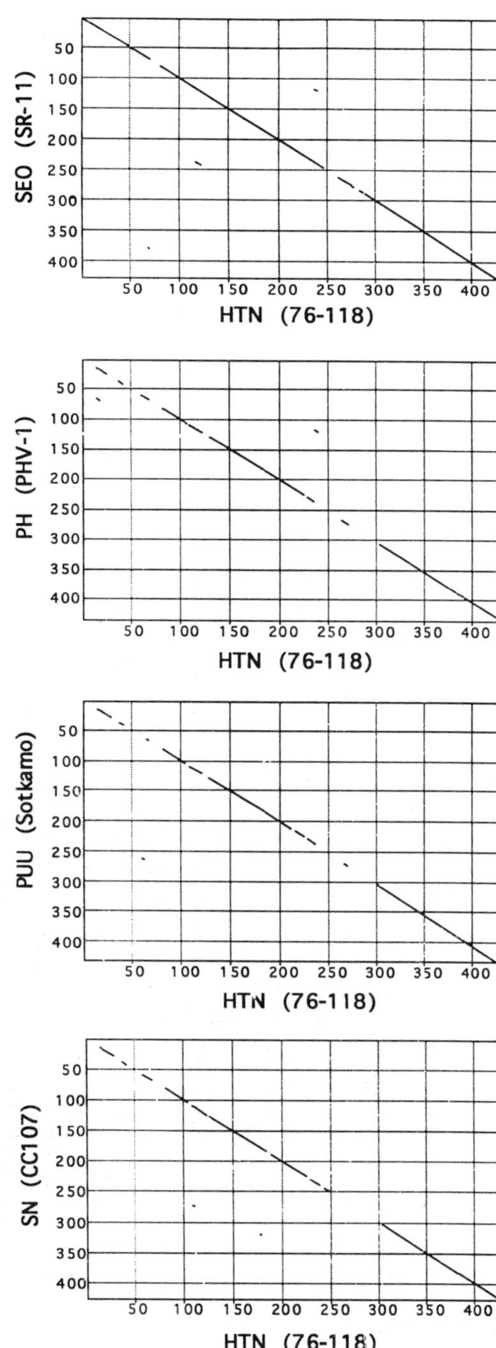

FIGURE 5. Dot matrix analyses of the N proteins of representative hantaviruses. Plots were generated using MacVector sequence analysis software. Window size = 8; minimum % score = 60; hash value 2. Solid diagonal lines indicate homology.

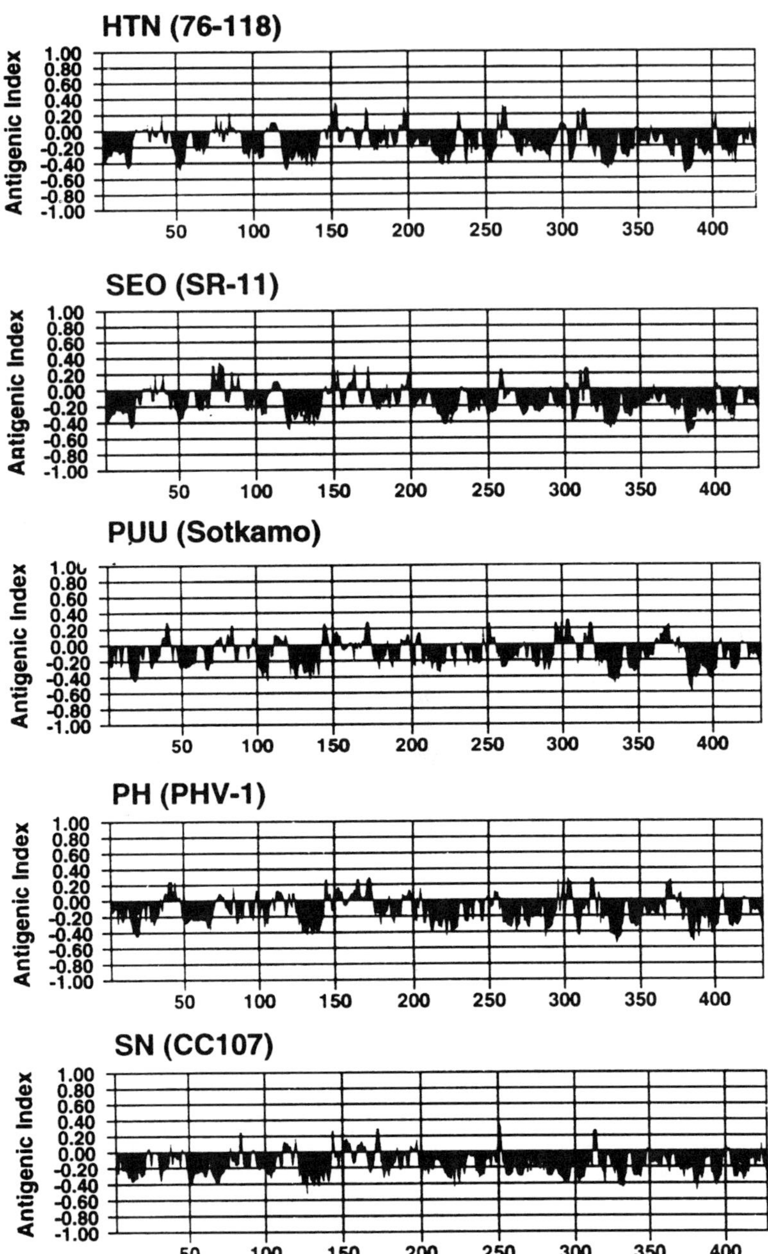

FIGURE 6. Antigenic indices of the N proteins of representative hantaviruses. Plots were generated using MacVector sequence analysis software. The analysis combines information from hydrophilicity, surface probability, backbone flexibility, and secondary structure predictions to produce a composite prediction of the surface profile of the protein. Regions that plot above the graph axis are predicted to be exposed at the protein's surface.

viruses, but not with serum from HTN virus-infected patients (Feldman *et al.*, 1993).

To generate antigen for specific identification of HTN, SEO, or PUU virus infections, truncated S segments, consisting only of two poorly conserved regions encoding amino acids 1–83 and 233–304 with respect to the amino terminus of HTN N, were expressed in *E. coli* (Wang *et al.*, 1993b). Although most epitopes recognized by panels of MAbs to N of HTN, SEO, and PUU viruses no longer reacted with the expressed truncated N proteins, at least one antibody maintained specific reactivity with the homologous protein, but not with the other N proteins. Using these expressed antigens, an ELISA was developed that was more specific than those employing complete antigen for differentiating sera from experimental or natural infections with HTN, SEO, or PUU viruses.

Although the antigenic characteristics of N have been studied extensively for various hantaviruses, little is known about their functional properties. The homology observed at the amino and carboxyl termini of the viral proteins may suggest a role for those regions in replication, perhaps in interaction with the viral RNA. Results of a single study on the RNA binding properties of the protein suggested that the carboxy-terminal portion of PUU virus N may be involved in RNA binding. In this study, however, a large amount of nonspecific binding was also observed (Gött *et al.*, 1993). Similarly, assembly of baculovirus expressed HTN N into nucleocapsidlike structures, and the association of these structures with cellular or baculovirus nucleic acid suggests that in the absence of the correct RNA, the protein will bind nonspecifically to available nucleic acid (Schmaljohn and Betenbaugh, 1993, and unpublished information). Other factors, such as those influencing the assembly of ribonucleocapsids, and the interactions of the nucleocapsids with the envelope proteins, have not yet been defined.

B. G1 and G2 Envelope Glycoproteins

1. Protein Properties

The G1 and G2 proteins of hantaviruses are similar to those of other viruses in the family in that they are glycosylated, rich in cysteine residues, and induce neutralizing antibody responses in animals (see below). Hydropathy plots of the deduced M segment expression products of HTN, DOB, SEO, THAI, PUU, PH, and SN viruses are nearly superimposable (Fig. 7), despite overall amino acid sequence identities of only 34% for G1 and 44% for G2 among all seven viruses. The amino-terminal sequences of G1 and G2 of HTN (76–118) and SEO (SR-11) viruses were determined by analysis of viral proteins (Schmaljohn *et al.*, 1987; Arikawa *et al.*, 1990). Amino acid sequences typical of signal sequences precede both G1 and G2 and both have predicted carboxy-terminal anchor sequences, suggesting G1 and G2 are type 1 membrane proteins.

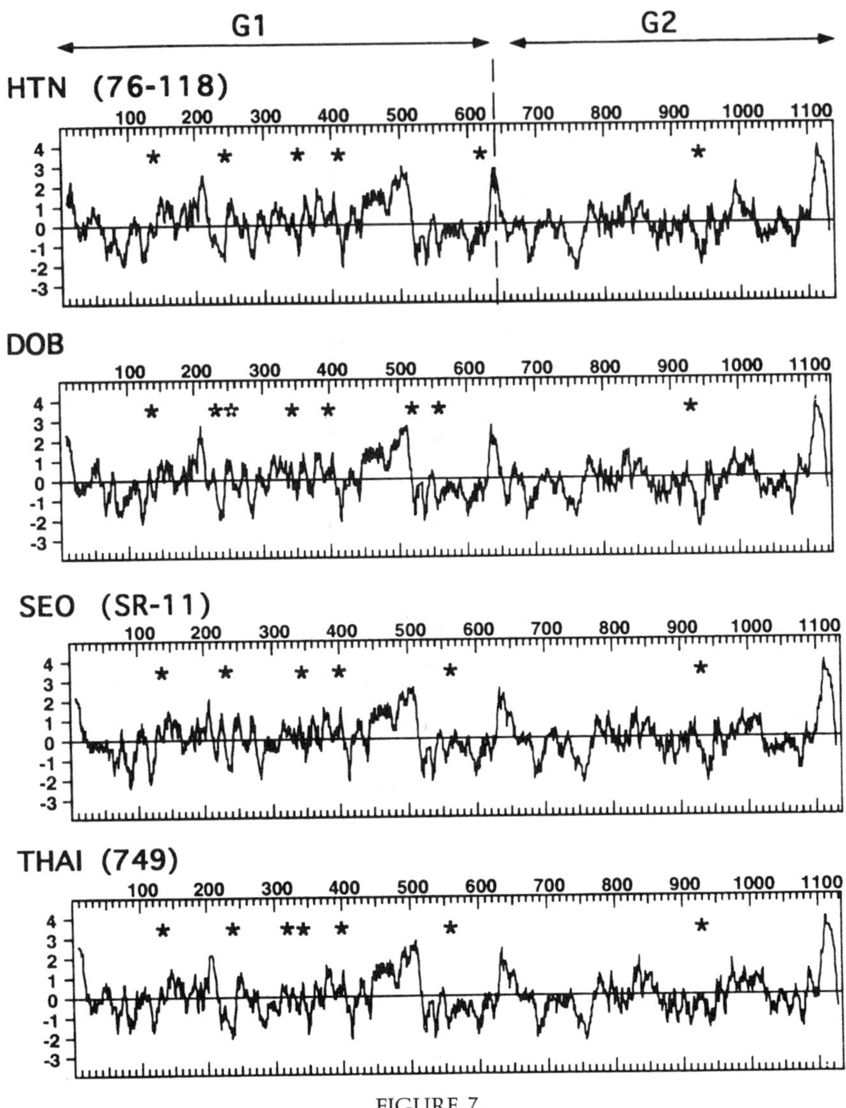

FIGURE 7.

The positions of the cysteine residues in the predicted G1 and G2 proteins are very highly conserved, with 27 of 30 residues in HTN virus G1, and 26 of 27 residues in G2 maintained among all other hantaviruses reported to date. These data suggest that the structures of the envelope proteins are similar among all of these viruses.

The glycans attached to both the G1 and G2 proteins of authentic HTN virus and also those expressed by vaccinia virus recombinants are largely

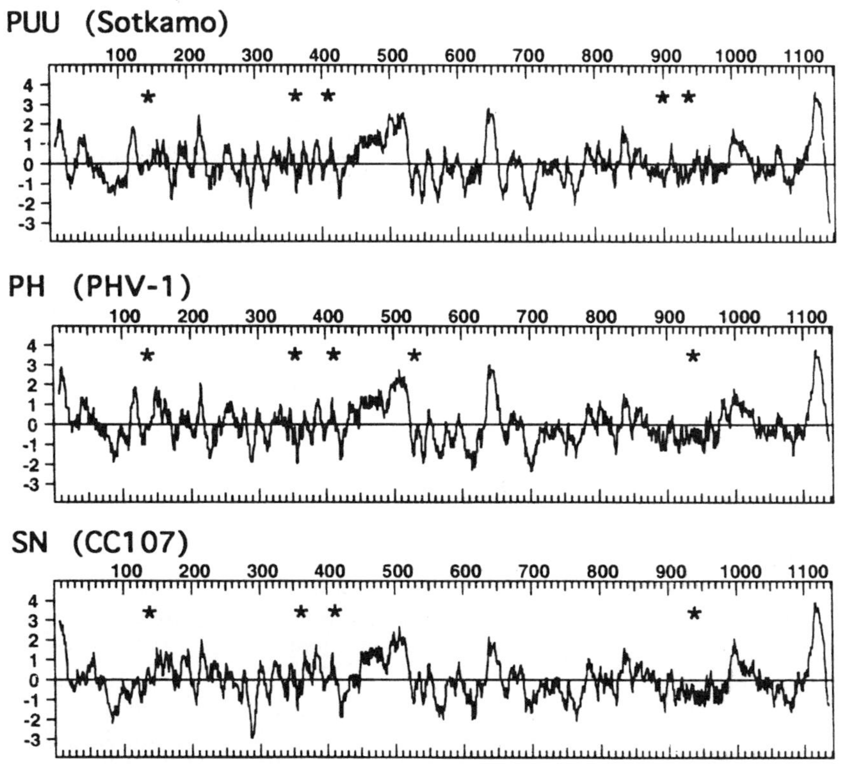

FIGURE 7. Hydrophobicity plots (Kyte Doolittle) of representative hantaviruses were prepared by using DNA Strider software. The regions representing G1 and G2 determined for HTN virus are shown with arrows above the HTN virus graph. The asterisks represent potential N-linked glycosylation sites having the amino acid sequence NXS/T where X is not P.

endoglycosidase H sensitive, indicating that they are N-linked and predominantly of the high-mannose type (Schmaljohn *et al.*, 1986a; Antic *et al.*, 1992; Ruusala *et al.*, 1992). The positions of potential N-linked glycosylation sites are conserved to some extent among hantaviruses studied so far. The G2 protein of all reported hantaviruses have one conserved potential N-linked glycosylation site. For all viruses except PUU virus, this is the only possible site (Fig. 6). Because it is known that the HTN virus G2 protein has N-linked glycans, this site is undoubtedly used (Schmaljohn *et al.*, 1986a), and it is presumably used by all of the others. PUU virus has an additional potential glycosylation site, located 39 amino acids nearer the amino terminus of G2, but studies to determine the extent of glycosylation for this virus have not been reported. The G1 protein of HTN virus has five possible N-linked glycosylation sites, three of which are conserved with all of the other viruses, and one other that is conserved only with SEO, THAI, and DOB viruses. The fifth possible site is very close to the coding sequence preceding G2, and may not even be a part of the G1 protein (Schmaljohn *et al.*, 1987) (Fig. 6). The potential glycosylation sites in DOB virus G1 are interesting, in that in addition to three possible sites conserved with all of the other hantaviruses, DOB virus has one unique site, one site conserved only with SEO and THAI viruses and another site adjacent to a unique site in PH virus (Fig. 6). These observations may correlate with the finding that several MAbs to the G1 protein of HTN virus do not react with the G1 protein of DOB virus (Van der Groen, 1990).

As with N, the predicted G1 and G2 amino acid sequences of HTN, DOB, SEO, and THAI viruses are more similar to each other than to the other viruses; and those of PUU, PH, and SN viruses more closely related to one another than to the former viruses (Table II). These data correlate with serological findings, and may suggest an early evolutionary split in hantaviruses into two distinct lineages.

2. Morphogenesis

The morphogenesis of hantaviruses has not yet been clearly defined. Intracellular assembly of viral particles has been relatively difficult to observe by electron microscopy, probably because of the low-titered, persistent mode of replication that these viruses exhibit in cell culture. Nevertheless, hantaviral antigen is generally not detected on the infected cell surface and viral particles have been observed to bud in the Golgi region, in a manner similar to that described for other viruses in the family (Hung, 1988). Also as described for other viruses in the *Bunyaviridae* family (Pettersson *et al.*, 1988; Pettersson, 1991), the G1 and G2 envelope proteins of HTN virus dimerize in the endoplasmic reticulum (Antic *et al.*, 1992). Although all factors dictating subcellular localization of viral proteins may not be known, it is clear that the envelope proteins possess whatever signal(s) are necessary to be dispatched to the Golgi without additional requirements for other viral proteins or nucleic acid. This has been demonstrated by following the transport of HTN virus G1 and G2 expressed with vaccinia virus recombinants in infected mammalian cell cultures. Such studies demonstrated that the proteins were efficiently transported from the ER to the Golgi only when both proteins were expressed. When individually expressed, both G1 and G2 remained predominantly in the ER (Ruusala *et al.*, 1992). In a separate study, HTN virus G1 expressed alone could exit the ER, and was suggested to contain the Golgi targeting signal (Pensiero and Hay, 1992). A possible explanation for these varying results might be differences in the expressed genes. That is, in the former study, a G1 gene was used that included the signal sequence and amino terminus of G2, thus permitting cotranslational cleavage of G1 and G2 analogous to authentic virus to occur. In the latter study, the G1 gene was expressed from cDNA that had been modified to introduce an artificial translation stop codon within the predicted signal sequence for

G2, thus the carboxy terminus of this G1 protein may not have been authentic. In both studies, G1 aggregated, probably because it could not achieve its correct conformation in the absence of G2, as indicated by the inability of certain G1 specific MAbs to recognize the protein (Ruusala *et al.*, 1992). In total, the findings can be interpreted to indicate that G1 and G2 must dimerize in the ER to achieve transport competency.

3. Immune Response to G1 and G2

The role of the envelope glycoproteins in protection from hantaviral infection is indisputable. Monoclonal antibodies to both the G1 and G2 proteins of HTN virus have been found to neutralize viral infectivity and to passively protect animals from infection (Dantas *et al.*, 1986; Schmaljohn *et al.*, 1990; Arikawa *et al.*, 1992). Such a humoral response is probably also important for cross-protection among hantaviruses, as evidenced by at least one neutralizing epitope that is shared by the G2 proteins of HTN and SR-11 viruses (Arikawa *et al.*, 1989). Whether one of the two envelope proteins induces a better protective immune response than the other is not known; however, protection studies performed with baculovirus or vaccinia virus recombinants that express G1 and G2 coding regions separately, or those that express the entire M segment ORF (i.e., both G1 and G2), demonstrated that both proteins are required to achieve efficient protection from HTN virus challenge (Schmaljohn *et al.*, 1990). These findings correlate with those listed above, in which it was determined that the viral proteins do not achieve correct antigenic conformation unless they dimerize prior to transport to the Golgi.

As described above for N-specific Mabs, G1 and G2-specific Mabs have also been used to define antigenic relationships among the viruses in the genus. For HTN virus, PRNT, HAI, and IFAT with G2-specific Mabs revealed a few sites conserved among the G2 proteins of most hantaviruses, while G1-specific MAbs recognized predominantly HTN-like viruses (Arikawa *et al.*, 1989; Chu *et al.*, 1994). Similar results were obtained with MAbs to the G1 and G2 proteins of a rat-borne hantavirus, B-1 virus (Dantas *et al.*, 1986)

Neutralizing epitopes are found on both the G1 and G2 proteins of HTN and PUU viruses (Arikawa *et al.*, 1989; Lundkvist and Niklasson, 1992). Specific gene regions encoding neutralizing epitopes were mapped for HTN virus by sequence analysis of antibody escape mutants and by reacting neutralizing Mabs with expressed, truncated G1 and G2 proteins (Wang *et al.*, 1993a). The neutralizing capabilities of three G1-specific and two G2-specific Mabs could be abolished in the escape mutants by nucleotide sequence substitutions resulting in single amino acid changes. For HTN virus, neutralization in vitro correlates with protection in animals, in that passive transfer of neutralizing, but not nonneutralizing G1 or G2 Mabs, will protect animals from infection with virulent HTN virus (Schmaljohn *et al.*, 1990; Arikawa *et al.*, 1992).

C. L Protein

The deduced amino acid sequences for the L proteins of HTN and SEO viruses are 85% homologous and those of HTN and PUU viruses are 70% homologous (Schmaljohn, 1990; Antic *et al.*, 1991b; Stohwasser *et al.*, 1991). Thus, the L proteins are more highly conserved than either the M or S segment gene products of these same viruses (Table II). Unlike results obtained by comparison of the N proteins (Fig. 5), dot matrix comparisons of the three known hantaviral L proteins indicate that the amino acid sequences are highly conserved throughout the protein rather than in specific regions (Fig. 8A). Evidence that the deduced amino acid sequence of the hantaviral L protein represents an RNA-dependent RNA polymerase is derived by comparing the sequence with other replicase sequences. Although very little direct sequence homology (<20%) occurs between hantaviral L proteins and those of other viruses in the family, clusters of more highly conserved amino acids are seen in the central regions of the L proteins of HTN virus, the bunyavirus, Bunyamwera (BUN) virus, and the tospovirus, tomato spotted wilt (TSW) virus (Fig. 8B). No homology could be detected between HTN virus L protein and those of the phleboviruses Rift Valley fever (RVF) virus (Fig. 8B), Toscana virus (Accardi *et al.*, 1993) or UUK virus (Elliott *et al.*, 1992) (not shown). Conserved, short-sequence motifs, indicative of RNA polymerases, were identified in the central conserved regions of HTN, BUN, TSW viruses, and also in the PB1 protein of influenza A virus, perhaps indicating an evolutionary relationship of these viruses (de Haan *et al.*, 1991).

Although not proven, by analogy to BUN virus, the hantaviral L protein is believed to carry out both transcription and replication of the genome. An RNA-dependent RNA polymerase has been described in association with detergent-disrupted HTN virions (Schmaljohn and Dalrymple, 1983). The enzyme activity was manganese-dependent and was enhanced by the presence of magnesium, 2-mercaptoethanol and sodium chloride. Although direct evidence of the ability of the hantaviral L protein to function without other viral or cellular factors in transcription and replication awaits development of an *in vitro* assay using expressed protein and a reverse genetic system, certain functional properties of the L protein can be inferred from properties of the viral mRNAs and genomic RNAs.

Like other viruses in the family, the L, M, and S mRNAs of HTN virus possess 5'-terminal nontemplated nucleotides, which are believed to be capped and to be appropriated from host cell mRNAs, probably through endonucleolytic activity of the L protein (Garcin *et al.*, 1995). Examination of the sequences of the terminal extensions revealed that although they were heterogeneous, there was a preponderance of G residues at the −1 position. Many of the extensions lacked one of the 3 AUG triplets (Fig. 2). Genomic RNAs, however, were found to begin with U at the +1 position, but only 5' monophosphates could be identified there, suggesting that initiation did not occur with UTP. A "prime-and-realign" mechanism for transcription was hypothesized to explain these findings. That is, it was suggested that mRNAs are initiated with a G-terminated host cell primer and genomes are initiated with GTP opposite the C at the +3 position. After synthesis of a few nucleotides, the nascent chain then realigns backwards by virtue of the terminal repeats and continues elongation. It was also hypothesized that for genome initiation, an endonuclease (probably the L protein) cleaves the 5' terminal extension leaving the 5' pU at position +1 (Garcin *et al.*, 1995).

V. RECOMBINANT VACCINES

As detailed above, the hantaviral envelope glycoproteins, G1 and G2, are presumed to be the major constituents involved in eliciting a protective immune response to HTN virus because: (1) MAbs to G1 or G2, but not to N, neutralize viral infectivity *in vitro*; and (2) passive transfer of neutralizing, but not nonneutralizing, MAbs specific for either G1 or G2, protected hamsters or mice from challenge with HTN virus (Schmaljohn *et al.*, 1990; Arikawa *et al.*, 1992). These findings suggest that a humoral response alone is sufficient for protection from hantaviral infection. The importance of the humoral response was also demonstrated in rats by passive transfer of immune sera and subsequent challenge with SEO virus. Similarly, maternal antibody, passively transferred to offspring, protected newborn rats from challenge with SEO virus (Zhang *et al.*, 1988). Likewise, both recombinant baculoviruses and vaccinia viruses expressing the M segment of HTN virus (Schmaljohn *et al.*, 1990) and vaccinia virus recombinants expressing the M segment of SEO virus (strain R22) (Xu *et al.*, 1992) have been found to induce neutralizing and protective antibody responses in animals. Interestingly, animals were also protected from HTN virus challenge when immunized with baculovirus-expressed N, despite the absence of a neutralizing antibody response (Schmaljohn *et al.*, 1990). These results suggest that a cell-mediated immune response to HTN N may be protective. This hypothesis is consistent with earlier findings (Asada *et al.*, 1989), which suggested that a cell-mediated immune response to HTN virus might also be protective. In those studies, HTN-specific cytotoxic T lymphocytes (CTL) were observed if immune mouse lymphocytes were restimulated *in vitro* with HTN antigen. Cross-reactive CTL could also be obtained by restimulating, with HTN virus, spleen cells from mice immunized with hantaviruses representing heterologous hantavirus serotypes, suggesting that the cell-mediated response is cross-reactive (Asada *et al.*, 1989). More recently, lymphocyte proliferation assays, performed with cells from HTN virus-immune humans, demonstrated that one of three individuals had a cross-reactive immune response to SN virus, strain CC107 (Schmaljohn *et al.*, 1995). Because of these findings, a recombinant vaccinia virus has been developed for use in humans

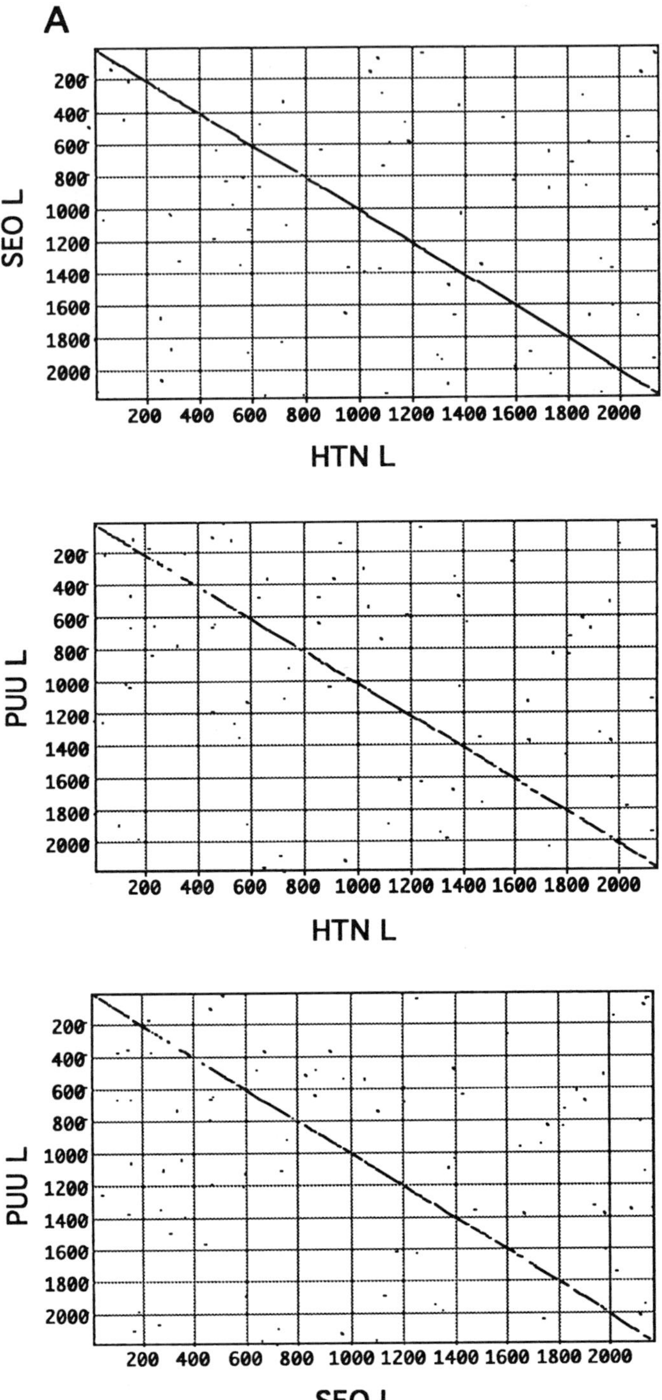

FIGURE 8. Dot matrix analyses of the L proteins of hantaviruses and other representative viruses in the family *Bunyaviridae*. Plots were generated using MacVector sequence analysis software. (A) The L proteins of HTN, SEO, and PUU viruses were compared with window size = 8; minimum % score = 60; hash value 2. The solid diagonal lines indicate homology. (B) The L

B

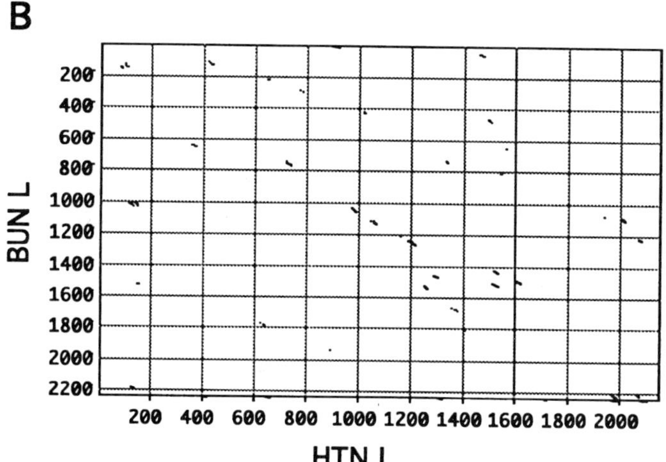

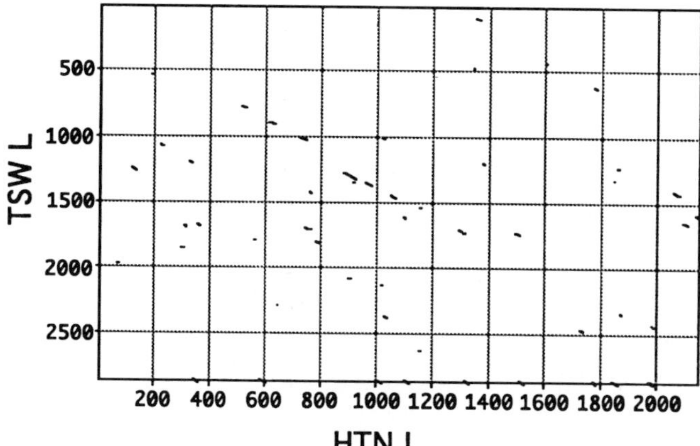

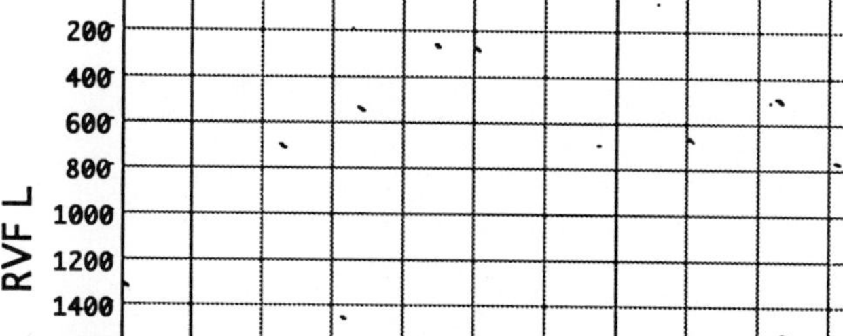

proteins of HTN virus compared at a lower stringency to those of the bunyavirus, BUN virus, the tospovirus, TSW virus, or the phlebovirus, RVF virus. Window size = 30; minimum % score = 20; hash value 2.

which expresses both the M and the S segments of HTN virus. This vaccine is currently being evaluated in human clinical trials (D. McClain and C. Schmaljohn, unpublished information).

VI. CONCLUSION

Hantaviruses are distributed throughout the world in numerous species of rodents. The viruses occasionally infect humans, sometimes causing devastating illnesses. Molecular characterization of hantaviruses has proven to be invaluable for assessing the genetic diversity among these viruses. Similarly, gene expression has been extremely useful for generating viral proteins for antigenic studies, developing diagnostics, and producing vaccines for hantaviral diseases. With the cognizance of these characteristics, the stage is set for expanding our understanding of the functional properties of the viral genes and gene products. Such knowledge may lead us to a means for manipulating the viruses to obtain less virulent, but still immunogenic viruses, or toward an understanding of potential ways to disrupt viral replication, thus mediating disease.

VII. REFERENCES

Accardi, L., Gro, M., Di Bonito, P., and Giorgi, C., 1993, Toscana virus genomic L segment: Molecular cloning, coding strategy and amino acid sequence in comparison with other negative strand RNA viruses, *Virus Res.* **27:**119.

Antic, D., Lim, B.-U., and Kang, C. Y., 1991a, Molecular characterization of the M genomic segment of the Seoul 80-39 virus: Nucleotide and amino acid sequence comparisons with other hantaviruses reveal the evolutionary pathway, *Virus Res.* **19:**47.

Antic, D., Lim, B.-U., and Kang, C. Y., 1991b, Nucleotide sequence and coding capacity of the large (L) genomic RNA segment of Seoul 80-39 virus, a member of the *Hantavirus* genus, *Virus Res.* **19:**59.

Antic, D., Wright, K. E., and Kang, C. Y., 1992, Maturation of Hantaan virus glycoproteins G1 and G2, *Virology* **189:**324.

Arikawa, J., Schmaljohn, A. L., Dalrymple, J. M., and Schmaljohn, C. S., 1989, Characterization of Hantaan virus envelope glycoprotein antigenic determinants defined by monoclonal antibodies, *J. Gen. Virol.* **70:**615.

Arikawa, J., Connors, L. C., Wang, M., and Schmaljohn, C. S., 1990, Coding properties of the S and the M genome segments of Sapporo rat virus: Comparison to other causative agents of hemorrhagic fever with renal syndrome, *Virology* **176:**114.

Arikawa, J., Yao, J. S., Yoshimatsu, K., Takashima, I., and Hashimoto, N., 1992, Protective role of antigenic sites on the envelope protein of Hantaan virus defined by monoclonal antibodies, *Arch. Virol.* **126:**271.

Asada, H., Balachandra, K., Tamura, M., Kondo, K., and Yamanishi, K., 1989, Cross-reactive immunity among different serotypes of virus causing haemorrhagic fever with renal syndrome, *J. Gen. Virol.* **70:**819.

Beaty, B., Sundin, D., and Chandler, L., 1985, Evolution of bunyaviruses by genome reassortment in dually infected mosquitoes (*Aedes triseriatus*), *Science* **230:**548.

Bishop, D. H., Beaty, B. J., and Shope, R. E., 1980, Recombination and gene coding assignments of bunyaviruses and arenaviruses, *Ann. N.Y. Acad. Sci.* **354:**84.

Carey, D. E., Reuben, R., Panicker, K. N., Shope, R. E., and Myers, R. M., 1971, Thottapalayam virus: A presumptive arbovirus isolated from a shrew in India, *Indian J. Med. Res.* **59**:1758.

Chandler, L. J., Hogge, G., Endres, M., Jacoby, D. R., Nathanson, N., and Beaty, B. J., 1991, Reassortment of La Crosse and Tahyna bunyaviruses in *Aedes triseriatus* mosquitoes, *Virus Res.* **20**:181.

Chapman, L. E. and Khabbaz, R. F., 1994, Etiology and epidemiology of the Four Corners hantavirus outbreak, *Infect. Agents Dis.* **3**:234.

Childs, J. E., Ksiazek, T. G., Spiropoulou, C. F., Krebs, J. W., Morzunov, S., Maupin, G. O., Gage, K. L., Rollin, P. E., Sarisky, J., Enscore, R. E., Frey, J. K., Peters, C. J., and Nichol, S. T., 1994, Serologic and genetic identification of *Peromyscus maniculatus* as the primary rodent reservoir for a new hantavirus in the southwestern United States, *J. Infect. Dis.* **169**:1271.

Chu, Y.-K., Rossi, C., LeDuc, J. W., Lee, H. W., Schmaljohn, C. S., and Dalrymple, J. M., 1994, Serological relationships among viruses in the *Hantavirus* genus, family Bunyaviridae, *Virology* **198**:196.

Dantas, J. R., Okuno, Y., Asada, H., Tamura, M., Takahashi, O., Takahashi, Y., Kurata, T., and Yamanishi, K., 1986, Characterization of glycoproteins of virus causing hemorrhagic fever with renal syndrome (HFRS) using monoclonal antibodies, *Virology* **151**:379.

Dantas, J. R., Jr., Okuno, Y., Tanishita, O., Takahashi, Y., Takahashi, M., Kurata, T., Lee, H. W., and Yamanishi, K., 1987, Viruses of hemorrhagic fever with renal syndrome (HFRS) grouped by immunoprecipitation and hemagglutination inhibition, *Intervirology* **27**:161.

de Haan, P., Kormelink, R., de Oliveira, R. R., van Poelwijk, F., Peters, D., and Goldbach, R., 1991, Tomato spotted wilt virus L RNA encodes a putative RNA polymerase, *J. Gen. Virol.* **72**:2207.

Elliott, R. M., Dunn, E., Simons, J. F., and Pettersson, F., 1992, Nucleotide sequence and coding strategy of the Uukuniemi virus L RNA segment, *J. Gen. Virol.* **73**:1745.

Elwell, M. R., Ward, G. S., Tingpalapong, M., and LeDuc, J. W., 1985, Serologic evidence of Hantaan-like virus in rodents and man in Thailand, *S.E. Asian J. Trop. Med. Public Health* **16**:349.

Feldman, H., Sanchez, A., Morzunov, S., Spiropoulou, C. F., Rollin, P. E., Ksiazek, T. G., Peters, C. J., and Nichol, S. T., 1993, Utilization of autopsy RNA for the synthesis of the nucleocapsid antigen of a newly recognized virus associated with hantavirus pulmonary syndrome, *Virus Res.* **30**:351.

Garcin, D., Lezzi, M., Dobbs, M., Elliott, R. M., Schmaljohn, C., Kang, C. Y. and Kolakofsky, D., 1995, The 5' ends of Hantaan virus (*Bunyaviridae*) RNAs suggest a prime-and-realign mechanism for the initiation of RNA synthesis, *J. Virol.* **69**:5754.

Giebel, L. B., Stohwasser, R., Zöller, L., Bautz, E. K. F., and Darai, G., 1989, Determination of coding capacity of the M genome segment of nephropathia epidemica virus strain Hällnäs B1 by molecular cloning and nucleotide sequence analysis, *Virology* **172**:498.

Gött, P., Zöller, L., Yang, S., Stohwasser, R., Bautz, E. K. F., and Darai, G., 1991, Antigenicity of hantavirus nucleocapsid proteins expressed in *E. coli*, *Virus Res.* **19**:1.

Gött, P., Stohwasser, R., Schnitzer, P., Darai, G., and Bautz, E. K. F., 1993, RNA binding of recombinant nucleocapsid proteins of hantaviruses, *Virology* **194**:332.

Hjelle, B., Anderson, B., Torrez-Martinez, N., Song, W., Gannon, W. F., and L., Y. T., 1995, Prevalence and geographic genetic variation of hantaviruses of New World harvest mice (*Reithrodontomys*): Identification of a divergent genotype from a Costa Rican *Reithrodontomys mexicanus*, *Virology* **207**:452.

Hjelle, B., Chavez-Giles, F., Torrez-Martinez, N., Yates, T., Sarisky, J., Webb, J. and Ascher, M., 1994, Genetic identification of a novel hantavirus of the harvest mouse *Reithrodontomys megalotis*. *J. Virol.* **68**:6751.

Hung, T., 1988, *Atlas of Hemorrhagic Fever with Renal Syndrome*, Science Press, Beijing, China.

Hung, T., Chou, Z., Zhao, T., Xia, S., and Hang, C., 1985, Morphology and morphogenesis of viruses of hemorrhagic fever with renal syndrome (HFRS), *Intervirology* **23**:97.

Kamrud, K. I., and Schmaljohn, C. S., 1994, Expression strategy of the M genome segment of Hantaan virus, *Virus Res.* **31**:109.

Kaplan, H. M., Brewer, N. R., and Blair, W. H., 1982, Physiology, in: *The Mouse in Biomedical Research*, Vol. III (H. L. Foster, J. D. Small, and J. G. Fox, eds.), pp. 260–292, Academic Press, New York.

Lee, H. W., and van der Groen, G., 1989, Hemorrhagic fever with renal syndrome, *Prog. Med. Virol.* **36**:62.

Lee, H. W., Lee, P. W., Baek, L. J., and Chu, Y. K., 1990, Geographical distribution of hemorrhagic fever with renal syndrome and hantaviruses, *Arch. Virol. Suppl.* **1**:5.

Liang, M., Li, D., and Schmaljohn, C. S., 1994, Antigenic and genetic diversity among hantavirus isolates from China, *Virus Res.* **31**:219.

Lundkvist, A., and Niklasson, B., 1992, Bank vole monoclonal antibodies against Puumala virus envelope glycoproteins: Identification of epitopes involved in neutralization, *Arch. Virol.* **126**:93.

Lundkvist, A., Fatouros, A., and Niklasson, B., 1991, Antigenic variation of European haemorrhagic fever with renal syndrome virus strains characterized using bank vole monoclonal antibodies, *J. Gen. Virol.* **72**:2097.

McCormick, J. B., Palmer, E. L., Sasso, D. R., and Kiley, M. P., 1982, Morphological identification of the agent of Korean haemorrhagic fever (Hantaan virus) as a member of the Bunyaviridae, *Lancet* **1**:765.

Martin, M. L., Regnery, H. L., Sasso, D. R., McCormick, J. B., and Palmer, E., 1985, Distinction between Bunyaviridae genera by surface structure and comparison with Hantaan virus using negative stain electron microscopy, *Arch. Virol.* **86**:17.

Murphy, F. A., Fauquet, C. M., Bishop, D. H. L., Ghabrial, S. A., Jarvis, A. W., Martelli, G. P., Mayo, M. A., and Summers, M. D., 1995, Classification and nomenclature of viruses. Sixth report of the International Committee on Taxonomy of Viruses, *Arch. Virol.* **Suppl 10**, in press.

Nerurkar, V. R., Song, K.-J., Song, J.-W., Nagle, J. W., Hjelle, B., Jenison, S., and Yanagihara, R., 1994, Genetic evidence for a hantavirus enzootic in deer mice (*Peromyscus maniculatus*) a decade before the recognition of hantaviral pulmonary syndrome, *Virology* **204**:563.

Parrington, M. A., and Kang, C. Y., 1990, Nucleotide sequence analysis of the S genomic segment of Prospect Hill virus comparison with the prototype hantavirus, *Virology* **175**:167.

Parrington, M. A., Lee, P. W., and Kang, C. Y., 1991, Molecular characterization of the Prospect Hill virus M RNA segment: A comparison with the M RNA segments of other hantaviruses, *J. Gen. Virol.* **72**:1845.

Pensiero, M. N., and Hay, J., 1992, The Hantaan virus M-segment glycoproteins G1 and G2 can be expressed independently, *J. Virol.* **66**:1907.

Pettersson, R. F., 1991, Protein localization and virus assembly at intracellular membranes, *Curr. Top. Microbiol. Immunol.* **170**:67.

Pettersson, R. F., Gahmberg, N., Kuismanen, E., Kaariainen, L., Ronnholm, R., and Saraste, J., 1988, Bunyavirus membrane glycoproteins as models for Golgi-specific proteins, *Mod. Cell. Biol.* **6**:65.

Plyusnin, A., Vapalahti, O., Lankinen, H., Lehvaslaiho, H., Apekina, N., Myasnikov, Y., Kallio-Kokko, H., Henttonen, H., Lundkvist, A., and Brummer-Korvenkontio, M., 1994, Tula virus: a newly detected hantavirus carried by European common voles, *J. Virol.* **68**:7833.

Pringle, C., Clark, W., Lees, J., and Elliott, R., 1984a, Restriction of subunit reassortment in the Bunyaviridae, in: *Segmented Negative Strand Viruses*. (R. Compans and D. Bishop, eds.), Academic Press, New York.

Pringle, C., Lees, J., Clark, W., and Elliott, R., 1984b, Genome subunit reassortment among bunyaviruses analysed by dot hybridization using molecularly cloned complementary DNA probes, *Virology* **135**:244.

Raju, R., and Kolakofsky, D., 1989, The ends of La Crosse virus genome and antigenome RNAs within nucleocapsids are base paired, *J. Virol.* **63**:122.

Ravkov, E. V., Rollin, P. E., Ksiazek, T. G., Peters, C. J. and Nichol, S. T., 1995, Genetic and serologic analysis of Black Creek Canal virus and its association with human disease and *Sigmodon hispidus* infection, *Virology* **210**:482.

Rossi, C. A., Schmaljohn, C. S., Meegan, J. M., and LeDuc, J. W., 1990, Diagnostic potential of a baculovirus-expressed nucleocapsid protein for hantaviruses, *Arch. Virol.* **115**:19.

Ruo, S. L., Sanchez, A., Elliott, L. H., Brammer, L. S., McCormick, J. B., and Fisher-Hoch, S. P., 1991, Monoclonal antibodies to three strains of hantaviruses: Hantaan, R22, and Puumala, *Arch. Virol.* **119**:1.

Ruusala, A., Persson, R., Schmaljohn, C. S., and Pettersson, R. F., 1992, Coexpression of the membrane glycoproteins G1 and G2 of Hantaan virus is required for targeting to the Golgi complex, *Virology* **186**:53.

Saluzzo, J. F., and Smith, J. F., 1990, Use of reassortant viruses to map attenuating and temperature-sensitive mutations of the Rift Valley fever virus MP-12 vaccine, *Vaccine* **8**:369.

Schmaljohn, A. L., Li, D., Negley, D. L., Bressler, D. S., Turell, M. J., Korch, G. W., Ascher, M. S., and Schmaljohn, C. S., 1995, Isolation and initial characterization of a new hantavirus from California, *Virology* **209**:963.

Schmaljohn, C. S., 1990, Nucleotide sequence of the L genome segment of Hantaan virus, *Nucleic Acids Res.* **18**:6728.

Schmaljohn, C. S., and Betenbaugh, M., 1993, Assembly of Hantaan virus-like particles from expressed gene products, *Abstract, IXth International Congress of Virology*, W6.

Schmaljohn, C. S., and Dalrymple, J. M., 1983, Analysis of Hantaan virus RNA: Evidence for a new genus of Bunyaviridae, *Virology* **131**:482.

Schmaljohn, C. S., Hasty, S. E., Harrison, S. A., and Dalrymple, J. M., 1983, Characterization of Hantaan virions, the prototype virus of hemorrhagic fever with renal syndrome, *J. Infect. Dis.* **148**:1005.

Schmaljohn, C. S., Hasty, S. E., Dalrymple, J. M., LeDuc, J. W., Lee, H. W., von Bonsdorff, C.-H., Brummer-Korvenkontio, M., Vaheri, A., Tsai, T. F., Regnery, H. L., Goldgaber, D., and Lee, P.-W., 1985, Antigenic and genetic properties of viruses linked to hemorrhagic fever with renal syndrome, *Science* **227**:1041.

Schmaljohn, C. S., Hasty, S. E., Rasmussen, L., and Dalrymple, J. M., 1986a, Hantaan virus replication: Effects of monensin, tunicamycin and endoglycosidases on the structural glycoproteins, *J. Gen. Virol.* **67**:707.

Schmaljohn, C. S., Jennings, C. B., Hay, J., and Dalrymple, J. M., 1986b, Coding strategy of the S genome of Hantaan virus, *Virology* **155**:633.

Schmaljohn, C. S., Schmaljohn, A. L., and Dalrymple, J. M., 1987, Hantaan virus M RNA: Coding strategy, nucleotide sequence, and gene order, *Virology* **157**:31.

Schmaljohn, C. S., Chu, Y. K., Schmaljohn, A. L., and Dalrymple, J. M., 1990, Antigenic subunits of Hantaan virus expressed by baculovirus and vaccinia virus recombinants, *J. Virol.* **64**:3162.

Smith, F. I., and Palese, P., 1989, Variation in influenza virus genes, in: *The Influenza Viruses* (R. M. Krug, ed.), pp. 319–359, Plenum Press, New York.

Spiropoulou, C. F., Morzunov, S., Feldmann, H., Sanchez, A., Peters, C. J., and Nichol, S. T., 1994, Genome structure and variability of a virus causing hantavirus pulmonary syndrome, *Virology* **200**:715.

Stohwasser, R., Giebel, L. B., Zoeller, L., Bautz, E. K. F., and Darai, G., 1990, Molecular characterization of the RNA S segment of nephropathia epidemica virus strain Hällnäs B1, *Virology* **174**:79.

Stohwasser, R., Raab, K., Darai, G., and Bautz, E. K. F., 1991, Primary structure of the large (L) RNA segment of nephropathia epidemica virus strain Hällnäs B1 coding for the viral RNA polymerase, *Virology* **183**:386.

Sugiyama, K., Morikawa, S., Matsuura, Y., Tkachenko, E. A., Morita, C., Komatsu, T., Akao, Y., and Kitamura, T., 1987, Four serotypes of haemorrhagic fever with renal syndrome viruses identified by polyclonal and monoclonal antibodies, *J. Gen. Virol.* **68**:979.

Turell, M. J., Korch, G. W., Rossi, C. A., Sesline, D., Enge, B. A., Dondero, D. V., Jay, M., Ludwig, G. V., Li, D., Schmaljohn, C. S., Jackson, R., and Ascher, M. S., 1995, Short report: prevalence of hantavirus infection in rodents associated with two fatal human infections in California, *Am. J. Trop. Med. Hyg.* **52**:180.

Urquidi, V. and Bishop, D. H. L., 1992, Non-random reassortment between the tripartite RNA genomes of La Crosse and snowshoe hare viruses, *J. Gen. Virol.* **73:**2255.

van der Groen, G., 1990, Hantavirus variation and disease distribution, in: *Applied Virology Research. Volume 2. Virus Variability, Epidemiology and Control* (E. Kurstak, R. G. Marusky, F. A. Murphy and M. V. Regenmortel, ed.), pp. 317, Plenum Press, New York.

van der Groen, G., Goyvaerts, D., Hoof, G., Xing, G. G., Gang, S., Leirs, H., Verhagen, R. and Yamanishi, K., 1989, Study of the antigenic relationship of hantaviruses using monoclonal antibodies, *JE & HFRS Bulletin* **3:**61.

Wang, M., Pennock, D. G., Spik, K. W., and Schmaljohn, C. S., 1993a, Epitope mapping studies with neutralizing monoclonal antibodies to the G1 and G2 envelope glycoproteins of Hantaan virus, *Virology* **197:**757.

Wang, M. W., Rossi, C. and Schmaljohn, C. S., 1993b, Expression of non-conserved regions of the S genome segments of three hantaviruses: evaluation of the expressed polypeptides for diagnosis of haemorrhagic fever with renal syndrome, *J. Gen. Virol.* **74:**1115.

White, J. D., Shirey, F. G., French, G. R., Huggins, J. W., Brand, O. M. and Lee, H. W., 1982, Hantaan virus, aetiological agent of Korean haemorrhagic fever, has Bunyaviridae-like morphology, *Lancet* **1:**768.

Xiao, S. Y., Chu, Y. K., Knauert, F. K., Lofts, R. S., Dalrymple, J. M., and LeDuc, J. W., 1992, Comparison of hantavirus isolates using a genus-reactive primer pair polymerase chain reaction, *J. Gen. Virol.* **73:**567.

Xiao, S. Y., Spik, K. W., Li, D., and Schmaljohn, C. S., 1993, Nucleotide and deduced amino acid sequences of the M and S genome segments of two Puumala virus isolates from Russia, *Virus Res.* **30:**97.

Xiao, S. Y., LeDuc, J. W., Chu, Y. K., and Schmaljohn, C. S., 1994, Phylogenetic analysis of virus isolates in the genus *Hantavirus*, family Bunyaviridae, *Virology* **198:**205.

Xu, X., Ruo, S. L., McCormick, J. B., and Fisher-Hoch, S., 1992, Immunity to hantavirus challenge in *Meriones unguiculatus* induced by vaccinia-vectored viral proteins, *Am. J. Trop. Med. Hyg.* **47:**397.

Yoo, D.-W., and Kang, C.-Y., 1987, Nucleotide sequence of the M segment of the genomic RNA of Hantaan virus 76-118, *Nucleic Acids Res.* **15:**6299.

Zaki, S. R., Albers, R. C., Greer, P. W., Coffield, L. M., Armstrong, L. R., Khan, A. S., Khabbaz, R. and Peters, C. J., 1994, Retrospective diagnosis of a 1983 case of fatal hantavirus pulmonary syndrome [letter], *Lancet* **343:**1037.

Zaki, S. R., Greer, P. W., Coffield, L. M., Goldsmith, C. S., Nolte, K. B., Foucar, K., Feddersen, R. M., Zumwalt, R. E., Miller, G. L., and Khan, A. S., 1995, Hantavirus pulmonary syndrome. Pathogenesis of an emerging infectious disease, *Am. J. Pathol.* **146:**552.

Zhang, X. K., Takashima, I., and Hashimoto, N., 1988, Role of maternal antibody in protection from hemorrhagic fever with renal syndrome virus infection in rats, *Arch. Virol.* **103:**253.

CHAPTER 4

Molecular Biology of Nairoviruses

A. C. MARRIOTT AND P. A. NUTTALL

I. DISTINGUISHING FEATURES OF NAIROVIRUSES

The tick-transmitted nairoviruses are serologically unrelated to other members of the *Bunyaviridae*. Relatively little is known about their comparative genetics and replication strategy. This is partly related to the problems of containment associated with handling the two most important members of the genus, Crimean-Congo hemorrhagic fever (CCHF) virus and Nairobi sheep disease (NSD) virus.

Membership of the genus was originally based on serological cross-reactivity, e.g., immunofluorescence, plaque neutralization, and hemagglutination inhibition (Casals and Tignor, 1980). Biochemical characteristics of the virus particles were later included to distinguish the group from other genera in the *Bunyaviridae* (Clerx *et al.*, 1981; Clerx-van Haaster *et al.*, 1982).

The following characteristics are shared by nairoviruses and distinguish them from members of other genera of the family *Bunyaviridae* (Clerx and Bishop, 1981; Clerx *et al.*, 1981; Marriott *et al.*, 1992):

1. The genome consists of three segments of single-stranded RNA of $0.6-0.7 \times 10^6$ Da (S), $1.5-1.9 \times 10^6$ Da (M), and $4.1-4.9 \times 10^6$ Da (L); these values correspond approximately to 1.7–2.0, 4.3–6.3, and 12–14 kb (Clerx *et al.*, 1981; Watret and Elliott, 1985; Ward *et al.*, 1990).

A. C. MARRIOTT AND P. A. NUTTALL • NERC Institute of Virology and Environmental Microbiology, Oxford OX1 3SR, England.

The Bunyaviridae, edited by Richard M. Elliott, Plenum Press, New York, 1996.

2. Virus particles contain three major proteins: the nucleocapsid protein N (49–54 kDa) and two glycoproteins G1 (72–84 kDa) and G2 (30–40 kDa) (Clerx *et al.*, 1981; El-Ghorr *et al.*, 1990). Additionally, nonstructural proteins are seen in extracts of infected cells, most notably glycoproteins of 92–115 and 78–85 kDa (Clerx *et al.*, 1981; Clerx and Bishop, 1981; Watret and Elliott, 1985; Cash, 1985; Marriott *et al.*, 1992). A further structural protein of 45 kDa was noted for Hazara virus (Foulke *et al.*, 1981).

3. The genomic RNAs have a unique consensus 3′ end sequence, namely 3′-AGAG(^A_U)UUCU ... for all three segments (Clerx-van Haaster *et al.*, 1982).

Relatively few studies have been made on the structure and replication of nairoviruses. Murphy *et al.* (1968, 1973) first described the morphology of CCHF virus in the brains of infected newborn mice and noted the similarity to members of the *Bunyaviridae* family. Subsequent studies on the morphogenesis and physicochemical properties of CCHF virus isolates showed them to be typical of the *Bunyaviridae* (Donets *et al.*, 1977). CCHF viral particles visualized by electron microscopy (EM) were 90–105 nm in diameter with surface spikes of 8–10 nm, and were located in the Golgi cisternae of infected cells. Filamentous internal nucleocapsids 9–10 nm wide were occasionally seen (Donets *et al.*, 1977).

David-West and Porterfield (1974) described an assay system for Dugbe (DUG) virus using a pig kidney (PS) cell line. Virus particles observed in infected PS cells resembled bunyaviruses in their morphology and morphogenesis. DUG virus was completely inactivated by sodium deoxycholate and trypsin, and by incubation at 37°C for up to 12 hr; it was stable at pH 9.0 but not at pH 3.0. Replication in PS cells was not affected by either actinomycin D or 5′-fluorodeoxyuridine. The G1 glycoprotein of DUG virus was localized to the virion surface, and N protein to the internal, filamentous nucleocapsids, by immunosorbent EM using monoclonal antibodies (El-Ghorr *et al.*, 1990). More detailed study of DUG virus showed that the 90-nm particles were bounded by a 5-nm-thick membrane, covered by 5- to 7-nm "spikes," presumably formed from one or both glycoproteins (Booth *et al.*, 1991). There is little cytopathic effect in either mammalian or tick cells. Electron microscopic studies of other nairoviruses show particles very similar to DUG virus particles, such as Erve virus (mean diameter 98 nm; Chastel *et al.*, 1989), CCHF virus (see above; Donets *et al.*, 1977), and Hazara, Bandia, and Hughes viruses (Clerx *et al.*, 1981). A unique structure was noted for Qalyub virus, in which the 90- to 120-nm particles with a fringe of 5- to 10-nm spikes were surrounded by spherical particles 10–20 nm in diameter (Clerx and Bishop, 1981). These spherical structures were removed by protease treatment, and have not been reported for any other nairovirus.

The receptor for attachment to vertebrate or tick cells is not known. Monoclonal antibodies (MAbs) against G1 of DUG virus are able to neutralize virus infection of SW13 (human) cells, and these neutralizing MAbs were able to passively protect mice from lethal intracerebral virus challenge (Buckley *et al.*, 1990). Further experiments using polyclonal antisera against DUG virus showed that neutralizing IgG did not block either attachment to mammalian cells, or subsequent fusion or internalization; consequently, neutralization was considered to involve a postinternalization step (Green *et al.*, 1992).

Experimental infection of ticks using capillary feeding methods indicates specificity in the virus–tick interaction. Virus acquired by a tick when it feeds on an infected vertebrate host must pass from the blood meal, through the gut wall to the salivary gland, in order to be transmitted in saliva when the tick takes its next blood meal (Nuttall *et al.*, 1994). When either *Amblyomma variegatum* or *Rhipicephalus appendiculatus* nymphs were allowed to feed on medium containing DUG virus, the virus persisted transstadially to the adult stage of *A. variegatum* but did not survive in *R. appendiculatus*. In *A. variegatum*, DUG virus persisted in the hemocytes, establishing an infection in the salivary glands after feeding commenced (Booth *et al.*, 1990). However, the virus persisted and was transmitted when either *A. variegatum* or *R. appendiculatus* was inoculated with DUG virus into the hemocoel, a route that bypasses the gut (Steele and Nuttall, 1989). The difference in response of *R. appendiculatus* to oral and intracoelomic routes of infection suggests that DUG virus is unable to disseminate from the gut of *R. appendiculatus* and, consequently, that this tick species is not a competent vector of DUG virus. DUG virus is apparently confined to Africa where most isolations have been from *A. variegatum*. In contrast, CCHF virus is one of the most widely distributed arboviruses and has been isolated from more than 20 tick species (Hoogstraal, 1979). Determination of the molecular basis of infection of the tick midgut may help explain the success of CCHF virus in terms of its extensive vector associations and broad geographical distribution.

Nairoviruses are currently divided into seven serogroups (see Chapter 1). Sequence data are available for members of the CCHF (CCHF and Hazara viruses) and NSD (DUG virus) serogroups. Cross-hybridization using labeled RNA probes derived from DUG virus showed reactions with the S RNAs of Ganjam (NSD serogroup), Hazara and CCHF S RNAs, indicating a closer relationship between the NSD and CCHF serogroups, than with other serogroups. Cross-reactivity with M-derived probes was limited to Ganjam M RNA, indicating larger sequence variation in M than in S RNA (Marriott *et al.*, 1990).

II. CODING STRATEGY OF THE S RNA SEGMENT

A. Nucleotide Sequence

The viral RNA of DUG virus was confirmed as negative sense by primer extension using synthetic oligonucleotides complementary to viral RNA

(Ward *et al.*, 1990). Four complete S segment sequences have been determined (Table I). Comparison of the end sequences of these viruses reveals conservation of the 21 nucleotides (nt) at the viral 3′ end and 28–30 nt at the viral 5′ end (Fig. 1), with 9 nt at each end being identical. For each virus, the 5′ and 3′ ends of the viral RNA can base pair (Fig. 2) for 11–13 nt, and there is a mismatch at position 10.

In all nairovirus S segments examined, the viral-complementary (vc) RNA strand contains a single long open reading frame (ORF). Following a 5′ untranslated region (UTR) of 49–82 nt, the ORF commences with the conserved sequence ARRAUGGAGA, which is a strong initiation sequence according to the rules of Kozak (1986). The 3′ UTR is longer at 140–337 nt. The 337-nt UTR of DUG S RNA resembles the long UTR of 370 nt in Hantaan virus S vcRNA, both of which are rich in U residues. Both show short potential ORFs of ~150 nt; however, these ORFs do not resemble each other in sequence, and neither is conserved in other members of their respective genera (Ward *et al.*, 1990; Chapter 3). There is no evidence that either the DUG or the Hantaan virus short potential ORFs are expressed.

The S mRNA of DUG virus is essentially full-length (Ward *et al.*, 1990), and contains host-derived primer sequences of 5–16 nt at the 5′ end (Jin and Elliott, 1993) indicating a "cap-snatching" mRNA priming mechanism similar to the *Orthomyxoviridae* and other members of the *Bunyaviridae*. The 5′ terminus of the vcRNA is UCUC. A preference for the sequence CUC at positions −1 to −3 of the nonviral sequence was taken as evidence of polymerase slippage analogous to that seen in the *Arenaviridae* and *Paramyxoviridae* (Jin and Elliott, 1993).

Comparison of the S RNA sequences of DUG, HAZ, and CCHF viruses indicates that the HAZ and DUG sequences show approximately the same percentage identity to each other as either does to the CCHF sequence (Table II). The variation found within CCHF virus is discussed in Section II.C. All of the sequences are relatively rich in U residues in the viral strand (Table III).

TABLE I. Summary of Sequence Data
for the S RNA Segment of Nairoviruses

Virus (isolate)	No. of nucleotides			No. of amino acids	GenBank accession number
	Total	5′ end[a]	3′ end[a]		
CCHF (C68031)[b]	1672	55	171	482	M86625
CCHF (AP92)[c]	1659	55	155	482	U04958
HAZ (JC280)[b]	1677	82	140	485	M86624
DUG (ArD 44313)[d]	1712	49	337	442	M25150

[a]Noncoding regions.
[b]Marriott and Nuttall (1992).
[c]Marriott *et al.* (1994).
[d]Ward *et al.* (1990).

```
3 ' ends
CCHF  AGAGUUUCUUUGUGCACGGCG...
HAZ   ********G*U**U******...
DUG   ********G*U********...

5 ' ends
CCHF  UCUCAAAGAUAUCGUUGCCGCACAGCCC...
HAZ   ****************************...
DUG   ********| |*****A**A***GC***...
              GAAA
```

FIGURE 1. Alignment of the S RNA ends of three nairoviruses. * represents identities.

B. Gene Products

The product of the ORF of DUG S RNA was shown to be the N protein
by expressing the cloned ORF as a β-galactosidase fusion protein in *E. coli*
and demonstrating that this protein was specifically bound by a DUG N-specific MAb (Ward *et al.*, 1990). When the ORF was expressed in insect cells, the
resulting protein was indistinguishable in size and reactivity to antibodies
from N protein derived from DUG virions (Ward *et al.*, 1992). Similarly,
CCHF S ORF expressed in insect cells reacted with a CCHF N-specific MAb
(Marriott *et al.*, 1994). Western blotting with polyclonal antisera raised to
whole virus showed that N was the most antigenic viral protein, at least in
laboratory animals (El-Ghorr *et al.*, 1990), and so expressed N proteins have
been used for immunodiagnosis (Ward *et al.*, 1992; Antoniadis *et al.*, 1992).

Sizes predicted from the sequence for DUG, HAZ, and CCHF N proteins
of 49.4, 54.2, and 54.0 kDa, respectively, agree with observations of SDS gel
electrophoresis of viral proteins. Alignment of the sequences of DUG, HAZ,
and CCHF N proteins reveals that DUG N lacks the 40 carboxyl-terminal
amino acids that are comparatively well conserved (80% identity) between

```
CCHF   5 ' UCUCAAAGAUA...
            | | | | | | | | | |  |
       3 ' AGAGUUUCUUU...

HAZ    5 ' UCUCAAAGAUA...
            | | | | | | | | | |  |
       3 ' AGAGUUUCUGU...

DUG    5 ' UCUCAAAGAGAAA...
            | | | | | | | | | |  | | |
       3 ' AGAGUUUCUGUUU...
```

FIGURE 2. Base-pairing between the 3' and
5' ends of the S RNAs of three nairoviruses.

TABLE II. Nucleotide and Amino Acid Sequence
Identities of Nairovirus S Segments

	% Sequence identity[a]			
Virus (isolate)	CCHF (C68031)	CCHF (AP92)	HAZ (JC280)	DUG (ArD44313)
CCHF (C68031)	100	81.6	61.5	57.8
CCHF (AP92)	91.7	100	61.1	58.3
HAZ (JC280)	60.0	59.3	100	57.3
DUG (ArD44313)	55.4	54.2	53.0	100

[a]Nucleotide identities are above the diagonal, amino acid identities below.

HAZ and CCHF N proteins (Marriott and Nuttall, 1992). Computer-generated translation of the S RNA of DUG beyond the stop codon at nt 1373–1375 shows a short reading frame with homology to the carboxyl-tails of HAZ and CCHF N proteins. This was taken as evidence that the truncation of DUG N relative to CCHF and HAZ N proteins was a recent event in evolutionary terms in which an amino acid codon had mutated to a UGA stop codon. Six isolates of DUG virus have N proteins indistinguishable in size from that of the sequenced isolate, ArD44313 (J. P. M. Clerx and A.C.M., unpublished observations), indicating that this truncation is not specific to one particular DUG isolate.

The N proteins have predicted positive charges (+8 to +10) as expected for proteins that bind to negatively charged RNA. Hydropathic plots for the proteins show conserved hydrophobic (residues 225–238) and hydrophilic peaks (residues 55–60 and 339–346), although these peaks do not correspond to conserved sequence motifs (Marriott and Nuttall, 1992). The overall similarity of these proteins (39.5% of amino acids are identical in all three viruses) indicates a common tertiary structure, despite the truncation in DUG N protein.

Although the coding strategy of the nairovirus S RNA resembles that of the hantavirus S RNA more than any other *Bunyaviridae* S segment (i.e.,

TABLE III. Nucleotide Composition
of Nairovirus RNA Segments

	RNA	% Nucleotide composition			
Virus (isolate)	segment	A	U	G	C
CCHF (C68031)	S	24.0	30.2	22.7	23.1
CCHF (AP92)	S	22.8	29.5	23.6	24.1
HAZ (JC280)	S	21.6	29.1	25.0	24.3
DUG (ArD44313)	S	25.9	30.6	20.1	23.4
	M	26.0	32.3	19.8	21.9
	L	26.7	34.2	17.6	21.5

long N protein ORF and no NSs protein), no homology can be detected between the nairovirus N protein sequences and the sequences of any other *Bunyaviridae* N or NSs proteins.

C. Genetic Variation in CCHF Virus

Analysis of the S RNA sequences of two isolates of CCHF virus showed that these isolates (C68031 from China and AP92 from Greece) were closer to each other than to the other nairovirus sequences, DUG and HAZ. To investigate a larger number of isolates, partial sequences were determined from PCR products derived from cell culture- or mouse brain-grown virus (A. C. Marriott, unpublished data). The amplified region of 499 bp corresponds to nt 135 to 670 of the published C68031 sequence. This region does not show significantly greater (or less) variability between C68031, AP92, DUG, and HAZ than the full-length coding regions of the S RNA sequences. Serologically, all isolates of CCHF virus are similar (Tignor *et al.*, 1980). However, the sequence data show a substantial amount of variability between CCHF virus isolates (Fig. 3). The 13 CCHF isolates cluster together, which is consistent with them comprising a single virus "species" that is distinct from HAZ and DUG viruses (which were used as outgroups to root the tree). Three major clusters of CCHF isolates are seen which differ from each other by ~20%: eight isolates from West and Central Africa, Madagascar, and China; four isolates from West Africa and Iran; and a single isolate from Greece. The clustering observed is mostly independent of the computer method used to construct the tree. From these data, it can be seen that at least three subtypes of CCHF N protein are in circulation, and at least two subtypes coexist in Senegal and Mauritania. Overall, clustering analysis revealed no obvious correlation with geographical distribution (with the possible exception of the Greek isolate) or time of isolation. This picture is consistent with gene flow between foci of infection rather than ecological isolation. As yet there are no corresponding data for variation in the M RNA of CCHF isolates. One potential route of gene flow is via virus dissemination by infected birds or infected ticks carried by birds (Hoogstraal, 1979). At least two species of African ground-feeding birds have been demonstrated to be competent (although nonviremic) hosts of CCHF virus (Zeller *et al.*, 1994).

III. CODING STRATEGY OF THE M RNA SEGMENT

A. Nucleotide Sequence

The size of the M segment of nairoviruses has been estimated in the range of 4.3 to 6.5 kb (Clerx *et al.*, 1981; Ward *et al.*, 1990) by gel electrophoretic methods. The only complete M RNA sequence published to date is that

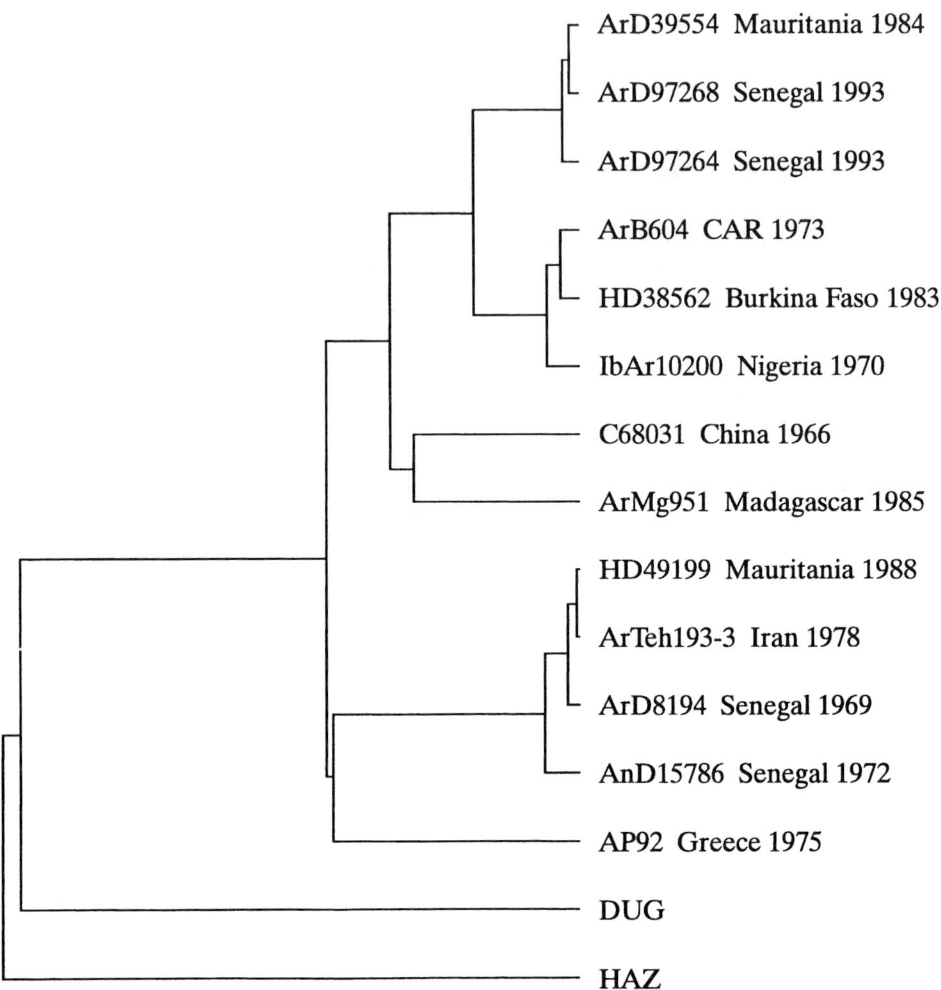

FIGURE 3. Dendrogram showing relatedness of partial S RNA sequences of 13 CCHF isolates. DUG and HAZ sequences serve as outgroups. The horizontal axis represents increasing similarity (from left to right). The isolates are labeled with country and year of isolation; CAR, Central African Republic.

of DUG virus (Marriott *et al.*, 1992; GenBank accession number M94133). This sequence has a total length of 4888 nt, similar in size to the tospovirus M RNA, and is considerably shorter than the value of 6.5 kb estimated from agarose gel electrophoresis (Ward *et al.*, 1990). However, synthetic RNA transcripts of 4.9 kb derived from the cloned cDNA also migrate at an apparent size of 6.5 kb in agarose gels.

The 5′ and 3′ ends of DUG M are complementary for 11 nt with a mismatch at position 10, as seen in the S RNA of DUG and other nairoviruses. The ends of DUG M and S RNAs match exactly for 9 (5′) or 11 (3′) nt, with similarity for 22 to 24 nt. As observed for the S RNAs of DUG, HAZ, and CCHF (Table III), the base composition shows that the viral strand is U-rich (32.3%).

Northern blotting of infected cell RNA using M-specific probes demonstrated that M-derived mRNA is essentially full-length. Nontemplated residues have been detected for 7–11 nt at the 5′ end of M mRNAs by primer extension with reverse transcriptase (A.C.M., unpublished observation), implying a cap-snatching mechanism as demonstrated for the S mRNA.

DUG M vcRNA contains a single long ORF encompassing nt 48–4703, with a coding capacity for 1551 amino acids. The 5′ and 3′ UTRs are 47 and 185 nt, respectively, thus similar in length to the corresponding regions of the S RNAs of the nairoviruses, although no conserved motifs have been detected (other than the ends, as described above). The start codon for the ORF is in a "weaker" context than the N ORF, i.e., ACA<u>AUG</u>U.

B. Processing of Gene Products

By analogy with other *Bunyaviridae*, the 1551-amino-acid M product of DUG virus was assumed to be the precursor to the glycoproteins, G1 and G2. The predicted nonglycosylated mass of the precursor is 173.3 kDa, much larger than the sum of G1 and G2 (nonglycosylated mass of 68 + 35 = 103 kDa; El-Ghorr *et al.*, 1990).

Nonstructural proteins have been reported in nairovirus-infected cells. Clerx and Bishop (1981) detected nonstructural glycoproteins of 115 and 85 kDa in Qalyub (QYB) virus-infected cells, and similar proteins have been reported for Hughes virus (115 and 78 kDa; Clerx *et al.*, 1981), DUG virus (92 and 82 kDa; Cash, 1985), and Clo Mor virus (115 kDa; Watret and Elliott, 1985). In the cases of QYB and Clo Mor viruses, pulse–chase experiments demonstrated that the larger proteins were cleaved to generate G1 protein. Marriott *et al.* (1992) showed that the nonstructural glycoproteins of 110 and 85 kDa were precipitated from DUG virus-infected cells by a G1-specific MAb as well as antivirus serum, and that the 85-kDa protein was apparently processed to G1 over a 4-hr chase. Amino-terminal sequencing of purified G1 protein mapped the end of the mature protein to residue 897 of the polyprotein. It was assumed that the 655 amino acids between residue 897 and the C-terminus of the polyprotein (nonglycosylated mass 73.2 kDa) corresponded to G1, with the hydrophobic region at residues 1448–1473 representing the membrane anchor of G1. This leaves three potential N-linked glycosylation sites in G1, compatible with the observation that glycosylation adds about 5 kDa to the apparent mass of G1 (El-Ghorr *et al.*, 1990). Unusually for the glycoproteins of members of the *Bunyaviridae*, the N-terminus of mature G1 is not immediately preceded by a hydrophobic signal sequence.

Antisera raised to fusion proteins derived from parts of the cloned M segment ORF of DUG virus demonstrated that the N-terminal part of the ORF encodes a nonstructural protein of 70 kDa (Marriott *et al.*, 1992). The C-terminal end of this protein overlaps the G2 coding region. Presumably the 70-kDa protein represents the precursor of G2, although processing has not yet been demonstrated. The part of the ORF between the presumed C-terminus of G2 and the N-terminus of the pre-G1 protein is strongly hydrophobic; no proteins derived from this region have yet been identified. Thus, the processing involved in the biosynthesis of DUG G1 and G2 shows more complexity than that involved in the processing of hantavirus, bunyavirus, or phlebovirus glycoproteins (Fig. 4).

IV. CODING STRATEGY OF THE L RNA SEGMENT

A. Nucleotide Sequence

Gel electrophoresis of nairovirus RNAs indicates a size of 12–14 kb for the L RNA. This is twice the length of hantavirus, bunyavirus, or phlebovirus L RNAs, 50% larger than the tospovirus L RNA, and accounts for over 60% of the total nairovirus genome. Recently a set of overlapping cDNA clones and PCR products of DUG L RNA have been sequenced, demonstrat-

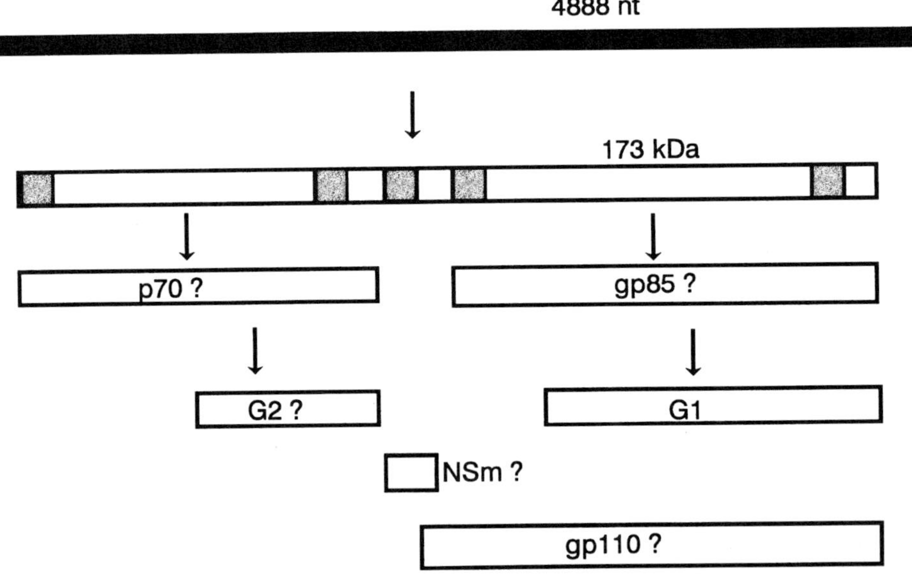

FIGURE 4. Schematic of the deduced pathway of processing of the M polyprotein of DUG virus. Black bar represents the M vcRNA, open bars represent proteins. Hydrophobic regions are shown as shaded boxes. No protein corresponding to the region labeled "NSm" has been detected experimentally.

ing that DUG L RNA has a total length of 12,255 nt (A.C. Marriott and P. Nuttall, unpublished data). This gives a total genome size of 18,855 nt for DUG virus, as compared to 16.6 kb for tospoviruses and 11.4–12.3 kb for other *Bunyaviridae* genomes.

The 3′ and 5′ ends of DUG L vRNA show sequence conservation with the M and S segments, with 11 nt at the 3′ end and 9 nt at the 5′ end being identical in all three segments (Fig. 5). The base composition of the L RNA is similar to that of the S and M segments, and is even richer in U residues (Table III). The first AUG is at nt 41–43 of the viral-complementary strand and is in a strong "Kozak" context, i.e., AACAUGG. Following this start codon is a predicted ORF of 4036 codons. No other long ORFs are present. The 3′ UTR of 104 nt is somewhat shorter than the corresponding regions of the other two segments. There is no conserved sequence (other than the termini mentioned above) in the 3′ UTRs of the three segments which could serve as a transcriptional termination sequence. The 3′ ends of the mRNAs have not been mapped.

B. Gene Products

An L RNA-encoded viral replicase is assumed to be present in nucleo-capsids by analogy with the other *Bunyaviridae*, but so far a nairovirus replicase protein has not been demonstrated either in virions or in infected cells. A large protein of ~210 kDa was detected in DUG virions by El-Ghorr *et al.* (1990), although it was not characterized further. The total coding capacity of DUG L RNA is 4036 amino acids (459 kDa). This is considerably greater than the L proteins of phleboviruses (2095–2104 amino acids), bunya-viruses (2238 amino acids), hantaviruses (2151–2156 amino acids), and tospo-viruses (2875 acids) (Accardi *et al.*, 1993; Elliott *et al.*, 1992; Elliott, 1989;

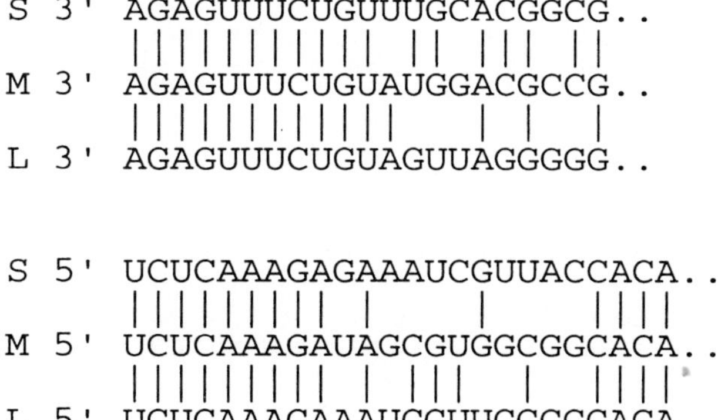

FIGURE 5. Alignment of the 3′ and 5′ ends of all three genomic RNAs of DUG virus.

Antic *et al.*, 1992; de Haan *et al.*, 1991). Analysis of the predicted amino acid sequence shows no homology to other proteins in the sequence databases, other than in a region of 308 amino acids (residues 2361–2668) which aligns with the core polymerase motif common to RNA-dependent RNA polymerases of segmented negative-strand viruses (Poch *et al.*, 1989). All six conserved modules of this polymerase motif (as defined by Muller *et al.*, 1994) are present including the Ser-Asp-Asp catalytic core tripeptide. The amino-terminal motifs described by Muller *et al.* (1994) for other segmented negative-strand viral replicases are not present in the DUG L sequence. The aligned region shows that the nairovirus L protein is distinct from the L proteins of other *Bunyaviridae*. The best alignment with other sequences available on the databases is with the phlebovirus L protein.

V. FUTURE RESEARCH

Full characterization of the nairoviruses at the molecular level will require one or more complete genome sequences. DUG virus has recently been completed, but sequences of other M and L RNAs will help identification of important conserved regions. The pathway of processing of the glycoproteins also requires elucidation. Expression of M-derived proteins in the absence of S and L products, e.g., in the baculovirus expression system, may facilitate such work.

More molecular data for CCHF field isolates are needed, especially M RNA sequences which would show whether reassortment of RNA segments is a major factor in CCHF epidemiology. There is certainly plenty of work to do for any researchers wishing to investigate further this important group of viruses!

VI. REFERENCES

Accardi, L., Gro, M. C., Di Bonito, P., and Giorgi, C., 1993, Toscana virus genomic L segment: Molecular cloning, coding strategy and amino acid sequence in comparison with other negative strand RNA viruses, *Virus Res.* **27**:119.

Antic, D., Kang, C. Y., Spik, K., Schmaljohn, C., Vapalahti, O., and Vaheri, A., 1992, Comparison of the deduced gene products of the L, M and S genome segments of hantaviruses, *Virus Res.* **24**:35.

Antoniadis, A., Polyzoni, T., Tomori, O., Williams, O., Marriott, A. C., and Nuttall, P. A., 1992, Evidence of previously unrecognised nairovirus infections of humans in Nigeria and Greece, *J. Viral Dis.* **1**:34.

Booth, T. F., Marriott, A. C., Steele, G. M., and Nuttall, P. A., 1990, Dugbe virus in ticks: Histological localisation studies using light and electron microscopy, *Arch. Virol.* [Suppl. 1]:207.

Booth, T. F., Gould, E. A., and Nuttall, P. A., 1991, Structure and morphogenesis of Dugbe virus (Bunyaviridae, Nairovirus) studied by immunogold electron microscopy of ultrathin cryosections, *Virus Res.* **21**:199.

Buckley, A., Higgs, S., and Gould, E. A., 1990, Dugbe virus susceptibility to neutralization by monoclonal antibodies as a marker of neurovirulence in mice, *Arch. Virol.* [Suppl. 1]:197.

Casals, J., and Tignor, G. H., 1980, The Nairovirus genus; serological relationships, *Intervirology* **14**:144.

Cash, P., 1985, Polypeptide synthesis of Dugbe virus, a member of the Nairovirus genus of the Bunyaviridae, *J. Gen. Virol.* **66**:141.

Chastel, C., Main, A. J., Richard, P., Le Lay, G., Legrand-Quillien, M. C., and Beaucournu, J. C., 1989, Erve virus, a probable member of Bunyaviridae family isolated from shrews (Crocidura russula) in France, *Acta Virol.* **33**:270.

Clerx, J. P. M., and Bishop, D. H. L., 1981, Qalyub virus, a member of the newly proposed Nairovirus genus (Bunyaviridae), *Virology* **108**:361.

Clerx, J. P. M., Casals, J., and Bishop, D. H. L., 1981, Structural characteristics of nairoviruses (genus Nairovirus, Bunyaviridae), *J. Gen. Virol.* **55**:165.

Clerx-van Haaster, C. M., Clerx, J. P. M., Ushijima, H., Akashi, H., Fuller, F., and Bishop, D. H. L., 1982, The 3' terminal RNA sequences of bunyaviruses and nairoviruses (Bunyaviridae): Evidence of end sequence generic differences within the virus family, *J. Gen. Virol.* **61**:289.

David-West, T. S., and Porterfield, J. S., 1974, Dugbe virus: A tick-borne arbovirus from Nigeria, *J. Gen. Virol.* **23**:297.

de Haan, P., Kormelink, R., Resende, R. D. O., Van Poelwijk, F., Peters, D., and Goldbach, R., 1991, Tomato spotted wilt virus L RNA encodes a putative RNA polymerase, *J. Gen. Virol.* **72**:2207.

Donets, M. A., Chumakov, M. P., Korolev, M. B., and Rubin, S. G., 1977, Physicochemical characteristics, morphology and morphogenesis of virions of the causative agent of Crimean hemorrhagic fever, *Intervirology* **8**:294.

El-Ghorr, A. A., Marriott, A. C., Ward, V. K., Booth, T. F., Higgs, S., Gould, E. A., and Nuttall, P. A., 1990, Characterisation of Dugbe virus (Nairovirus, Bunyaviridae) by biochemical and immunochemical procedures using monoclonal antibodies, *Arch. Virol.* [Suppl. 1]:169.

Elliott, R. M., 1989, Nucleotide sequence analysis of the large (L) genomic RNA segment of Bunyamwera virus, the prototype of the family Bunyaviridae, *Virology* **173**:426.

Elliott, R. M., Dunn, E., Simons, J. F., and Pettersson, F., 1992, Nucleotide sequence and coding strategy of the Uukuniemi virus L RNA segment, *J. Gen. Virol.* **73**:1745.

Foulke, R. W., Rosata, R. R., and French, G. R., 1981, Structural polypeptides of Hazara virus, *J. Gen. Virol.* **53**:169.

Green, E. M., Armstrong, S. J., and Dimmock, N. J., 1992, Mechanisms of neutralisation of a nairovirus (Dugbe virus) by polyclonal IgG and IgM, *J. Gen. Virol.* **73**:1995.

Hoogstraal, H., 1979, The epidemiology of tick-borne Crimean-Congo hemorrhagic fever in Asia, Europe, and Africa, *J. Med. Entomol.* **15**:307.

Jin, H., and Elliott, R. M., 1993, Non-viral sequences at the 5' ends of Dugbe nairovirus S mRNAs, *J. Gen. Virol.* **74**:2293.

Kozak, M., 1986, Point mutations define a sequence flanking the AUG initiator codon that modulates translation by eukaryotic ribosomes, *Cell* **44**:283.

Marriott, A. C., and Nuttall, P. A., 1992, Comparison of the S RNA segments and nucleoprotein sequences of Crimean-Congo hemorrhagic fever, Hazara and Dugbe viruses, *Virology* **189**:795.

Marriott, A. C., Ward, V. K., Higgs, S., and Nuttall, P. A., 1990, RNA probes detect nucleotide sequence homology between members of two different nairovirus serogroups, *Virus Res.* **16**:77.

Marriott, A. C., El-Ghorr, A. A., and Nuttall, P. A., 1992, Dugbe nairovirus M RNA: Nucleotide sequence and coding strategy, *Virology* **190**:606.

Marriott, A. C., Polyzoni, T., Antoniadis, A., and Nuttall, P. A. 1994, Detection of human antibodies to Crimean-Congo haemorrhagic fever virus using expressed viral nucleocapsid protein, *J. Gen. Virol.* **75**:2157.

Muller, R., Poch, O., Delarue, M., Bishop, D. H. L., and Bouloy, M., 1994, Rift Valley fever virus L segment: Correction of the sequence and possible functional role of newly identified regions conserved in RNA-dependent polymerases, *J. Gen. Virol.* **75**:1345.

Murphy, F. A., Harrison, A. K., and Tzianabos, T., 1968, Electron microscopic observations of mouse brain infected with Bunyamwera group arboviruses, *J. Virol.* **2**:1315.

Murphy, F. A., Harrison, A. K., and Whitfield, S. G., 1973, Bunyaviridae: Morphologic and morphogenetic similarities of Bunyamwera serological supergroup viruses and several other arthropod-borne viruses, *Intervirology* **1**:297.

Nuttall, P. A., Jones, L. D., Labuda, M., and Kaufman, W. R., 1994, Adaptations of arboviruses to ticks, *J. Med. Entomol.* **31**:1.

Poch, O., Sauvaget, I., Delarue, M., and Tordo, N., 1989, Identification of four conserved motifs among the RNA-dependent polymerase encoding elements, *EMBO J.* **8**:3867.

Steele, G. M., and Nuttall, P. A., 1989, Difference in vector competence of two species of sympatric ticks, Amblyomma variegatum and Rhipicephalus appendiculatus, for Dugbe virus (Nairovirus, Bunyaviridae), *Virus Res.* **14**:73.

Tignor, G. H., Smith, A. L., Casals, J., Ezeokoli, C. D., and Okoli, J., 1980, Close relationship of Crimean hemorrhagic fever-Congo (CHF-C) virus strains by neutralising antibody assays, *Am. J. Trop. Med. Hyg.* **29**:676.

Ward, V. K., Marriott, A. C., El-Ghorr, A. A., and Nuttall, P. A., 1990, Coding strategy of the S RNA segment of Dugbe virus (Nairovirus: Bunyaviridae), *Virology* **175**:518.

Ward, V. K., Marriott, A. C., Polyzoni, T., El-Ghorr, A. A., Antoniadis, A., and Nuttall, P. A., 1992, Expression of the nucleocapsid protein of Dugbe virus and antigenic cross-reactions with other nairoviruses, *Virus Res.* **24**:223.

Watret, G. E., and Elliott, R. M., 1985, The proteins and RNAs specified by Clo Mor virus, a Scottish nairovirus, *J. Gen. Virol.* **66**:2513.

Zeller, H. G., Cornet, J.-P., and Camicas, J.-L., 1994, Experimental transmission of Crimean-Congo hemorrhagic fever virus by West African wild ground-feeding birds to Hyalomma marginatum rufipes ticks, *Am. J. Trop. Med. Hyg.* **50**:676.

CHAPTER 5

Molecular Biology of Phleboviruses

COLOMBA GIORGI

I. INTRODUCTION

The *Phlebovirus* genus encompasses two groups of viruses, the phlebotomus or sandfly fever group (PHL) and the Uukuniemi group (Calisher, 1991). In a previous taxonomic organization the two groups, based on serological relationships, were considered as different genera in the *Bunyaviridae* family, but because of the similarities in biochemical properties and in genome organization and expression, the two groups were combined into one genus, *Phlebovirus*. However, the biological properties of the two groups are distinct: the phlebotomus fever viruses are transmitted by sandflies and gnats whereas the uukuviruses are transmitted by ticks (Bishop and Shope, 1979). Several PHL viruses, but none of the uukuviruses, have been associated with disease in humans, although antibodies to Uukuniemi virus have been detected in humans (Tesh, 1988; Beaty and Calisher, 1991). This chapter focuses on phlebovirus genome organization and expression, and the data obtained in the past few years will be discussed to provide an update on earlier reviews (Elliott, 1990; Bouloy, 1991; Elliott *et al.*, 1991). The *Phlebovirus* genus is composed of at least 51 different viruses (39 within the sandfly fever group and 12 within the Uukuniemi group; Calisher, 1991; see Chapter 1), but the information on the molecular biology of these viruses comes from the analysis of only a few of them, namely Punta Toro (PT), Rift Valley fever (RVF), sandfly fever Sicilian (SFS), Toscana (TOS), and Uukuniemi (UUK) viruses.

COLOMBA GIORGI • Laboratory of Virology, Istituto Superiore di Sanità, 00161 Rome, Italy.

The Bunyaviridae, edited by Richard M. Elliott, Plenum Press, New York, 1996.

II. STRUCTURAL COMPONENTS OF THE VIRION

Early observations on the morphology of phleboviruses were made by electron microscopy of the negatively stained viral particles. UUK virions appeared to be spherical and approximately 90–100 nm in diameter. The surface of viral particles showed a regular hexagonal arrangement of subunits which resembled hollow capsomers with projections measuring about 9 nm (Saikku and Von Bonsdorff, 1968; Saikku et al., 1970). Similar capsomer-like projections had been observed previously on the surface of the RFV virus particles (Levitt et al., 1963), and subsequently similar morphological features were described for two other phleboviruses, PT and Karimabad viruses (Smith and Pifat, 1982). It was also evident from these studies that assembly of viral particles occurred exclusively at smooth membrane vesicles and predominantly at the membranes in, or adjacent to, the Golgi complex.

Disruption of UUK virions released a coiled inner component, a loosely wound helix composed of one strand of ribonucleoprotein (Saikku et al., 1970). A 25-kDa protein (N) was shown to be associated with the ribonucleoprotein strand whereas two similarly sized glycosylated proteins (G1 and G2, 65–75 kDa) were associated with the surface projections (Pettersson et al., 1971; Von Bonsdorff and Pettersson, 1975). Polypeptides of similar sizes were found in virions of several PHL viruses (Robeson et al., 1979; Rice et al., 1980). A fourth viral protein with a mass greater than 200 kDa was found in purified virions of RVF virus (Collett et al., 1985), and is thought to be a component, at least, of the viral transcriptase (Bishop et al., 1980).

Besides the N, G1, G2, and L structural proteins, all phleboviruses express one nonstructural protein, termed NSs, while a second nonstructural protein, designated NSm, seems to be expressed by the sandfly fever group viruses but not from the Uukuniemi group viruses.

Like all other viruses in the *Bunyaviridae* family, the genome of phleboviruses was shown to comprise three single-stranded RNA segments called L, M, and S. Each RNA segment appeared to contain a unique primary sequence with no overlap between the segments (Pettersson et al., 1977; Robeson et al., 1979). The sequences at the 3' ends of each RNA segment were shown to be conserved in the same virus and to be very similar to those of other phleboviruses (Parker and Hewlett, 1981; Clerx-van Haaster et al., 1982; Ihara et al., 1984; Collett et al., 1985).

III. GENOME STRUCTURE AND ORGANIZATION

Sequence data obtained from molecular clones of the genomic segments of different phleboviruses (Ihara et al., 1984, 1985b; Collett et al., 1985; Rönnholm and Pettersson, 1987; Marriott et al., 1989; Giorgi et al., 1991; Muller et al., 1991; Elliott et al., 1992; Accardi et al., 1993; Giorgi et al., unpublished data) have shown that the 3' and 5' ends are conserved not only within the segments of the same virus but also among the different viruses (Fig. 1). The consensus sequences consist of eight nucleotides at the 3' end and 11 nucleotides at the 5' end. The bases at the 3' and 5' termini appear to be complementary and able to form stable base-paired panhandles. The lengths of the most stable structures predicted from the terminal sequences vary depending on the segment and on the virus.

These analyses has been made without considering that the complete sequence of the RNA segment may influence the possible secondary structure at the termini. A secondary structure model of the complete M RNA segment of RVF virus was suggested by Collett *et al.* (1985). This model clearly reveals an energetically stable base-paired stem structure involving 51 nucleotides from each terminus of the RNA segment. This model, however, should also be considered theoretical, because it does not take into account that the genomic RNA, *in vivo*, is never naked but is complexed with the nucleocapsid protein, N, which could strongly influence the RNA structure. Nevertheless, the circular nature of isolated ribonucleoprotein seen by electron microscopy (Pettersson and Von Bonsdorff, 1975; Hewlett *et al.*, 1977) suggests that the terminally paired RNA structure can exist *in vivo*.

The base-paired structure and the sequences at the termini of the genomic segments may play different but essential roles in viral RNA transcription and replication. The structure may contain the recognition site for the viral polymerase, and the signal for encapsidation by N protein. In this regard it should be noted that viral mRNAs, which are truncated at the 3' end and possess nonviral sequences at the 5' end, are not encapsidated. Moreover, the pattern of possible base-pairing structures seems to be unique for each genomic segment of the same virus (Elliott *et al.*, 1991), suggesting a role in the selective packaging of the genomic segments in the viral particles.

A. S RNA Segment

The gene organization and expression of the S segment are the major features that distinguish the *Phlebovirus* genus from bunyaviruses, hantaviruses, and nairoviruses, but are similar to those of tospoviruses.

Complete nucleotide sequences have been reported for the S segment of PT (Ihara *et al.*, 1984), SFS (Marriott *et al.*, 1989), UUK (Simons *et al.*, 1990), TOS and RVF (Giorgi *et al.*, 1991) viruses. All possess two open reading frames (ORF): one in the viral-complementary sense and the other in the viral sense. This coding strategy is referred to as "ambisense" (Fig. 2) (Ihara *et al.*, 1984).

The ORFs extend over the two halves of the genomic segment at the 3' end and at the 5' end of the S RNA but do not overlap each other. The intergenic region (the region extending between the two stop codons of the ORFs) has a different length and sequence in the different viruses. The intergenic sequence of PT S RNA is the longest of all of the viruses analyzed so far (356 nt) and has many A-rich and U-rich regions that may potentially

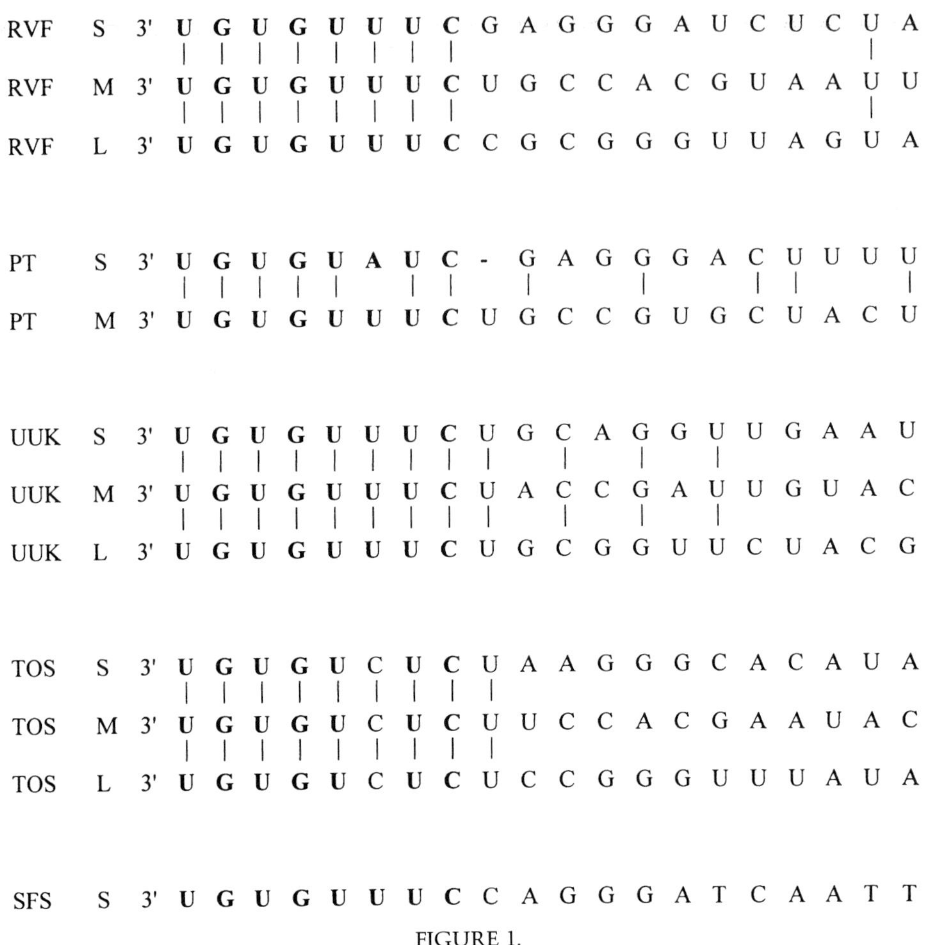

FIGURE 1.

form an energetically favored hairpin structure (Emery and Bishop, 1987). In UUK virus the S segment intergenic region is shorter (Simons *et al.*, 1990) but could also potentially form a stem-loop structure. In contrast, the intergenic regions of the TOS (62 nt), RVF (82 nt), and SFS (141 nt) viruses have G-rich sequences, in the viral-complementary sense, with several homopolymeric G tracts that range up to six residues in length. Other motifs are conserved in the three sequences, including the pentanucleotide GCUGC followed by G-tracts (in the viral-complementary sense), or C-tracts (in the viral-sense RNA). The structure and sequences of the intergenic regions are thought to play some role in the transcription termination mechanism (see below).

The product of the gene at the 3′ end of the S genomic segment has been identified as the nucleocapsid protein, N, in PT, UUK, and TOS viruses (Overton *et al.*, 1987; Simons *et al.*, 1990; Giorgi *et al.*, 1991). Evidence that the product of the ORF in the viral-sense sequence is the NSs protein has

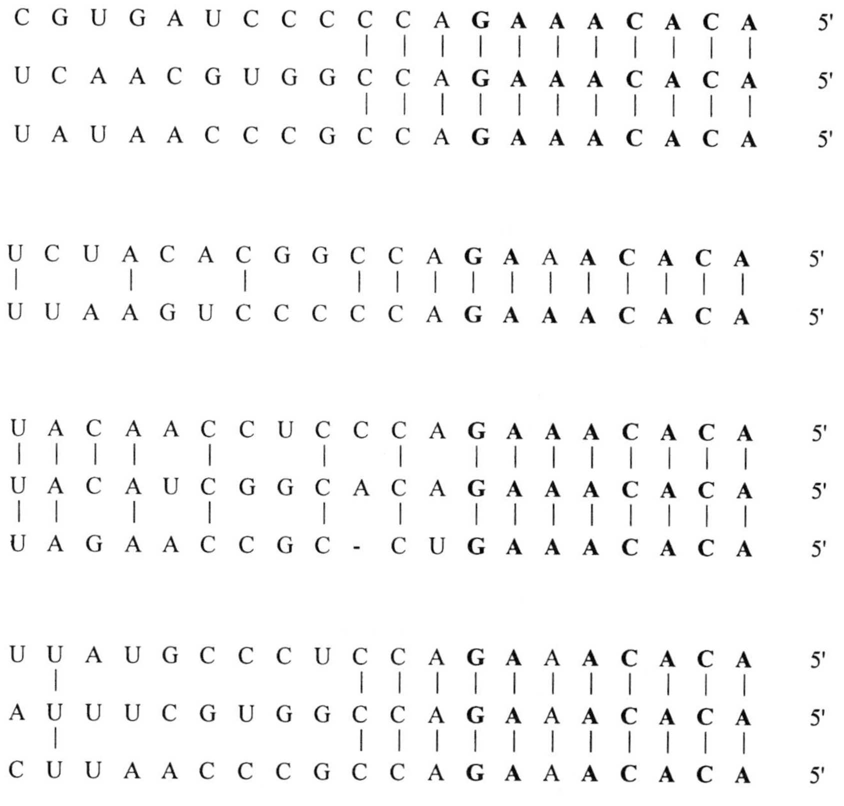

FIGURE 1. Conservation of the terminal sequences of the genomic segments of RVF (Giorgi *et al.*, 1991; Collett *et al.*, 1985; Muller *et al.*, 1991), PT (Ihara *et al.*, 1984, 1985b), UUK (Simons *et al.*, 1990; Rönnholm and Pettersson, 1987; Elliott *et al.*, 1992), TOS (Giorgi *et al.*, 1991; Giorgi *et al.*, unpublished data; Accardi *et al.*, 1993), and SFS (Marriott *et al.*, 1989) viruses. The sequences conserved at the 3' and 5' ends among the segments of the same virus are indicated with a vertical bar. The complementary nucleotides at the termini of each segment are indicated in boldface.

been obtained for the PT and TOS viruses (Overton *et al.*, 1987; Giorgi *et al.*, unpublished data).

B. M RNA Segment

The M segments of UUK virus and of the phlebotomus fever group viruses PT, RVF, TOS have been cloned and sequenced. In the case of the RVF virus, the M sequences of three strains have been reported: two wild-type

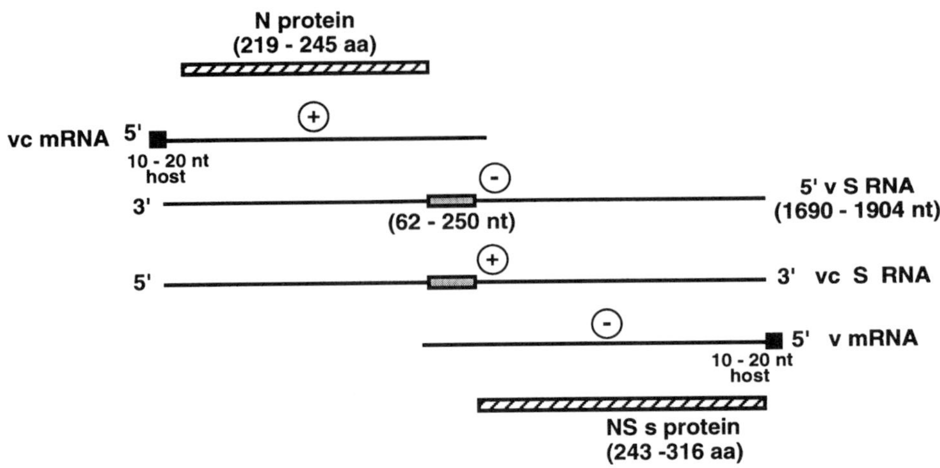

FIGURE 2. Schematic representation of the ambisense expression of the phlebovirus S segment. The viral S RNA segment (v S RNA) is transcribed to give a subgenomic viral-complementary messenger RNA (vc mRNA) which expresses the N protein. The v S RNA is also replicated to give a full-length viral-complementary RNA (vc S RNA). The vc S RNA is transcribed to give a viral-sense subgenomic mRNA (v mRNA) which expresses the NSs protein. The two mRNAs possess nontemplated sequences at their 5' ends (black boxes). The shaded boxes in the v S RNA and vc S RNA indicate the intergenic region. The numbers in parentheses represent the variation in nucleotides or amino acids among RVF, PT, TOS, and UUK viruses.

strains, ZH501 (Collett *et al.*, 1985) and ZH548, and a derivative, attenuated strain ZH548M12 (Takehara *et al.*, 1989). The latter virus is a candidate vaccine strain, and was derived from the wild-type ZH548 strain by consecutive passage in the presence of the mutagen 5-fluorouracil (Caplen *et al.*, 1985). The sequence data show that all viruses possess a single large ORF extending the length of the M segment in the viral complementary RNA (Ihara *et al.*, 1985b; Collett *et al.*, 1985; Rönnholm and Pettersson, 1987; Giorgi, unpublished data). This ORF codes for a polypeptide precursor of the viral glycoproteins G1 and G2 (Fig. 3). The nomenclature of the two proteins refers only to their electrophoretic mobility on gels, the slower-migrating glycoprotein being designated G1, and does not reflect differences in gene arrangement. The precursor of the glycoproteins has never been found in infected cells. However, a 110-kDa protein (p110) was synthesized when UUK mRNA, which has the same size as the M RNA genomic segment, was translated *in vitro*. In the presence of microsomal membranes, p110 was processed to give two glycoproteins which migrated similarly to UUK virus glycoproteins (Ulmanen *et al.*, 1981).

The gene order and the potential cleavage sites in the glycoprotein precursor of PT and RVF viruses have been determined by limited amino-terminal sequence analysis of the viral glycoproteins. For PT and RVF viruses the coding capacity of the entire ORF is greater than that estimated to be re-

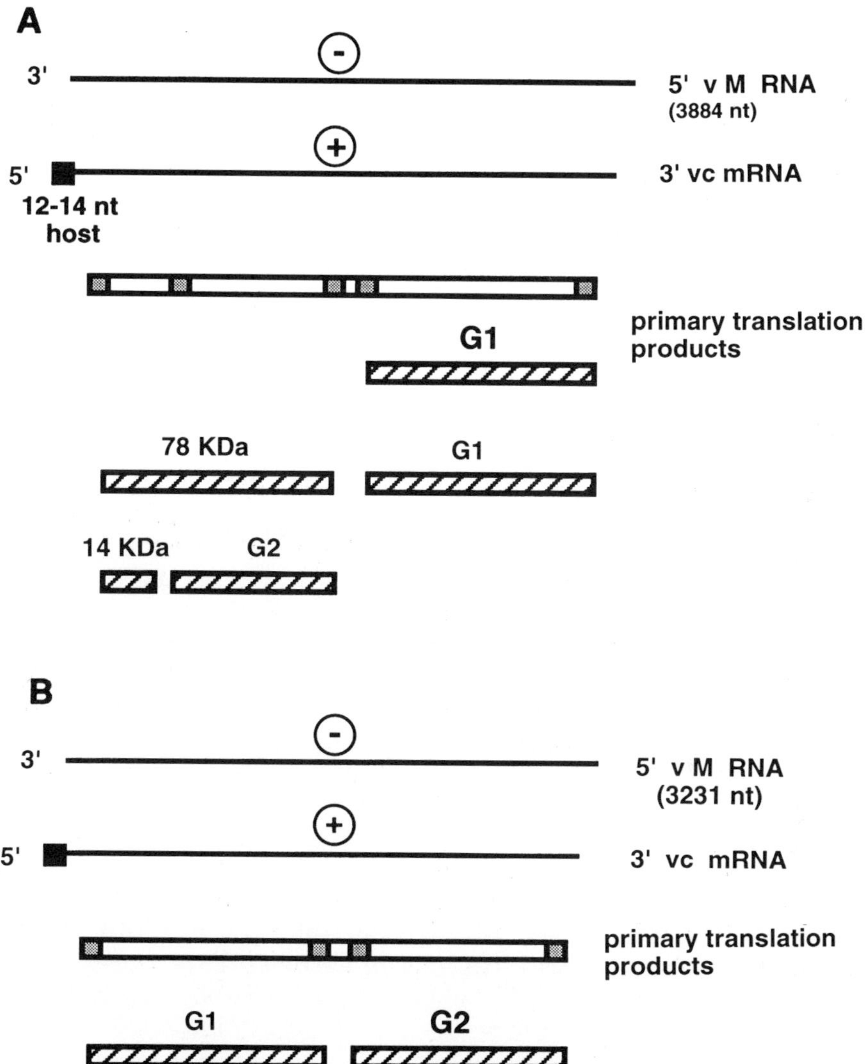

FIGURE 3. Schematic representation of the expression strategy of the M segments of RVF (panel A) and UUK (panel B) viruses. The M genomic segment is transcribed to give a viral-complementary sense mRNA (vc M RNA), which possesses host primers at the 5' end (black boxes) and is truncated at the 3' end. The primary translation product is a polyprotein, which has not been found in virus-infected cells. For RVF virus, some of the G1 protein synthesized may also be a primary translation product (see text).

   The shaded boxes in the precursor represent stretches of hydrophobic amino acids. In the case of RVF the mature products are the G1, G2, and 14-kDa (NSm) proteins, in the case of UUK only G1 and G2 are produced. Based on features of the encoded ORFs, expression of the M segments of PT and TOS viruses probably resembles that of RVF virus.

quired to encode the two glycoproteins, indicating that there are 30 and 14 kDa of protein information respectively preceding the first glycoprotein. Thus, the gene order of the PT virus M segment is 5′ NSm–G1–G2 3′, whereas that of RVF virus is 5′ NSm–G2–G1 3′. The coding capacity of the TOS M segment ORF also suggests that a nonstructural protein in addition to the structural glycoproteins is encoded, with the gene order 5′ NSm–G2–G1 3′ (Giorgi, unpublished data).

The gene order of the M segment of UUK virus was determined by synchronized initiation of protein synthesis and pulse labeling (Kuismanen, 1984), and later confirmed by terminal sequencing of the G1 and G2 glycoproteins (Rönnholm and Pettersson, 1987), to be 5′ G1–G2 3′. There is no evidence for an equivalent of the NSm protein.

C. L RNA Segment

In the past few years the nucleotide sequences of the L RNA segment of the three phleboviruses (RVF, UUK, TOS) have been reported (Muller *et al.*, 1991; Elliott *et al.*, 1992; Accardi *et al.*, 1993), thus completing the entire genome sequences of RVF and UUK viruses.

The three L RNA segments are similar in length (6404–6423 nt) and present the same characteristics; there is a single ORF in the viral-complementary RNA that could code for a high-molecular-weight protein. Since in bunyaviruses, using reassortants between viruses of the California serogroup (Endres *et al.*, 1989) or viruses of the Bunyamwera serogroup (Elliott, 1989), it had been shown that the L segment encodes the L protein, the same was predicted for phleboviruses. Recently, the L protein of RVF virus was expressed via a recombinant vaccinia virus (Lopez *et al.*, 1995) and a protein with molecular weight >200 kD was detected in TOS virus infected cells, using monospecific antisera raised to different portions of the TOS L segment encoded protein (Di Bonito *et al.*, unpublished results), confirming the coding assignment of the genomic L segment in phleboviruses.

IV. GENE EXPRESSION

A. S Segment

The S segment of phleboviruses contains the genetic information for two proteins, one coded in the viral-complementary RNA and the other in the viral (genomic)-sense RNA.

Transcriptional analysis has revealed that the two ORFs are expressed via two subgenomic mRNAs of opposite polarity (Ihara *et al.*, 1984, 1985a; Marriott *et al.*, 1989; Simons *et al.*, 1990): the viral-complementary mRNA codes for the N protein and the viral-sense mRNA codes for the NSs protein (Ihara *et al.*, 1985a; Emery and Bishop, 1987; Fig. 2). Analysis of viral mRNA synthesis in PT virus-infected cells, in the presence of inhibitors of protein synthesis, has revealed that mRNA transcription and genome replication have the same features as the other negative-strand virus genomes (Ihara *et al.*, 1985a). In fact, while in PT-infected cells full-length S viral RNA, S viral-complementary RNA, and subgenomic N and NSs mRNAs were detected, only newly synthesized N mRNA was identified in the presence of protein synthesis inhibitors. These results indicate that transcription of the NSs mRNA takes place on the antigenome only after genome replication has begun.

The 5′ and 3′ ends of these mRNAs have been analyzed for PT (Ihara *et al.*, 1985a), UUK (Simons and Pettersson, 1991), and TOS (Grò *et al.*, 1993) viruses. For all three viruses the 5′ ends were shown to contain heterogeneous, nonviral sequences, about 10 to 15 nucleotides in length. Determination of the sequences of individual 5′ ends of UUK mRNAs showed a wider length distribution (7–25 nucleotides), and that the first virus-encoded base (A in the viral-complementary sense) was frequently missing from the mRNAs, and replaced by a G residue. A similar situation has been described for influenza (Krug, 1981), snowshoe hare (Bishop *et al.*, 1983), and Germiston (Bouloy *et al.*, 1990) virus mRNAs. These data are taken as evidence that phleboviruses, like other viruses in the family *Bunyaviridae*, initiate mRNA synthesis by a mechanism similar to "cap-snatching" originally described for influenza virus mRNA production (Krug *et al.*, 1989).

The subgenomic N and NSs mRNAs of the PT, UUK, and TOS viruses have been shown to terminate in the noncoding intergenic region. In the case of PT virus, the 3′ termini of these mRNAs were mapped at either side of the tip of the large hairpin structure predicted for folding of the AU-rich intergenic region (Emery and Bishop, 1987). In the case of UUK virus, the last 100 residues at the 3′ ends of the two mRNAs overlap each other and contain a palindromic sequence that can be folded into a stem-loop structure. However, comparison of the UUK and PT S RNA sequences around the 3′ ends of the two mRNAs did not reveal any similarities (Simons and Pettersson, 1991).

In contrast, the 3′ ends of N and NSs mRNAs of the TOS virus terminate in a sequence motif (described above; Grò *et al.*, 1993) which is conserved in the RVF and SFS viral intergenic regions (Giorgi *et al.*, 1991). In these latter two viruses, the 3′ ends of N and NSs mRNAs have not been mapped directly, but the conservation of the motif indicates its possible role in the termination of transcription. Like the UUK N and NSs mRNAs, those of TOS virus cover the intergenic region and their 3′ ends overlap by about 80 nt. Since the viral mRNAs are not encapsidated in virus-infected cells, the 3′ complementary regions could anneal and in some way control the stability or translation of the N and NSs mRNAs.

From the above results no clear picture can be obtained for the mechanism of transcription termination of the mRNAs derived from the S segment of phleboviruses and perhaps different mechanisms exist for the different viruses.

B. M Segment

The phlebovirus M segments examined to date all possess a single ORF in the viral-complementary sense RNA which is expressed via an mRNA similar in size to the respective genomic segment (Parker *et al.*, 1984; Pettersson *et al.*, 1985; Collett, 1986; Giorgi *et al.*, unpublished data).

Examination of the termini of RVF M segment mRNA showed that approximately 112 nucleotides at the 3' end were missing relative to the genomic template, and that the 5' end was extended by about 12–14 nucleotides of heterogeneous nonviral sequence (Collett, 1986). We found the same characteristics for the mRNA expressed from the M genomic segment of TOS virus: it lacks about 100 nucleotides at the 3' end and possesses an extension of 10–15 nucleotides at the 5' end (Giorgi *et al.*, unpublished data). Collett (1986) mapped the 3' end of RVF M mRNA near a symmetrical sequence rich in purines, and inspection of the sequences around the predicted 3' end of TOS M segment mRNA revealed the presence of similar sequences. However, there is no direct experimental evidence yet for the involvement of these apparently conserved sequences in transcription termination.

Although the ORF of the phlebovirus M segment is expressed by a single mRNA, it encodes at least two proteins, the virion glycoproteins G1 and G2. As the product corresponding to the complete ORF has never been observed in virus-infected cells, it is likely that the putative precursor is processed cotranslationally. Gene expression of the M segment has been studied extensively in the case of RVF virus (Fig. 3). A primary translation product of 133 kDa, the size expected for a polypeptide encompassing the entire ORF of the M segment, was obtained in a cell-free translation system using mRNA-like transcripts synthesized from M segment cDNA-containing plasmids (Suzich and Collett, 1988). In the presence of microsomal membranes, this primary protein was cotranslationally processed to yield the two glycoproteins G1 (63–65 kDa) and G2 (56 kDa), as well as three other proteins of 78, 21, and 14 kDa; the nonglycosylated 14-kDa and the glycosylated 78-kDa proteins co-migrated with M segment-encoded proteins found in RVF virus-infected cell lysates, but the 21-kDa protein is apparently not synthesized *in vivo*.

A protein pattern similar to that seen in authentic RVF virus-infected cells was obtained by expressing M segment cDNAs using a recombinant vaccinia virus (Kakach *et al.*, 1988). By manipulation of the sequences expressed, in conjunction with monospecific antisera, it was shown that the M segment expression strategy involved aspects not only of protein processing, but also of translational initiation. Within the preglycoprotein region, which precedes the amino terminus of the G2 protein, there are five in-frame AUGs which could potentially initiate protein translation. The utilization of these codons was investigated using vaccinia virus recombinants in which selected AUG codons in the preglycoprotein region were eliminated (Suzich *et al.*, 1990). It was shown that the 78-kDa protein, which encompasses the preglycoprotein as well as the G2 coding sequences, was initiated at the first

AUG (Kakach *et al.*, 1988; Suzich *et al.*, 1990). However, the 78-kDa protein does not seem to be the precursor of G2 or of the 14-kDa protein; the 14-kDa protein is initiated independently at the second in-frame AUG codon. Interestingly, synthesis and correct processing of G1 and G2 glycoproteins occurred if the first two AUGs were removed, indicating that the 78- and 14-kDa proteins are not functionally required in the biogenesis of the mature G1 and G2 glycoproteins. However, no G2 synthesis was observed if the 22 amino acids preceding the amino terminus of the protein were removed. The same vaccinia virus M recombinant produced correctly processed G1, albeit at significantly reduced levels. Synthesis of G2 appeared to require an internal initiation codon in the preglycoprotein region, i.e., not the first AUG but probably the second. In contrast, only a portion of the G1 synthesized seemed to involve the AUGs in the preglycoprotein region.

These data together suggest that the expression of the 78-kDa, 14-kDa, G2 proteins, and a proportion of G1 involves at least two primary polyprotein translation products and their proteolytic processing. One putative precursor, starting at the first AUG codon and encompassing all of the M segment ORF, would yield the 78-kDa protein, and possibly some of the G1 made in the cell (Fig. 3). The putative second translation product, beginning at the second AUG and including all of the ORF, would serve as a precursor for the 14-kDa protein and for the mature glycoproteins G1 and G2. However, about half of the G1 synthesized appears to follow a biosynthetic pathway distinct from that just discussed, and is suggested to originate via an internal translational initiation at a specific initiation codon just upstream of the G1 region (Suzich *et al.*, 1990).

Studies on the utilization of N-linked glycosylation sites within the gene products of the M segment revealed that the site predicted in the preglycoprotein region (Collett *et al.*, 1985), which would be present in both 78- and 14-kDa species, was utilized only in the 78-kDa protein (Kakach *et al.*, 1989). Thus, the use of the first or second AUG codons seems to determine not only the proteolytic processing of the primary polyprotein products but also the utilization of the N-linked glycosylation site within the preglycoprotein region (Kakach *et al.*, 1989). Since proteolytic cleavage at the preglycoprotein-G2 junction is independent of glycosylation at the preglycoprotein site (Kakach *et al.*, 1989), it appears it is the amino acid sequence between the first and second AUG codons that influences the glycosylation and processing. Thus, it has been proposed (Suzich *et al.*, 1990) that the use of two different sites for the initiation of translation might serve to control posttranslational protein modification.

C. L Segment

The sequence data indicate that the L segment employs a negative-sense coding strategy, and evidence for a viral-complementary L RNA, approximately the size of the genomic segment, was obtained in RVF and UUK virus-infected cells (Parker *et al.*, 1984; Pettersson *et al.*, 1985). Using L-specific probes, the presence of an L mRNA was shown in TOS virus-infected cells (Accardi *et al.*, 1993), proving that the L segment ORF is expressed via a viral-complementary mRNA (Fig. 4). The TOS L mRNA also possesses non-viral sequences at its 5′ end, 9–14 nt in length, as the other viral mRNAs.

V. GENE PRODUCTS AND THEIR FUNCTION

A. N and NSs Proteins

The S genomic segment of the phlebovirus codes for two proteins, N and NSs. The sizes of the N proteins of PT (Ihara *et al.*, 1985a; Overton *et al.*, 1987), SFS (Marriott *et al.*, 1989), UUK (Simons *et al.*, 1990), RVF and TOS (Giorgi *et al.*, 1991) viruses are comparable, ranging from 27 kDa to 28.5 kDa (243–254 amino acids).

Comparison of the amino acid sequences shows the N proteins of phlebotomous fever group viruses to be highly related (Simons *et al.*, 1990; Giorgi *et al.*, 1991), with UUK being more distantly related. The percentage of amino acid identity ranges from 53% (PT versus RVF) to 31% (TOS versus UUK). Comparison at the amino acid level identifies some regions of exact identity among the N proteins of PT, SFS, RVF, and TOS viruses. The same regions can also be detected in the UUK N amino acid sequence but they are less extended. Whether these regions play a role in protein–nucleic acid and/or in protein–protein interaction is not known. In contrast, the analysis of NSs amino acid sequences shows low homology indicating a higher degree of variability among these proteins (Giorgi *et al.*, 1991).

The N protein is the most abundant viral protein found in the virion and in virus-infected cells; it is associated with the genomic RNA to constitute the viral nucleocapsids. Synthesis of N protein can be detected early after

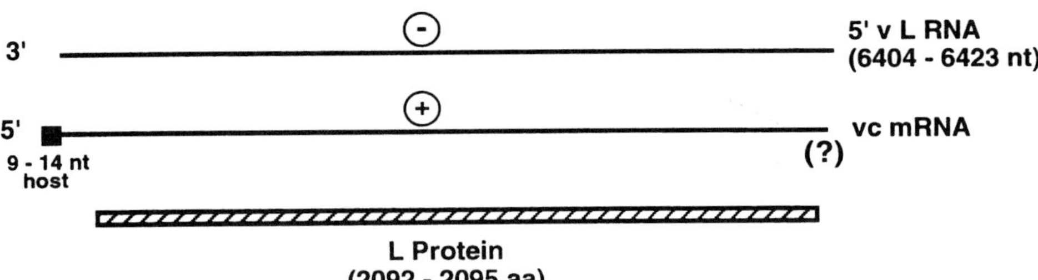

FIGURE 4. Schematic representation of the expression strategy of the phlebovirus L segment. The viral L RNA segment (v L RNA) is transcribed to give a viral-complementary mRNA (vc mRNA) which expresses a large-molecular-weight protein (L protein). The 5′ end of the mRNA possesses host nucleotides (black box). The 3′ end of the mRNA has not yet been mapped. The numbers in parentheses indicate the size range among RVF, TOS, and UUK viruses.

infection (2 hr postinfection) and the protein has a half-life of several hours (Ulmanen *et al.*, 1981; Simons *et al.*, 1992).

In phlebovirus-infected cells the N protein seems to accumulate in the Golgi region during later stages of infection (Smith and Pifat, 1982; Kuismanen *et al.*, 1982, 1984; Simons *et al.*, 1992), whereas N protein expressed by recombinant vaccinia virus (in absence of other viral proteins) localizes only in the cytoplasm (Simons *et al.*, 1992). The Golgi accumulation is thought to be caused by the association of the cytoplasmic ribonucleoprotein with the transmembranal sequences of the viral envelope glycoproteins, which reside in the Golgi complex. The direct association seems to represent the first step in the assembly process for viruses in the *Bunyavirdae*, which lack an equivalent to the matrix protein present in other negative-strand RNA virus groups.

The nucleocapsid assembly process for phleboviruses has not yet been elucidated, but it is thought to be controlled at the level of initiation by interaction of the N protein with a specific sequence on its target RNA. The encapsidation process would then proceed by addition of further N protein molecules to the preexisting complex by RNA–N and N–N interaction. As the genome and antigenome are always found encapsidated (in contrast to the mRNAs which are not encapsidated) the sequence required for nucleocapsid assembly must be present at the 3' end of both the genome and antigenome. It is likely, therefore, that these sequences are contained in the conserved regions found at the termini of each viral genomic segment (Raju and Kolakofsky, 1987; Kolakofsky and Hacker, 1991).

There are no reports on the RNA binding properties of the phlebovirus N protein. Recently, Gött *et al.* (1993) described binding *in vitro* of recombinant N protein of Hantaan and Puumala viruses (*Hantavirus* genus) to S RNA, and showed that the interaction between N protein and RNA occurs via a domain near the carboxy terminus which is well conserved among the hantaviruses. In the case of phleboviruses the carboxy-terminal domain of the N proteins is variable and hence interaction would probably be mediated by the conserved regions found in other parts of the N protein (Giorgi *et al.*, 1991).

Studies on the NSs protein of a number of phleboviruses have generated data that are difficult to reconcile if one assumes a common function for the protein. In RVF virus-infected cells NSs was found to be phosphorylated and accumulated in the nucleus (Struthers and Swanpoel, 1982; Struthers *et al.*, 1984). In contrast, the NSs of Karimabad (Smith and Pifat, 1982) and UUK viruses (Simons *et al.*, 1992) was found dispersed in the cytoplasm, whereas in TOS virus-infected cells, NSs accumulates in the Golgi complex during the later stages of the infection (Giorgi *et al.*, unpublished data). Moreover, the NSs protein has been found associated with virions and nucleocapsids in the case of PT (Overton *et al.*, 1987) and TOS viruses (Giorgi *et al.*, unpublished data), and with the 40 S ribosomal subunits in UUK (Simons *et al.*, 1992) and PT (Watkins and Jones, 1993) virus-infected cells. Given these contrasting results, the function of the NSs protein remains a matter of speculation.

As a consequence of the ambisense expression strategy of the S RNA, the synthesis of NSs protein requires prior genome replication (Ihara *et al.*, 1985a). In fact, it is detected in virus-infected cells some 2 hr later than the N protein (Ulmanen *et al.*, 1981; Simons *et al.*, 1992; Giorgi *et al.*, unpublished data). However, its association with the virion suggests a role in an early event during infection, possibly in transcription and/or replication. For UUK virus, the NSs protein has not been detected in the virion, but it could be synthesized in infected cells as early as the N protein; in fact UUK virions carry both full-length virus-sense and viral-complementary sense S RNAs (Simons *et al.*, 1990), templates for N mRNA and NSs mRNA, respectively. The association with the nucleocapsid, and the accumulation in the Golgi complex late in the infection also suggests, for some viruses, a possible role for NSs in the interaction with the viral glycoproteins during virion morphogenesis.

B. Glycoproteins

Amino acid sequence comparison of the large ORF of the phlebovirus M segments reveals a clear homology among RVF, PT, and TOS viruses (about 40% identity), especially in the carboxy terminus of the precursor protein. Between the glycoproteins of UUK virus and the other three phleboviruses the homology is lower (about 25%), but significant considering that the position of about 70% of the cysteine residues are conserved (Rönnholm and Pettersson, 1987; Giorgi *et al.*, unpublished data). The precursor proteins are rich in cysteine residues (about 5%) and their conservation strongly suggests a conserved three-dimensional structure of G1 and G2. In the M segment of RVF and PT viruses the G1 and G2 glycoprotein coding regions are preceded by a region coding for a nonstructural protein. These regions do not show any obvious homology (Ihara *et al.*, 1985b).

The hydropathy profiles of the M segment ORFs of the four phleboviruses exhibit striking similarities. At the carboxy termini of G1 and G2, there are hydrophobic stretches of amino acids (about 20–30 residues in length, depending on the virus) followed by some charged amino acids, similar to the "stop transfer" sequences seen in the transmembrane domains of other viral envelope proteins (Garoff *et al.*, 1980; Rose *et al.*, 1980), which probably anchor the proteins in the lipid bilayer. The amino termini of the mature G1 and G2 proteins are preceded by stretches of predominantly hydrophobic amino acids which may function as signal sequences (Fig. 3; Rönnholm and Pettersson, 1987).

There are few potential N-linked glycosylation sites in these glycoproteins: four sites are present in both G1 and G2 of UUK, in G2 of PT, and in G1 of RVF and TOS viruses; of these only two are conserved between G2 of PT virus and G1 of RVF virus (i.e., the second proteins in the ORF; Ihara *et al.*, 1985b). Only one site is predicted in G1 of PT and in G2 of RVF. Evidence that these sites are utilized comes from the analysis of the glycoproteins in virus-infected cells in the presence of tunicamycin, a drug that inhibits N-linked glycosylation and thus reduces the electrophoretic mobility of the un-glycosylated glycoproteins (Kuismanen, 1984; Collett *et al.*, 1985; Robeson *et al.*, 1979). Using recombinant vaccinia viruses which express regions of the RVF virus M segment, it was shown that G2 was glycosylated at its single site and G1 at at least three sites out of four. As described above, the N-linked glycosylation site present in the RVF preglycoprotein region was utilized only in 78-kDa protein and not in 14-kDa protein (Kakach *et al.*, 1989).

The detailed structure of the glycans associated with these glycoproteins has only been analyzed for UUK virus (Pesonen *et al.*, 1982). Two types of structures were found: a high-mannose form (endo H-sensitive), and an inter-mediate type (endo H-resistant) which probably represents an intermediate phase in the maturation from high-mannose to complex glycans. Uukuniemi virus G1 contains only the endo H-resistant glycans, whereas G2 contains mainly endo H-sensitive ones.

A characteristic feature of the members of the *Bunyaviridae* is their intracellular maturation by budding at smooth-surfaced membranes in the Golgi complex (Murphy *et al.*, 1973; Kuismanen *et al.*, 1982; Smith and Pifat, 1982). Budding at the plasma membrane has been observed, however, in rat hepatocytes infected with RVF virus (Anderson and Smith, 1987), suggesting that the factors that mediate post-Golgi transport may vary in different cell types. A major factor determining the site of virus maturation seems to be the accumulation of the two glycoproteins G1 and G2 in the Golgi complex. In cells infected with a temperature-sensitive (ts) mutant of UUK virus (Gahmberg, 1984), it was shown that the glycoproteins were transported to the Golgi complex in the absence of virus maturation, and were retained there for several hours when protein synthesis was inhibited (Gahmberg *et al.*, 1986a). In this system, only a small amount, if any, of the glycoproteins was transported to the cell membrane. The accumulation of the glycopro-teins induces vacuolization and expansion of the Golgi complex which, however, retains its functional integrity. In fact, in cells coinfected with UUK and Semliki Forest viruses, the glycoproteins of the latter virus were properly glycosylated and transported to the plasma membrane (Gahmberg *et al.*, 1986b). The retention of glycoproteins in the Golgi, even in the absence of virus maturation, suggests that the glycoproteins themselves may have a signal for retention in this organelle.

The possible role of the amino-terminal sequences (preglycoprotein re-gion) of the RVF virus M segment ORF in intracellular transport and target-ting not only of the products they encode (78- and 14-kDa proteins) but also of G1 and G2, has been investigated. Vaccinia virus recombinants which ex-press the 14-kDa protein and G1 and G2 glycoproteins, or G1, G2, and pos-sibly a 12.5-kDa preglycoprotein, or only G1 and G2 were constructed by inserting M segment coding sequences progressively deleted at the 5' region. By immunofluorescence, the proteins were shown to localize to the Golgi complex, suggesting that the phlebovirus NSm was not involved in this process. Results from expression of the G1 and G2 coding regions of the PT virus M segment by recombinant vaccinia viruses lead to the same conclusion (Matsuoka *et al.*, 1988).

The role in anchorage and retention in the Golgi complex of the hydrophobic regions, present at the carboxy and amino termini of the G2 glycoprotein (the second protein on the ORF) of PT virus, was investigated by expression of some G2 constructs in a vaccinia-based expression system (Chen *et al.*, 1991). A vaccinia virus containing G2 coding sequences, with a partial deletion in the presumptive signal peptide, did not express the protein. In contrast, this protein was stably produced and glycosylated using a construct containing the complete hydrophobic region preceding G2. This result indicates that the hydrophobic sequences at the amino terminus of G2 act as a signal peptide mediating the translocation of G2 in the endoplasmic reticulum (ER).

Using a similar approach, it was shown that a recombinant vaccinia virus containing G2 sequences and the hydrophobic domains at the amino and carboxy termini expressed a G2 protein which localized to the cell membrane. If the hydrophobic domain at the carboxy terminus was removed, the G2 protein was secreted into the medium. These results indicate that the carboxy terminus of G2 serves as the membrane anchor, and that the G2 protein does not contain a specific signal for the Golgi retention; thus, the Golgi localization of G2 observed in cells that coexpress G1 must be related to its interaction with the G1 glycoprotein.

Similar conclusions were reached when G1 and G2 of UUK virus were expressed from simian virus 40-based vectors (Rönnholm, 1992). In fact, when expressed together G1 and G2 were correctly translocated, processed, and targetted to the Golgi. When expressed separately, G1 was transported to the Golgi, whereas G2 could not be detected in the Golgi; probably G2 was retained and finally degraded in the ER.

From further studies (Persson and Pettersson, 1991; Chen and Compans, 1991; Chen *et al.*, 1992), it appears that the maturation of G1 and G2 of both PT and UUK viruses results in the formation of heterodimers. Heterodimerization occurs rapidly, probably in the ER, between newly synthesized G1 (which appears to acquire its mature form within a few minutes) and mature G2. In PT virus-infected cells a fraction of the G2 protein is also assembled into G2 homodimers (Chen and Compans, 1991). Both the G1–G2 heterodimers and the G2 homodimer are able to exit from the ER and be transported to the Golgi, where they are retained, the former possibly by a specific retention mechanism mediated by the G1 protein and the latter by interaction with the G1–G2 heterodimer. Taken together these results indicate that the signal for retention in the Golgi must reside in the first glycoprotein coded by the M segment. Preliminary results obtained by expression of UUK G1 mutants, constructed by progressive deletion or insertion of specific sequences, in BHK 21 cells using the vaccina virus–T7 RNA polymerase expression system have suggested that the retention signal is not contained in the transmembrane domain or in the carboxy-terminal tail of the protein (Melin *et al.*, 1993).

The glycoproteins of phleboviruses, and of members of the *Bunyaviridae* family in general, are important for viral infection and pathogenesis. They have been shown to induce and interact with virus-neutralizing antibodies (Gentsch *et al.*, 1980; Dalrymple *et al.*, 1981). The amino acid sequences of some epitopes, defined by monoclonal antibodies (MAbs) able to neutralize virus infectivity, have been identified on the RVF virus G2 glycoprotein (Keegan and Collett, 1986). Competitive binding assays with a panel of MAbs have defined four antigenic regions on RVF virus G1 and G2, with domains I, II, and IV on G1 and domain I on G2 being involved in neutralization and hemagglutination (Besselaar and Blackburn, 1991). Synergistic neutralization was observed with different pairs of these MAbs (Besselaar and Blackburn, 1992). Distinct antigenic determinants have been also defined by MAbs on both glycoproteins of PT virus. MAbs to both G1 and G2 glycoproteins were able to neutralize virus infectivity *in vitro* and inhibit hemagglutination, but only G1-specific MAbs provided efficient protection in mice (Pifat *et al.*, 1988; Pifat and Smith, 1987).

The G1 and G2 proteins of RVF virus, expressed either in a bacterial system, in vaccinia recombinant virus (Collett *et al.*, 1987; Dalrymple *et al.*, 1989), or in recombinant baculovirus (Schmaljohn *et al.*, 1989), have been shown to elicit a good immune response in mice. When used separately to immunize mice, G2, but not G1, induced neutralizing antibodies that protect animals against a challenge with infectious virus.

C. L Protein

The product of the L segment ORF is a protein with a molecular weight greater than 200 kDa, as predicted by the amino acid sequences (Muller *et al.*, 1991, 1994; Elliott *et al.*, 1992; Accardi *et al.*, 1993). In TOS virus infected cells, the L protein is synthesized early in infection (2h p.i.) and it is present in mature virions (C. Giorgi *et al.*, unpublished data).

Recently, the complete L segment of RVF virus was expressed via a recombinant vaccinia virus (Lopez *et al.*, 1995). The expressed L protein was shown to restore the RNA polymerase activity in natural ribonucleoproteins, depleted of this activity after purification in CsCl gradient, and to achieve the transcription of a synthetic template in cells coexpressing the N protein. These results clearly indicate that the L protein is an essential component of the transcriptase complex.

Until now the L segment nucleotide sequences of three phleboviruses (TOS, RVF, and UUK viruses) are available. The percentage of overall homology of the predicted amino acid sequences ranges between 52% (RVF-TOS) and 37% (RVF-UUK). The homology is not uniformly distributed along the sequences but is more pronounced in the central part of the proteins. Some conserved regions were observed in the middle third of the three sequences (Elliott *et al.*, 1992; Accardi *et al.*, 1993; Muller *et al.*, 1994). These regions correspond to the polymerase motifs identified by Poch *et al.* (1989) in all

RNA-dependent RNA polymerases examined. The possible involvement of these regions in the activity of the proteins has been tested on the L protein of Bunyamwera virus (Bunyavirus genus). This protein, expressed by recombinant vaccinia virus, is functional in terms of RNA synthesis activity (Jin and Elliott, 1991) and possesses both transcriptase and replicase activities (Jin and Elliott, 1992). Substitution of amino acids strictly conserved among the L proteins of different viruses in the *Bunyaviridae* family abolished the RNA synthesis activity of the L protein (Jin and Elliott, 1993), thus suggesting that the conserved regions may represent the catalytic sites of the L proteins.

VI. PERSISTENT INFECTIONS

Like other arboviruses, members of the *Bunyaviridae* can replicate both in permissive vertebrate cells and in insect cells. The infection of cultured vertebrate cells is highly cytopathic and leads to cell death, whereas that in cultured mosquito cells is asymptomatic and becomes persistent. This situation may be related to the inapparent infection in mosquitoes *in vivo*.

Mosquito cells, persistently infected with bunya- and phleboviruses (Newton *et al.*, 1981; Nicoletti and Verani, 1985; Carvalho *et al.*, 1986; Elliott and Wilkie, 1986; Rossier *et al.*, 1988; Nicoletti *et al.*, 1989; Delord *et al.*, 1989), can be maintained for long periods with subculturing, and have been shown to produce virus, although at lower levels than from lytically infected cells.

During the early stages of the persistence in *Aedes albopictus* C6/36 cells infected with La Crosse (Rossier *et al.*, 1988) and Bunyamwera (Scallan and Elliott, 1992) viruses, the virus replication has a self-limiting nature, due to translational control by the N protein. This control is mediated by the encapsidation of the N mRNA by N protein, thereby preventing its translation (Hacker *et al.*, 1989). At the later passages of persistently infected cells, the presence of subgenomic RNAs derived from the L RNA segment was observed, but these shorter RNAs were not packaged into particles (Elliott and Wilkie, 1986; Scallan and Elliott, 1992). It is not clear if they play some role in the maintenance of the persistent infection in mosquito cells.

Even though infection of mammalian cells is generally lytic, persistent infections have also been established: a persistent infection of Vero cells was established by serial undiluted passages of Toscana phlebovirus, a virus normally lytic in these cells (Verani *et al.*, 1984). The persistently infected cells were morphologically similar to the parental Vero cells, released variable amounts of infectious virus, and were resistant to superinfection with homologous virus. The virus from these cells caused interference with the multiplication of the standard virus. These characteristics are compatible with a mechanism mediated by defective interfering (DI) particles (Holland, 1990).

Recently, subgenomic RNAs derived from the L segment have been found in TOS virus-infected cells after sequential passages of the virus at high multiplicity (Marchi, Nicoletti, and Giorgi, unpublished data). These RNAs are encapsidated and replicated, suggesting they behave as DI particles. The subgenomic L RNAs contain large internal deletions, representing approximately 87–93% of the L RNA segment. Interestingly, short nucleotide sequences (5–7 nt) are repeated at either side of the junction sites. This feature has also been reported for defective L segment-derived RNAs for tomato spotted wilt virus (*Tospovirus* genus), a plant-infecting member of the *Bunyaviridae* (de Oliveira Resende *et al.*, 1992). DI particles, with large deletions in the L RNA segment, have also been described in BHK and L cells infected with Bunyamwera bunyavirus (Patel and Elliott, 1992).

The mechanisms involved in the establishment and maintenance of a persistent infection in mosquito and mammalian cells appear to be very different. The role of DI particles and cellular interaction in determining the different outcome of infection in the two systems remains to be elucidated.

VII. CONCLUSION

Over the past few years, a number of advances have been made in our knowledge of the organization and gene expression of different members of the Phlebovirus genus. However, there are still many gaps in our understanding of the mechanisms involved in each phase of the viral replication. With the application of the experimental approach employing genetic manipulation techniques, developed for the negative strand viruses, the research is changing from a predominantly descriptive phase to one that can surely help answer the many important questions on the molecular biology of these viruses.

ACKNOWLEDGMENTS. I am indebted to Mrs. Giulia Pacetto for excellent preparation of the manuscript.

VIII. REFERENCES

Accardi, L., Grò, M. C., Di Bonito, P., and Giorgi, C., 1993, Toscana virus L segment: Molecular cloning, coding strategy and amino acid sequence in comparison with other negative strand RNA viruses, *Virus Res.* **27**:119.

Anderson, G. W., Jr., and Smith, J. F., 1987, Immunoelectron microscopy of Rift Valley fever viral morphogenesis in primary rat hepatocytes, *Virology* **161**:91.

Beaty, B. J., and Calisher, C. H., 1991, Bunyaviridae—natural history, *Curr. Top. Microbiol. Immunol.* **169**:27.

Besselaar, T. G., and Blackburn, N. K., 1991, Topological mapping of antigenic sites on the Rift Valley fever virus envelope glycoproteins using monoclonal antibodies, *Arch. Virol.* **99**:75.

Besselaar, T. G., and Blackburn, N. K., 1992, The synergistic neutralization of Rift Valley fever virus by monoclonal antibodies to the envelope glycoproteins, *Arch. Virol.* **125**:239.

Bishop, D. H. L., and Shope, R. E., 1979, Bunyaviridae, in: *Comprehensive Virology* (H. Fraenkel-Conrat and R. A. Wagner, eds.), Vol. 14, pp. 1–156, Plenum Press, New York.

Bishop, D. H. L., Calisher, C., Casals, J., Chumakov, M. P., Gaidomovich, S. Y. A., Hannoun, C., Luou, D. K., Marshall, I. D., Oker-Blom, N., Pettersson, R. F., Porterfield, J. S., Russel, P. K., Shope, R. E., and Westaway, E. G., 1980, Bunyaviridae, *Intervirology* **14**:125.

Bishop, D. H. L., Gay, M. E., and Matsuoka, Y., 1983, Non viral heterogeneous sequences are present at the 5' ends of one species of snowshoe hare bunyavirus S complementary RNA, *Nucleic Acids Res.* **11**:6409.

Bouloy, M., 1991, Bunyaviridae: Genome organization and replication strategies, *Adv. Virus Res.* **40**:235.

Bouloy, M., Pardigon, N., Vialat, P., Gerbaud, S., and Girard, M., 1990, Characterization of the 5' and 3' end of viral messenger RNAs isolated from BHK 21 cells infected with Germiston virus (Bunyavirus), *Virology* **175**:50.

Calisher, C. H., 1991, Bunyaviridae, in: *Classification and nomenclature of viruses, Arch. Virol.* (Suppl. 2):273.

Caplen, H., Peters, C. J., and Bishop, D. H. L., 1985, Mutagen-directed attenuation of Rift Valley fever virus as a method for vaccine development, *J. Gen. Virol.* **66**:2271.

Carvalho, M. G., Frugulhetti, I. C., and Revello, M. A., 1986, Marituba (Bunyaviridae) virus replication in cultured *Aedes albopictus* cells and in L-A9 cells, *Arch. Virol.* **90**:325.

Chen, S.-Y., and Compans, R. W., 1991, Oligomerization, transport and Golgi retention of Punta Toro virus glycoproteins, *J. Virol.* **65**:5902.

Chen, S.-Y., Matsuoka, Y., and Compans, R. W., 1991, Golgi complex localization of the Punta Toro virus G2 protein requires its association with the G1 protein, *Virology* **183**:351.

Chen, S.-Y., Matsuoka, Y., and Compans, R. W., 1992, Assembly of G1 and G2 glycoprotein oligomers in Punta Toro virus-infected cells, *Virus Res.* **22**:215.

Clerx-van Haaster, C. M., Clerx, J. P. M., Ushijima, H., Akashi, H., Fuller, F., and Bishop, D. H. L., 1982, The 3' terminal RNA sequences of Bunyaviruses and Nairoviruses (Bunyaviridae). Evidence of end sequence generic differences within the virus family, *J. Gen. Virol.* **61**:289.

Collett, M. S., 1986, Messenger RNA of the M segment RNA of Rift Valley fever virus, *Virology* **151**:151.

Collett, M. S., Purchio, A. F., Keegan, K., Frazier, S., Hays, W., Anderson, D. K., Parker, M. D., Schmaljohn, C., Schmidt, J., and Dalrymple, J. M., 1985, Complete nucleotide sequence of the M RNA segment of Rift Valley fever virus, *Virology* **144**:228.

Collett, M. S., Keegan, K., Hu, S.-L., Sridhar, P., Purchio, A. F., Ennis, W. H., and Dalrymple, J. M., 1987, Protective subunit immunogens to Rift Valley fever virus from bacteria and recombinant vaccinia virus, in: *The Biology of Negative Strand Viruses* (B. W. J. Mahy and D. Kolakofsky, eds.), pp. 321–329, Elsevier, Amsterdam.

Dalrymple, J. M., Peters, C. J., Smith, J. F., and Gentry, M. K., 1981, Antigenic components of Punta Toro virus, in: *Replications of Negative Strand Viruses* (D. H. L. Bishop and R. W. Compans, eds.), pp. 167–172, Elsevier, Amsterdam.

Dalrymple, J. M., Hasty, S. E., Kakach, L. T., and Collett, M. S., 1989, Mapping of protective determinants of Rift Valley fever virus using recombinant vaccinia viruses, in: *Vaccines '89* (R. A. Lerner, H. Ginsberg, R. Chanock, and F. Brown, eds.), pp. 371–375, Cold Spring Harbor Laboratory, Cold Spring Harbor, NY.

Delord, B., Poveda, J. D., Astier-Gein, T., Gerbaud, S., and Fleury, H. J. A., 1989, Detection of the bunyavirus Germiston in Vero and *Aedes albopictus* C6/36 cells by *in situ* hybridization using cDNA and asymmetric RNA probes, *J. Virol. Methods* **24**:253.

de Oliveira Resende, R., de Haan, P., van de Vossen, E., de Àvila, A. C., Goldbach, R., and Peters, D., 1992, Defective interfering L RNA segments of tomato spotted wilt virus retain both virus genome termini and have extensive internal deletions, *J. Gen. Virol.* **73**:2509.

Elliott, R. M., 1989, Nucleotide sequence analysis of the large (L) genomic RNA segment of Bunyamwera virus, the prototype of the family Bunyaviridae, *Virology* **173**:426.

Elliott, R. M., 1990, Molecular biology of Bunyaviridae, *J. Gen. Virol.* **71**:501.

Elliott, R. M., and Wilkie, M. L., 1986, Persistent infection of *Aedes albopictus* C6/36 cells by Bunyamwera virus, *Virology* **150**:21.

Elliott, R. M., Schmaljohn, C. S., and Collett, M. S., 1991, Bunyaviridae genome structure and gene expression, *Curr. Top. Microbiol. Immunol.* **169**:91.

Elliott, R. M., Dunn, E., Simons, J. F., and Pettersson, R. F., 1992, Nucleotide sequence and coding strategy of the Uukuniemi virus L RNA segment, *J. Gen. Virol.* **73**:1745.

Emery, V. C., and Bishop, D. H. L., 1987, Characterization of Punta Toro S mRNA species and identification of an inverted complementary sequence in the intergenic region of Punta Toro phlebovirus ambisense S RNA that is involved in mRNA transcription termination, *Virology* **156**:1.

Endres, M. J., Jacoby, D. R., Janssen, R. S., Gonzalez-Scarano, F., and Nathanson, N., 1989, The large viral RNA segment of California serogroup bunyaviruses encodes the large viral protein, *J. Gen. Virol.* **70**:223.

Gahmberg, N., 1984, Characterization of two recombination-complementation groups of Uukuniemi virus temperature-sensitive mutants, *J. Gen. Virol.* **65**:1079.

Gahmberg, N., Kuismanen, E., Keranen, S., and Pettersson, R. F., 1986a, Uukuniemi virus glycoproteins accumulate in and cause morphological changes of the Golgi complex in the absence of virus maturation, *J. Virol.* **57**:899.

Gahmberg, N., Pettersson, R. F., and Kääriäinen, L., 1986b, Efficient transport of Semliki Forest virus glycoproteins through a Golgi complex morphologically altered by Uukuniemi virus glycoproteins, *EMBO J.* **5**:3111.

Garoff, H., Frishaus, A.-M., Simon K., Lehrach, H., and Delius, H., 1980, Nucleotide sequence of cDNA coding for Semliki Forest virus membrane glycoproteins, *Nature* **288**:236.

Gentsch, J. E., Rozhom, E. J., Klimas, R. A., El-Said, L. H., Shope, R. E., and Bishop, D. H. L., 1980, Evidence from recombinant Bunyavirus studies that the M RNA gene product elicit neutralizing antibodies, *Virology* **102**:190.

Giorgi, C., Accardi, L., Nicoletti, L., Grò, M. C., Takehara, K., Hilditch, C., Morikawa, S., and Bishop, D. H. L., 1991, Sequences and coding strategies of the S RNAs of Toscana and Rift Valley fever viruses compared to those of Punta Toro, Sicilian sandfly fever and Uukuniemi viruses, *Virology* **180**:733.

Gött, P., Stohwasser, R., Schnitzler, P., Darai, G., and Bautz, E. K. F., 1993, RNA binding of recombinant nucleocapsid proteins of Hantaaviruses, *Virology* **194**:332.

Grò, M. C., Di Bonito, P., Accardi, L., and Giorgi, C., 1992, Analysis of 3' and 5' ends of N and NSs messenger RNAs of Toscana phlebovirus, *Virology* **191**:435.

Hacker, D., Raju, R., and Kolakofsky, D., 1989, La Crosse virus nucleocapsid protein controls its own synthesis in mosquito cells by encapsidating its mRNA, *J. Virol.* **63**:5166.

Hewlett, M. J., Pettersson, R. F., and Baltimore, D., 1977, Circular forms of Uukuniemi virion RNA: An electron microscopy study, *J. Virol.* **21**:1085.

Holland, J. J., 1990, Defective viral genomes, in: *Virology* (N. Fields and D. M. Knipe, eds.), pp. 151–165, Raven Press, New York.

Ihara, T., Akashi, H., and Bishop, D. H. L., 1984, Novel coding strategy (ambisense genomic RNA) revealed by sequence analysis of Punta Toro phlebovirus S RNA, *Virology* **136**:293.

Ihara, T., Matsuura, Y., and Bishop, D. H. L., 1985a, Analyses of the mRNA transcription processes of Punta Toro phlebovirus (Bunyaviridae), *Virology* **147**:317.

Ihara, T., Smith, J., Dalrymple, J. M., and Bishop, D. H. L., 1985b, Complete sequences of the glycoprotein and M RNA of Punta Toro phlebovirus compared to those of Rift Valley fever virus, *Virology* **144**:246.

Jin, H., and Elliott, R. M., 1991, Expression of functional Bunyamwera virus L protein by recombinant vaccinia viruses, *J. Virol.* **65**:4182.

Jin, H., and Elliott, R. M., 1992, Mutagenesis of the L protein encoded by Bunyamwera virus and production of monospecific antibodies, *J. Gen. Virol.* **73**:2235.

Jin, H., and Elliott, R. M., 1993, Characterization of Bunyamwera virus S RNA that is transcribed and replicated by the L protein expressed from recombinant vaccinia virus, *J. Virol.* **67**:1369.

Kakach, L. T., Wasmoen, T. L., and Collett, M. S., 1988, Rift Valley fever virus M segment: Use of recombinant vaccinia viruses to study phlebovirus gene expression, *J. Virol.* **62**:826.

Kakach, L. T., Suzich, J. A., and Collett, M. S., 1989, Rift Valley fever virus M segment: Phlebovirus expression strategy and protein glycosylation, *Virology* **170**:505.

Keegan, K., and Collett, M. S., 1986, Use of bacterial expression cloning to define the amino acid sequences of antigenic determinants on the G2 glycoprotein of Rift Valley fever virus, *J. Virol.* **58**:263.

Kolakofsky, D., and Hacker, D., 1991, Bunyavirus RNA synthesis: Genome transcription and replication, *Curr. Top. Microbiol. Immunol.* **169**:143.

Krug, R. M., 1981, Priming of influenza viral RNA transcription by capped heterologous RNAs, *Curr. Top. Microbiol. Immunol.* **93**:125.

Krug, R. M., Alons-Caplen, F. V., Julknem, I., and Katze, M. G., 1989, Expression and replication of the influenza virus genome, in: *The Influenza Viruses* (R. M. Krug, ed.), pp. 89–152, Plenum Press, New York.

Kuismanen, E., 1984, Post-translational processing of Uukuniemi virus glycoproteins G1 and G2, *J. Virol.* **51**:806.

Kuismanen, E., Hedman, K., Saraste, J., and Pettersson, R. F., 1982, Uukuniemi virus maturation, accumulation of virus particles and viral antigens in the Golgi complex, *Mol. Cell. Biol.* **2**:1444.

Kuismanen, E., Bang, B., Hurme, M., and Pettersson, R. F., 1984, Uukuniemi virus maturation: Immunofluorescence microscopy with monoclonal glycoprotein-specific antibodies, *J. Virol.* **51**:137.

Levitt, J., Nandè, W., Du, T., and Polson, A., 1963, Purification and electron microscopy of pantropic Rift Valley fever virus, *Virology* **20**:530.

Lopez, N., Muller, R., Prehaud, C., and Bouloy, M., 1995, The L protein of Rift Valley fever Virus can rescue viral ribonucleoproteins and transcribe synthetic genome-like RNA molecules, *J. Virol.* **69**:3972.

Marriott, A. C., Ward, V. K., and Nuttall, P. A., 1989, The S RNA segment of sandfly fever Sicilian virus: Evidence for an ambisense genome, *Virology* **169**:341.

Matsuoka, Y., Ihara, T., Bishop, D. H. L., and Compans, R. C., 1988, Intracellular accumulation of Punta Toro virus glycoproteins expressed from cloned cDNA, *Virology* **167**:251.

Melin, L., Bean, A., Bergquist, A., Bergström, A., and Pettersson, R. F., 1993, In search of a Golgi complex retention signal in the membrane glycoproteins G1 and G2 of Uukuniemi virus, Abstracts of IXth International Congress of Virology, August 1993, Glasgow, Scotland, p. 36.

Muller, R., Poch, O., Delarue, M., Bishop, D. H. L., Bouloy, M., 1994, Rift Valley fever virus L segment: correction of the sequence and possible functional role of newly identified regions conserved in RNA-dependent polymerases, *J. Gen. Virol.* **75**:1345.

Muller, R., Argentini, C., Bouloy, M., Prehaud, C., and Bishop, D. H. L., 1991, Completion of the genome sequence of Rift Valley fever phlebovirus indicates that the L RNA is negative sense and codes for a putative transcriptase-replicase, *Nucleic Acids Res.* **19**:5433.

Murphy, F. A., Harrison, A. K., and Whitefield, S. G., 1973, Bunyaviridae: Morphologic and morphogenetic similarities of Bunyamwera serologic supergroup viruses and several other arthropod-borne viruses, *Intervirology* **1**:297.

Newton, S. E., Short, N. J., and Dalgarno, L., 1981, Bunyamwera virus in replication in cultured *Aedes albopictus* (mosquito) cells: Establishment of a persistent infection, *J. Virol.* **38**:1015.

Nicoletti, L., and Verani, P., 1985, Growth of phlebovirus Toscana in a mosquito (*Aedes pseudoscutellaris*) cell line (AP-61): Establishment of a persistent infection, *Arch. Virol.* **85**:35.

Nicoletti, L., Marchi, A., Bellisario, L., and Verani, P., 1989, Toscana virus (Phlebovirus) replication in cultured arthropod cells: Establishment of persistent infections, in: *Invertebrate Cell System Application* (J. Mitsuashi, ed.), Vol. II, pp. 169–176, CRC Press, Boca Raton, FL.

Overton, H. A., Ihara, T., and Bishop, D. H. L., 1987, Identification of the N and NSs proteins coded by the ambisense S RNA of Punta Toro phlebovirus using monospecific antisera raised to baculovirus expressed N and NSs proteins, *Virology* **157**:338.

Parker, M. D., and Hewlett, M. J., 1981, The 3' terminal sequences of Uukuniemi and Inkoo virus RNA genome segments, in: *Replication of Negative Strand Viruses* (D. H. L. Bishop and R. W. Compans, eds.), pp. 125–145, Elsevier, Amsterdam.

Parker, M. D., Smith, J. F., and Dalrymple, J. M., 1984, Rift Valley fever virus intracellular RNA: A functional analysis, in: *Segmented Negative Strand Viruses* (R. W. Compans and D. H. L. Bishop, eds.), pp. 21–28, Academic Press, New York.

Patel, A. H., and Elliott, R. M., 1992, Characterization of Bunyamwera virus defective interfering particles, *J. Gen. Virol.* **73:**389.

Persson, R., and Pettersson, R. F., 1991, Formation and intracellular transport of a heterodimeric viral spike protein complex, *J. Cell Biol.* **112:**257.

Pesonen, M., Kuismanen, E., and Pettersson, R. F., 1982, Monosaccharide sequence of protein-bound glycans of Uukuniemi virus, *J. Virol.* **41:**390.

Pettersson, R. F., and Von Bonsdorff, C.-H., 1975, Ribonucleoproteins of Uukuniemi virus are circular, *J. Virol.* **15:**386.

Pettersson, R. F., Kääriäinen, L., Von Bonsdorff, C.-H., and Oker-Blom, N., 1971, Structural components of Uukuniemi virus, a non-cubical tick-borne arbovirus, *Virology* **46:**721.

Pettersson, R. F., Hewlett, M. J., Baltimore, D., and Coffin, J. M., 1977, The genome of Uukuniemi virus consists of three unique RNA segments, *Cell* **11:**51.

Pettersson, R. F., Kuismanen, E., Rönnholm, R., and Ulmanen, I., 1985, mRNAs of Uukuniemi virus, a bunyavirus, in: *Viral Messenger RNA* (Y. Becker, ed.), pp. 283–300, Nijhoff, The Hague.

Pifat, D. Y., and Smith, J. F., 1987, Punta Toro virus infection of C57B1-6J mice: A model of phlebovirus-induced disease, *Microb. Pathog.* **3:**409.

Pifat, D. Y., Osterling, M. C., and Smith, J. J., 1988, Antigenic analysis of Punta Toro virus and identification of protective determinate with monoclonal antibodies, *Virology* **167:**442.

Poch, O., Sauvaget I., Delarne, M., and Tordo, N., 1989, Identification of four conserved motifs among the RNA-dependent polymerase encoding elements, *EMBO J.* **8:**3867.

Raju, R., and Kolakofsky, D., 1987, Unusual transcripts in La Crosse virus-infected cells and the site for nucleocapsid assembly, *J. Virol.* **61:**667.

Rice, R. M., Erlick, B. J., Rosato, R. R., Eddy, G. A., and Mohanty, S. B., 1980, Biochemical characterization of Rift Valley fever virus, *Virology* **105:**256.

Robeson, G., El-Said, L. H., Brandt, W., Dalrymple, J., and Bishop, D. H. L., 1979, Biochemical studies on the Phlebotomus fever group viruses (Bunyaviridae family), *J. Virol.* **30:**339.

Rönnholm, R., 1992, Localization to the Golgi complex of Uukuniemi virus glycoproteins G1 and G2 expressed from cloned cDNAs, *J. Virol.* **66:**4525.

Rönnholm, R., and Pettersson, R. F., 1987, Complete nucleotide sequence of the M RNA segment of Uukuniemi virus encoding the membrane glycoproteins G1 and G2, *Virology* **160:**191.

Rose, J. K., Welch, W. J., Sefton, B. M., Esch, F. S., and Ling, N., 1980, Vesicular stomatitis virus glycoprotein is anchored in viral membrane by a hydrophobic domain near the COOH terminus, *Proc. Natl. Acad. Sci. USA* **77:**3884.

Rossier, C., Raju, R., and Kolakofsky, D., 1988, La Crosse virus gene expression in mammalian and mosquito cells. *Virology* **165:**539.

Saikku, P., and Von Bonsdorff, C.-H., 1968, Electron microscopy of the Uukuniemi virus, an ungrouped arbovirus, *Virology* **34:**804.

Saikku, P., Von Bonsdorff, C.-H., and Oker-Blom, N., 1970, The structure of Uukuniemi virus, *Acta Virol.* **14:**103.

Scallan, M. F., and Elliott, R. M., 1992, Defective RNAs in mosquito cells persistently infected with Bunyamwera virus, *J. Gen. Virol.* **73:**53.

Schmaljohn, C. S., Parker, M. D., Ennis, W. H., Dalrymple, J. M., Collett, M. S., Suzich, J. A., and Schmaljohn, A. L., 1989, Baculovirus expression of the M genome segment of Rift Valley fever virus and examination of antigenic and immunogenic properties of the expressed proteins, *Virology* **170:**184.

Simons, J. F., and Pettersson, R. F., 1991, Host-derived 5′ends and overlapping complementary 3′ends of the two mRNAs transcribed from the ambisense S segment of Uukuniemi virus, *J. Virol.* **65:**4741.

Simons, J. F., Hellman, U., and Pettersson, R. F., 1990, Uukuniemi virus S segment: Ambisense coding strategy, packaging of complementary strands into virions and homology to members of the genus Phlebovirus, *J. Virol.* **64**:247.

Simons, J. F., Persson, R., and Pettersson, R. F., 1992, Association of the nonstructural protein NSs of Uukuniemi virus with the 40S ribosomal subunit, *J. Virol.* **66**:4233.

Smith, J. F., and Pifat, D. Y., 1982, Morphogenesis of sandfly fever viruses (Bunyaviridae family), *Virology* **121**:61.

Struthers, J. K., and Swanepoel, R., 1982, Identification of a major non-structural protein in the nuclei of Rift Valley virus-infected cells, *J. Gen. Virol.* **60**:381.

Struthers, J. K., Swanepoel, R., and Shepard, S. P., 1984, Protein synthesis in Rift Valley fever virus-infected cell, *Virology* **134**:118.

Suzich, J. A., and Collett, M. S., 1988, Rift Valley virus M segment: Cell free transcription and translation of virus-complementary RNA, *Virology* **164**:478.

Suzich, J. A., Kakach, L. T., and Collett, M. S., 1990, Expression strategy of a phlebovirus: Biogenesis of proteins from the Rift Valley fever virus M segment, *J. Virol.* **64**:1549.

Takehara, K., Min, M. K., Battles, J. K., Sujiyama, K., Emery, V. C., Dalrymple, J. F., and Bishop, D. H. L., 1989, Identification of mutations in the M RNA of a candidate vaccine strain of Rift Valley fever virus, *Virology* **169**:452.

Tesh, R. B., 1988, Phlebotomus fever, in: *The Arboviruses: Epidemiology and Ecology* (T. P. Monath, ed.), Vol. IV, pp. 181–195, CRC Press, Boca Raton, FL.

Ulmanen, I., Seppälä, P., and Pettersson, R. F., 1981, In vitro translation of Uukuniemi virus-specific RNAs: Identification of a nonstructural protein and a precursor to the membrane glycoproteins, *J. Virol.* **37**:72.

Verani, P., Nicoletti, L., and Marchi, A., 1984, Establishment and maintenance of persistent infection by the phlebovirus Toscana in Vero cells, *J. Gen. Virol.* **65**:367.

Von Bonsdorff, C.-H., and Pettersson, R. F., 1975, Surface structure of Uukuniemi virus, *J. Virol.* **16**:1296.

Watkins, C. A., and Jones, I. M., 1993, Association of the 40S ribosomal subunit with the NSs nonstructural protein of Punta Toro virus, Abstracts of IXth Congress of Virology, August 1993, Glasgow, Scotland, p. 136.

CHAPTER 6

Molecular and Biological Aspects of Tospoviruses

ROB GOLDBACH AND DICK PETERS

I. INTRODUCTION

A. Historical Aspects

Most members of the *Bunyaviridae* infect animals but some are able to infect plants. The plant-infecting viruses, which are propagatively transmitted by thrips species, are classified into a separate genus, *Tospovirus*, named after the type species *tomato spotted wilt virus* (Francki *et al.*, 1991). Since the late 1980s, when increased knowledge in the molecular biology of tomato spotted wilt virus (TSWV) allowed it to be included in the *Bunyaviridae*, a number of distinct tospovirus species are still being identified and classified within the same family.

The first report on the "spotted wilt" disease of tomatoes dates back to 1915 when Brittlebank described this plant disease in Australia (Brittlebank, 1919). Later it was demonstrated that the causal agent of the infectious disease was a virus (Samuel *et al.*, 1930) for which the name *tomato spotted wilt virus* was coined. The virus was shown to be spread by thrips (Pittman, 1927), a large family of minute insects (order *Thysanoptera*; family *Thripidae*). These discoveries were followed by an increasing number of publications on the occurrence of TSWV in many countries on a large number of host plants. These reports also led to some confusion since the virus was

ROB GOLDBACH AND DICK PETERS • Department of Virology, Agricultural University, 6709 PD Wageningen, The Netherlands.

The Bunyaviridae, edited by Richard M. Elliott, Plenum Press, New York, 1996.

reported under many different names, e.g., tomato bronzing virus, Kromnek virus, pineapple yellow potyvirus, makhorka tip chlorosis virus, vira cabeça virus, and others (Best, 1968; Smith, 1972; Sakimura, 1962). This variation in nomenclature reflects the wide variation of disease symptoms, depending on virus isolate, host species, and the different regions where the disease was found. After World War II the occurrence of tomato spotted wilt disease declined in Western Europe and the United States, whereas in Eastern Europe, South America, and South Africa the virus remained a serious problem. The decline in Western Europe and the United States has been explained by the effective chemical control of the onion thrips, *Thrips tabaci*, which at that time were probably the most important vector of TSWV.

From the early 1980s on, a rapid emergence and marked geographic spread of TSWV has taken place (Barker, 1989; Marchoux *et al.*, 1991; Vaira *et al.*, 1993) which was preceded by a rapid expansion of another efficient vector, the Western flower thrips (*Frankliniella occidentalis*) (Mantel and Van de Vrie, 1988; Brødsgaard, 1989). Reports of new hosts accumulated at an exceptionally high rate (reviewed by German *et al.*, 1992) and, to date, TSWV is known to infect more than 650 different plant species, monocots as well as dicots, belonging to more than 70 distinct botanical families. Susceptible hosts include many important agricultural crops such as pepper, peanut, pea, potato, tobacco, and tomato, vegetables such as celery and lettuce, and ornamentals such as dahlia, chrysanthemum, gerbera, impatiens, and iris, and many weeds. TSWV ranks among the ten economically most important plant viruses, causing crop losses worldwide of more than $1 billion (Goldbach and Peters, 1994). The disease is widespread in many agricultural production areas on all continents, mainly in the warmer climate zones, while in areas with a temperate climate the virus is prevalent in greenhouse cultivations. Furthermore, surveys have revealed the existence of a number of other tospoviruses in addition to TSWV, the economic impact of which still remains to be assessed.

B. Classification of Tospoviruses

For a long period TSWV was classified as the single representative of a monotopic plant virus group, the "tomato spotted wilt virus group" (Matthews, 1979). With the unraveling of its molecular biology, this virus was shown to have properties in common with members of the *Bunyaviridae*. Therefore, TSWV has recently been classified as the prototype of a newly created genus within the *Bunyaviridae*, the genus *Tospovirus*. Using techniques like ELISA and sequence determination, it has been found that TSWV itself (i.e., as a single species) is in many instances responsible for infections worldwide of a wide range of plant species (de Àvila *et al.*, 1990). However, these studies also revealed the existence of a number of other, distinct tospovirus species (Table I). Impatiens necrotic spot virus (INSV) was the first virus recognized as a separate tospovirus (Law and Moyer, 1990). This virus has a distinct host range, infecting mainly ornamental plants such as impa-

TABLE I. The Genus *Tospovirus*

	Acronym	Homology of N protein[a]
Established species		
Tomato spotted wilt	TSWV	100%
Groundnut ringspot	GRSV	78%
Tomato chlorotic spot	TCSV	76%
Impatiens necrotic spot	INSV	55%
Watermelon silver leaf mottle	WSMV	29%
Tentative species		
Groundnut bud necrosis	GBNV	N.D.[b]
Groundnut yellow spot	GYSV	N.D.

[a]Compared to TSWV.
[b]N.D., not determined.

tiens, gerbera, alstroemeria, and cineraria, and does not occur in solanaceous crops. INSV was first reported in the United States in 1987 (Law and Moyer, 1990), and in Europe in 1990 from the Netherlands (de Àvila *et al.*, 1992). Currently, it is the predominant tospovirus in the commercial ornamental plant industry in the United States (German *et al.*, 1992). Two further isolates, which share great overlap in host range with TSWV but display no clear biological differences, were recognized only after serological and genome sequence analysis. These species have been denoted *tomato chlorotic spot virus* (TCSV) and *groundnut ringspot virus* (GRSV) (de Àvila *et al.*, 1993a). They seem to occur mainly in South America and South Africa, respectively, though their economic impact still remains to be assessed. Serological and sequence studies have demonstrated the occurrence of yet another tospovirus, infecting watermelon in Taiwan and Japan. For this virus, names such as *watermelon silver leaf mottle virus* and *watermelon spotted wilt virus* have been suggested (Yeh *et al.*, 1992).

de Àvila *et al.* (1993a,b) made a proposal to classify tospoviruses based on the nucleotide sequence of the nucleoprotein (N) gene. Comparison of the various, distinct isolates reveals that there are different levels of similarity between the N genes, which appear to vary between 29% (TSWV–WSMV) and 80–82% (TSWV–GRSV–TCSV). Sequence data are needed for two other groundnut-infecting tospoviruses (GBNV, GYSV; see Table I) to verify their status as distinct species on this basis.

II. ORGANIZATION AND EXPRESSION OF THE TOSPOVIRAL GENOME

A. Genetic Organization of the Tospoviral Genome

Typical of members of the *Bunyaviridae*, TSWV particles have a lipid envelope which contains two types of glycoproteins termed G1 and G2. The envelope encompasses a tripartite RNA genome which is tightly packaged by numerous copies of nucleoprotein (N) subunits, and 10–20 copies of a large (L) protein (Van Poelwijk *et al.*, 1993) which is the putative viral polymerase (Fig. 1). All three genomic segments have complementary termini over a length of approximately 65 nucleotides, which may hybridize into a panhandle structure and account for the circular appearance of isolated TSWV nucleocapsids (Peters *et al.*, 1991). Only the terminal 8 nucleotides (sequence: 3′ UCUCGUUA...) are strictly conserved among the three genomic segments. The genome of TSWV contains five genes which specify at least six mature, functional proteins. The L RNA is completely negative-stranded and encodes the L protein of size 331.5 kDa (de Haan *et al.*, 1991). The S RNA is ambisense (de Haan *et al.*, 1990; Maiss *et al.*, 1991), a property that TSWV shares with viruses in the genus *Phlebovirus*. Similar to the phleboviral S RNAs, the S RNA of TSWV encodes the nucleoprotein (N) in the viral-complementary (vc) strand, and a nonstructural protein (NSs) in the viral (v) strand (Figs. 1 and 2). A notable difference to all of the animal-infecting members of the family is that the M RNA of TSWV is also ambisense, encoding the glycoprotein precursor protein (G2/G1) in the vcRNA and a nonstructural protein (NSm) in vRNA (Kormelink *et al.*, 1992a, and Fig. 1). Elucidation of the sequences of S and M RNA segments of another tospovirus, INSV, indicates that the possession of two ambisense RNA segments is an apparently genus-specific characteristic (Law *et al.*, 1992; de Haan *et al.*, 1992a), setting tospoviruses apart from the other *Bunyaviridae* (Fig. 2). The possible function of the NSm cistron in the tospoviral M RNA will be discussed in Section IV.C.

Analysis of purified TSWV particles revealed that for the M RNA and the S RNA genomic segments both complementary strands are encapsidated, in a ratio of approximately 10 (v strand) to 1 (vc strand) (Kormelink *et al.*, 1992b). Since the vc strand of the L RNA was not detected, it is tempting to assume that the presence of both M and S complements is somehow related to the ambisense character of these RNA segments. In this context it is worthwhile mentioning that with Uukuniemi phlebovirus both complements of its ambisense S RNA are also encapsidated (Simons *et al.*, 1990).

Analysis of virus-specific RNA species in TSWV-infected *Nicotiana benthamiana* plants revealed the presence of two subgenomic mRNAs each for the S and M RNAs, which correspond to the four cistrons in these genome segments (de Haan *et al.*, 1990; Kormelink *et al.*, 1992a). The size of these mRNAs fits with the idea that the long, stable AU-rich hairpins present in the intercistronic region of both S and M RNA act as transcription terminator signals (de Haan *et al.*, 1990).

Primer extension analysis of the two, partially purified S RNA-specific mRNAs demonstrated the presence of 12 to 20 extra, nontemplated nucleotides at the 5′ end, indicating that, similar to the other *Bunyaviridae*, tospoviruses, in their plant cell background, also utilize a "cap-snatching" mechanism for initiation of transcription (Kormelink *et al.*, 1992c).

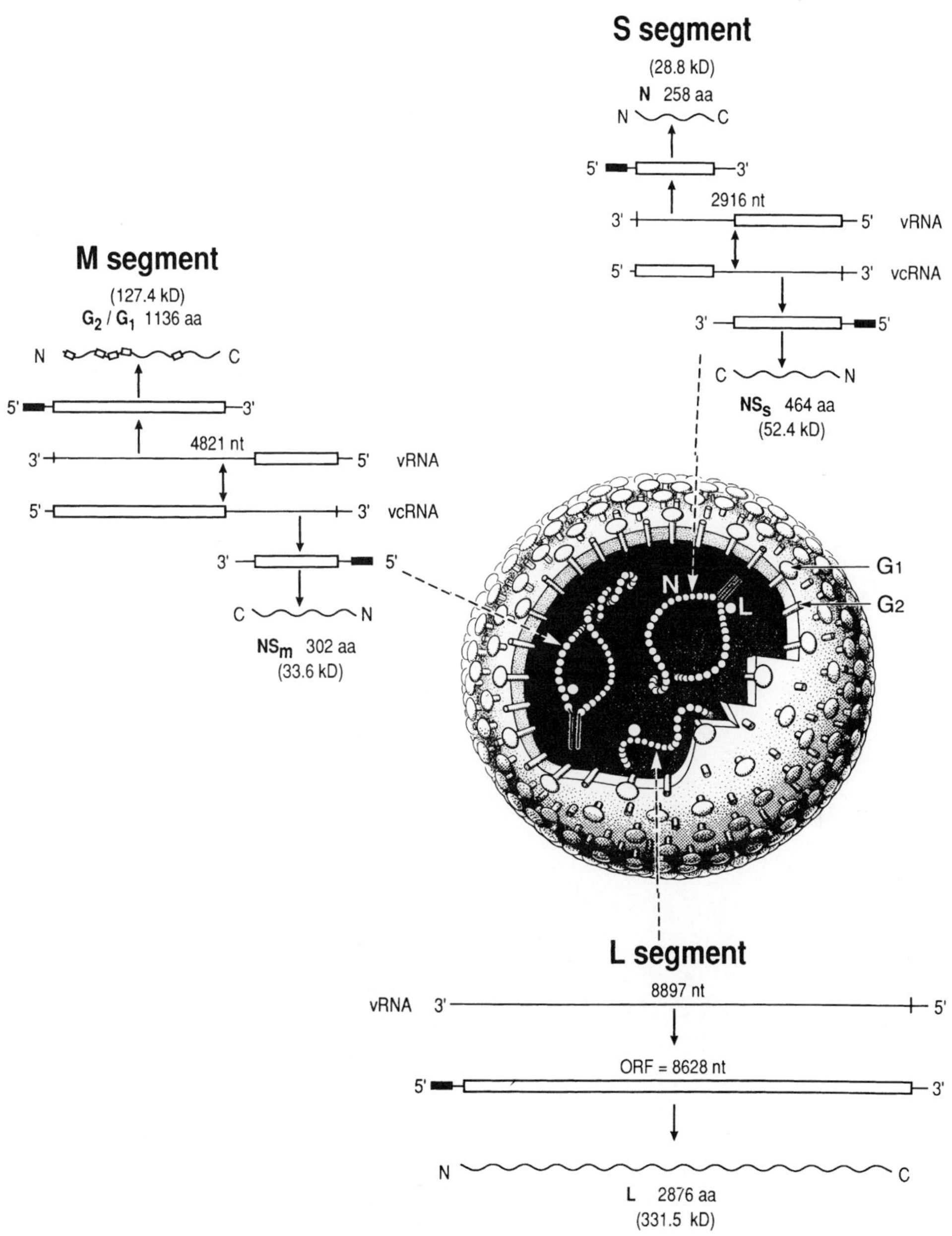

FIGURE 1. Morphological and genetic organization of tomato spotted wilt virus (TSWV), type species of the genus *Tospovirus*.

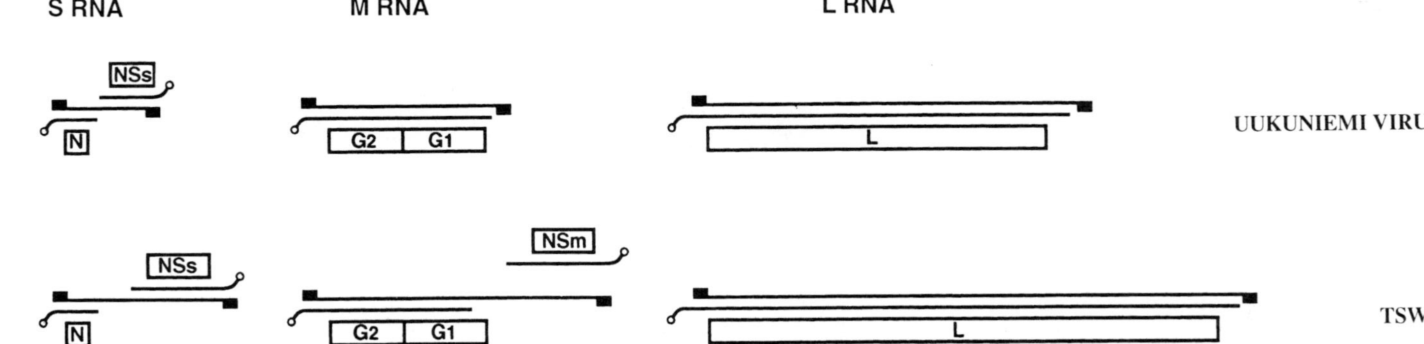

FIGURE 2. Comparison of the genetic maps of Uukuniemi virus (genus *Phlebovirus*) and TSWV (genus *Tospovirus*). Terminal inverted repeats are indicated by black squares, capped cellular leader sequences at the 5' ends of mRNAs as open circles, and gene products (proteins) as large, open boxes.

TABLE II. Sizes and Coding Properties of Tospoviral RNAs

Virus	RNA species	Size of RNA (nt)	Database accession number	Translation products
TSWV	L	8897	D10066	L protein (331.5 kDa)
	M	4821	S48091	G2/G1 (127.4 kDa) NSm (33.6 kDa)
	S	2916	D00645	N (28.8 kDa) NSs (52.4 kDa)
INSV	L	N.D.[a]	—	N.D.
	M	4972	M74904	G2/G1 (124.9 kDa) NSm (34.1 kDa)
	S	2992	X66972	N (28.7 kDa) NSs (51.2 kDa)

[a]N.D., not determined.

B. The Structural Proteins

The apparent size of the nucleoprotein of various tospoviruses, as judged by SDS–PAGE, varies between 29 kDa (TSWV) and 32 kDa (WSMV), which fits with the theoretical values based on sequence information (de Haan *et al.*, 1990; Yeh *et al.*, 1992). There is no evidence for glycosylation of tospoviral N proteins and asparagine-linked glycosylation sites are lacking. The N proteins of tospoviruses are good antigens and most antisera used for detection and diagnosis are based on antisera raised against these protein species (de Àvila *et al.*, 1990, 1993b; Resende *et al.*, 1991a; Law and Moyer, 1990). Nothing is known about the nature of the interaction of the N protein with the viral RNA.

Fractionation of purified enveloped TSWV particles by SDS–PAGE revealed the presence of distinct electrophoretic classes of glycoproteins (Mohamed *et al.*, 1973; Tas *et al.*, 1977). Further analysis indicated that the band with the lowest mobility represents the large glycoprotein G1 (apparent size 78 kDa) while the faster-migrating bands have been suggested to represent two forms of the second glycoprotein G2 (G2a about 58 kDa, and G2b about 52 kDa) (Mohamed *et al.*, 1973; Tas *et al.*, 1977). The resolution of G2 into two size classes seems typical for tospoviruses. G2b is believed to be a stable degradation product of the intact G2 (G2a) glycoprotein. The glycoprotein precursor (G2/G1) protein has been cloned into a recombinant baculovirus and expressed in insect cells (Kormelink, 1994); cleavage into G1 and G2 (the G2a form) occurred faithfully. By treating infected cells with tunicamycin it was established that both G1 and G2 underwent asparagine-linked glycosylation, and that glycosylation took place at the stage of the uncleaved G2/G1 precursor protein.

Computer-assisted alignments revealed homology between the G2/G1 protein sequence of TSWV and those of Bunyamwera and snowshoe hare (SSH) bunyavirus (Kormelink *et al.*, 1992a). Homology was mainly restricted to G1, with 22% identity and 45% similarity, in a 485-amino acid region. Based on this homology the tospovirus G2 could be mapped to the N-terminal part of the precursor protein, and G1 to the C-terminal part (Fig. 3). This

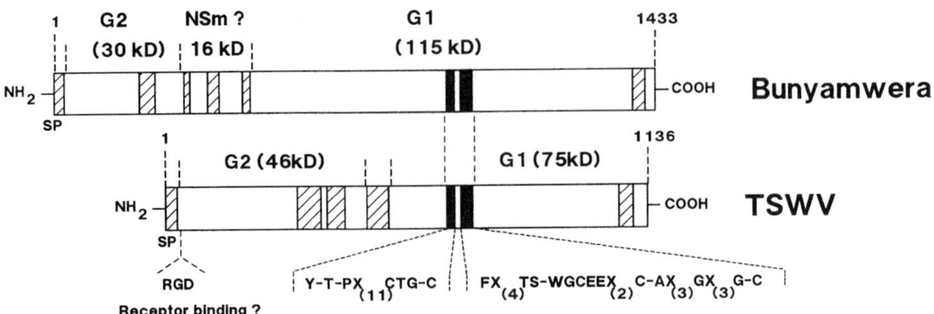

FIGURE 3. Topological comparison of the glycoprotein precursors of Bunyamwera virus (genus *Bunyavirus*) and TSWV (genus *Tospovirus*). Sequences conserved between the two genera are indicated in black, and hydrophobic domains by diagonal lines. X and hyphens refer to nonidentical amino acids, SP to the signal peptide sequence.

orientation was experimentally confirmed by expression of appropriate parts of the M segment cDNA in *E. coli* and subsequent immunological analysis of the products obtained. A similar topography of the glycoprotein precursor cistron has also been reported for the M segment of INSV (Law *et al.*, 1992).

Analysis of the hydropathy plot of the TSWV glycoprotein precursor shows a predicted hydrophobic N-terminus of some 35 residues that probably corresponds to a signal sequence for translocation across the endoplasmic reticulum (ER) membrane (Kormelink *et al.*, 1992a). This sequence may be cleaved from the glycoprotein precursor at amino acid residue 35 from the N-terminus (VLLAFLIFRATDAˆKV) according to the algorithms of Von Heijne (1986). Similarly the hydrophobic domain found between amino acid residues 400 and 500 of the precursor probably functions as a separate signal sequence for the G1 protein. Hydrophobic domains predicted between amino acid positions 300–400 and 1000–1100 may represent anchoring domains of the two glycoproteins in the membrane (Fig. 3). Based on these assumptions, the sizes of G2 and G1 can be estimated at about 46 and 75 kDa, respectively. The latter is in good agreement with the experimentally estimated size of G1 (78 kDa). However, the calculated size of G2 is considerably smaller than the apparent size from SDS–PAGE gels (58 kDa), the difference presumably relating to glycosylation.

The glycoprotein precursor contains an RGD motif (Fig. 3) at its amino terminus, immediately downstream of the hydrophobic signal sequence. Furthermore, this motif, and its position, is conserved between TSWV and INSV. RGD motifs are typically found in glycoproteins of the extracellular matrix of animal and plant cells, and are thought to be involved in adhesion of cells to the extracellular matrix (Pierschbacher and Ruoslahti, 1984; Ruoslahti and Pierschbacher, 1986, 1987; D'Souza *et al.*, 1988; Sanders *et al.*, 1991). The RGD sequence in such adhesive proteins is crucial for recognition by cell surface receptors (referred to as integrins). For tospoviruses it is tempting to speculate that this sequence is only involved in the insect part of the virus multiplication cycle, possibly in binding of virus particles to cell receptors in the midgut of the vector. Indeed, morphological defective isolates, i.e., virus mutants defective in glycoprotein synthesis, are still able to infect plants but are deficient in thrips transmission (see Section III). The relevance of the RGD motif in the glycoprotein precursor of TSWV, however, remains to be tested, as analysis of the glycoprotein precursor proteins of other members of the *Bunyaviridae* revealed this motif only in those encoded by Germiston and SSH bunyaviruses and Punta Toro phlebovirus.

For the animal-infecting *Bunyaviridae* it is known that, prior to maturation, viral RNPs condense at the cytoplasmic face of the Golgi complex only where the viral glycoproteins are found to accumulate. This indicates that an interaction between the nucleocapsid protein and a cytoplasmic domain of the glycoproteins triggers the process of budding of RNPs into the lumen of the Golgi. For TSWV no information is available on the process of morphogenesis. However, if an interaction of the N protein with either of the two glycoproteins occurs, then it is most likely to be with G2, as this protein has a large cytoplasmic domain (Fig. 3). Indeed, antiserum raised against purified nucleocapsids often cross-reacts in Western blots with G2, but not with G1, indicating a close interaction between the N and G2 protein species, leading to copurification (M. Kikkert and D. Peters, unpublished results).

A comparative Western blot study, including TSWV and INSV isolates, indicated that among different tospoviruses the glycoproteins are more conserved than the N proteins (Law and Moyer, 1990; de Àvila *et al.*, 1993b). This finding is supported by sequence analyses which show that the N proteins of TSWV and INSV exhibit 69.7% homology (55.4% identity), and the glycoproteins 79.1% homology (with 64.4% identity) (de Haan *et al.*, 1992a; Law *et al.*, 1992; Kormelink *et al.*, 1992a).

The fourth protein species present in mature virions is the L RNA-coded viral polymerase (also denoted L protein). This protein shares, together with the other *Bunyaviridae* L proteins, sequence motifs characteristic for polymerases of negative-strand RNA viruses (Poch *et al.*1990; Tordo *et al.*, 1992). The tospoviral enzyme is significantly larger than the L proteins of bunya- and phleboviruses. However, using antibodies to the N- and C-terminal parts of the primary translation product of TSWV L RNA, it was established that this protein does not undergo any proteolytic cleavages and is present at 10–20 copies per mature virion (Van Poelwijk *et al.*, 1993).

III. DEFECTIVE ISOLATES OF TOSPOVIRUSES

For many positive-strand RNA viruses, collections of natural mutants and/or designed mutants (generated from infectious cDNA clones) are available which have been very useful in unraveling the functions of the various virally encoded proteins during the infection cycle. This does not hold for tospoviruses, which, like all members of the *Bunyaviridae*, cannot be studied yet by reverse genetics. In the early days, Best (Best, 1968; Best and Gallus, 1953, 1955) tried to distinguish TSWV strains on the basis of symptom severity and conducted some reassortment studies using these strains. In mixed infections intermediate symptoms were obtained which were assumed to be caused by reassortant viruses, but little insight was thus obtained in the genetics of the virus. Currently, two types of defective isolates of tospoviruses have been characterized which may be useful for studying various steps in the infection cycle, i.e., envelope-deficient isolates and isolates with defective interfering particles.

A. Envelope-Deficient Isolates

Tospoviruses tend to lose their envelope when maintained solely by serial mechanical inoculations of plants, without the aid of their insect vector. Thus, "morphological defective" isolates (Ie, 1982) are obtained, lacking the lipid envelope but still being infectious. Such envelope-deficient (env⁻) isolates of TSWV fail to produce detectable amounts of the G1 and G2 glycoproteins (Verkleij and Peters, 1983; Resende *et al.*, 1991b,c). After elucidation of the nucleotide sequence of the M RNA of two tospoviruses, i.e., TSWV and INSV (Kormelink *et al.*, 1992a; Law *et al.*, 1992), the genetic nature of the morphological defectiveness of the env⁻ isolates could be identified by comparing M RNA sequences with those of wild-type isolates (Resende, 1993). Comparison of (partial) M RNA sequences of several env⁻ isolates of TSWV and INSV revealed an accumulation of point mutations in the G2/G1 ORF. Thus, it was found that an env⁻ isolate (US-01) of INSV had acquired an extra nucleotide in this gene, causing an early frameshift and consequently the loss of the putative signal peptide of the glycoprotein precursor (Law *et al.*, 1992; Resende, 1993). In this case, the env⁻ phenotype may be explained by a block in transmembrane transport, and hence further maturation of the glycoprotein precursor. Apparently, on serial passage of tospoviruses in plants, without the involvement of an insect stage, the G2/G1 gene is no longer under selective pressure, implying that its products are not required for virus multiplication in plants. The characterization of an env⁻ isolate of TSWV, NL-04 (Resende, 1993), has already led to two important conclusions, with respect to the possible involvement of the glycoproteins during the tospovirus infection cycle. First, it was observed that this defective isolate is still able to infect leaf tissues at a similar rate to that of wild-type infection. Thus, the glycoproteins (and lipid envelope) are neither essential for cell-to-cell movement of the virus, nor for long-distance transport in the plant. The conclusion seems justified, therefore, that tospoviruses, during the development of infection, are transported mainly as free nucleocapsid structures. Second, the presence of the lipid envelope (and glycoproteins) appeared to be essential for successful thrips transmission since the env$^-$ isolate of NL-04 so far fails to show vector transmissibility (unpublished results). This finding strongly suggests a role for G1 and/or G2 glycoproteins in virus–vector interactions. It is noteworthy in this context that G2 contains a so-called "cell attachment" site (RGD sequence motif) at its N-terminus (Law *et al.*, 1992; Kormelink *et al.*, 1992a) which may be involved in the recognition of a receptor in the thrips midgut.

B. Defective Interfering RNAs

A second class of defective mutants are represented by the defective interfering (DI) RNAs, which are generated during sequential inoculations on plants at high multiplicity of infection. RNA analysis revealed that for TSWV, DI RNAs were exclusively generated from the (polymerase-encoding) L RNA segment (Resende *et al.*, 1991b). These shortened L RNA molecules accumulate at the expense of the wild-type genomic RNA segments, and their appearance is associated with attenuation of the disease symptoms, both phenomena being diagnostic for true DI RNAs. Characterization of a number of different DI RNAs arising in TSWV and GRSV revealed that they have all retained their original 5' and 3' ends, and must have been generated by internal deletion events (Resende *et al.*, 1992). Each DI RNA studied contained a single internal deletion of approximately 60 to 80% of the L RNA and all of them maintained a large ORF, suggesting that these defective molecules might encode protein. PCR cloning and sequence analysis demonstrated that the deletions occur between short sequence repeats (up to five nucleotides), one of which is lost during the generation of the DI RNA (Resende *et al.*, 1992).

In view of their replicative advantage over wild-type RNAs, the TSWV DI RNAs may constitute a powerful tool to study the genomic information required for viral replication, encapsidation, and packaging into virus particles, and to unravel the RNA replication process.

IV. CYTOLOGY OF TOSPOVIRAL INFECTIONS

A. Cytopathological Effects

Cytopathological studies on ultrathin sections of cells from infected plants reveal the presence of numerous virus particles and distinct cytopathic structures (Kitajima, 1965; Francki and Grivell, 1970; Milne, 1970; Ie, 1971). Mature particles occur invariably and exclusively in the cisternae of the ER system (Fig. 4A). Several to many particles occur, usually in a single cavity, whereas the particles of a few isolates are singly bounded by distinct membranes of the ER. The way the particles accumulate may be a stable property of an isolate, but this is not so for all members of a species. The

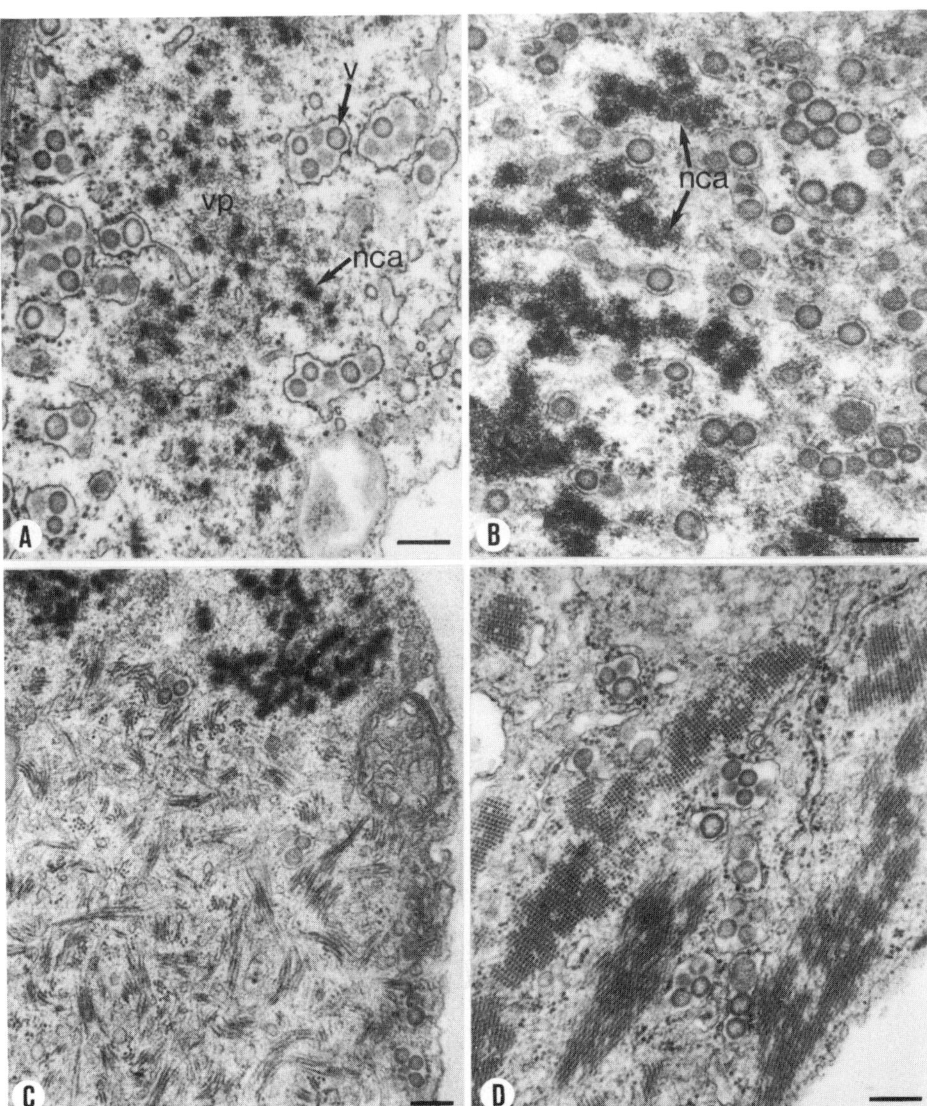

FIGURE 4. Cytopathology of tospoviruses. Osmium-fixed ultrathin sections of TSWV-infected *N. rustica* plants reveal the following cytopathological structures: (A) mature virion particles (v) in the ER system, viroplasm (vp), and nucleocapsid aggregates (nca); (B) nucleocapsid aggregates; (C, D) fibrillar structures in TSWV- and INSV-infected plant cells, respectively. Bars = 200 nm.

particles of TCSV and GRSV are often arranged in small crystalline arrays within a single cavity.

In addition to the membrane-bound clusters of enveloped particles, several distinct structures are produced. Moderately dense, amorphous masses, which will be referred to here as viroplasms, are found in each cell (Ie, 1982). They occur close to the regions where the virus particles accumulate (Fig. 4A). Immunostaining shows that these viroplasms contain N protein whereas the glycoproteins are virtually absent. Further studies (Ie, 1982; Kitajima *et al.*, 1992) disclosed that the viroplasms often contain, especially after some mechanical transfers of the viruses, numerous small complexes of material of higher electron density, often either loosely arranged in apparent chains or strings or grouped together in clusters (Fig. 4B). These complexes have cubic, circular, or elliptic profiles with diameters ranging in size between 30 and 120 nm. A striated arrangement with a periodicity of approximately 5 nm can often be discerned in these complexes. They react intensively with antiserum to the N protein and are believed to be aggregates of nucleocapsids (Kitajima *et al.*, 1992), which are either not yet enveloped or fail to acquire a membrane during the replication of the virus. The occurrence of these N protein-containing complexes may suggest that they are precursors for the nucleocapsids to be incorporated into the virus particles. However, it is more likely that they will not be enveloped. Their sizes are usually larger than the space to be filled within the membrane and their electron density is more intense than that of the core of the virus particles. Their existence may be caused by an imbalance between nucleocapsid and glycoprotein production resulting from the synthesis of defective glycoproteins, or even the absence of glycoproteins caused by mutations in the M RNA (see Section III.A) on serial passage in plants (Verkleij and Peters, 1983; Resende *et al.*, 1991b,c).

A second type of inclusion found in tospovirus-infected plant cells consists of loose aggregates or bundles of filamentous material consisting of NSs protein (Kormelink *et al.*, 1991). The filaments found in TSWV-infected cells are loosely aggregated (Fig. 4C). In contrast, the filaments in INSV inclusions are ordered in a crystalline structure of alternating rows of threads and points (Fig. 4D). These structures actually consist of layers in which the direction of the filaments in one layer is perpendicular to that in the adjacent one. The arrangement of the filaments is a feature by which TSWV can be distinguished from INSV. These inclusions differ in amount relative to the viroplasms and the nucleocapsid aggregates. Large numbers of NSs filaments are found in cells infected with one isolate, while only few occur in cells infected by another isolate. Their development might be also time and host dependent. For TSWV a slight, but positive correlation has been observed between the severity of symptoms and the amount of NSs protein in infected plants (Kormelink *et al.*, 1991). The function of this protein and the need to accumulate into filamentous structures remained thus far unknown. Among different isolates studied, NSs is the least conserved protein (50% homology between TSWV and INSV), indicating that this protein is not involved in a primary process such as RNA replication.

The NSs protein is also expressed in viruliferous thrips (Wijkamp *et al.*, 1993), and occurs abundantly in salivary gland tissue. This might indicate that the NSs protein, besides having a function in the plant, may play a role in thrips infection. Its presence in the salivary glands might suggest a role in transmission.

B. Assembly of the Virus

The process of virus assembly is one of the most enigmatic features of tospoviruses. Their accumulation in the cisternae of the ER suggests that the assembly involves an interaction of the nucleocapsid produced in the cytoplasm and the glycoproteins at the membrane. Involvement of the Golgi system or of vesicles required to transport the virus to the cell surface has not been observed or demonstrated. The symplastic mechanism by which the virus moves from one plant cell to another (see below) and the way thrips acquire the virus, i.e., ingesting the cell's content, does not require the Golgi apparatus as a steering body to release the virus from the infected cell.

Some remarkable structures have only rarely been found which could be interpreted as intermediate stages of particle morphogenesis. Flattened and curved double membrane structures with dense material in the concave side were found in the cytoplasm and the Golgi bodies (Kitajima *et al.*, 1992). When such structures are present, the Golgi body contains a few single particles surrounded by a closely juxtaposed membrane ("double enveloped particles"). The presence of these structures suggests the involvement of the Golgi body in a maturation process as depicted by Kitajima *et al.* (1992). However, the possibility that these particles are aberrant products of an uncontrolled mechanism in a late phase of the infection cycle in the cell cannot be ruled out. This is even more apparent in view of the massive accumulation of virus particles in the cisternae of the ER and the absence of any essential intermediate function of the Golgi body in intracellular virus transport.

C. Cell-to-Cell Movement

Systemic infection of plants by viruses proceeds only when the virus is translocated from one cell to another. It is generally accepted that plant-infecting viruses pass the rigid cell wall, which surrounds each cell, through the plasmodesmata—cytoplasmic channels that bridge the walls between the cells. However, the sizes of all viruses (diameter 25 nm or more) or even their genome (diameter of naked, single-stranded RNA 12 nm) greatly exceed the functional diameter of the plasmodesmata (Roberts and Lucas, 1990;

Lucas *et al.*, 1993). For a number of positive-strand viruses, it has been shown that they specify a cell-to-cell "movement" protein which modifies the plasmodesmata in such a way to allow passage of the viral nucleic acid (e.g., tobamovirus TMV) or virus particles (e.g., comovirus CPMV). Comparison of the genome of tospoviruses with those of the animal-infecting *Bunyaviridae* reveals the presence of an extra gene located on the tospoviral M RNA segment. This gene encodes a nonstructural protein of 34 kDa, designated NSm, which may act as the movement protein. Analysis of samples taken from systemically TSWV-infected plants demonstrated that the NSm protein is expressed early and transiently between 6 and 9 days after infection (Kormelink *et al.*, 1994).

Immunogold analysis of infected plant tissues demonstrated that the NSm protein specifically associates with free, nonenveloped nucleocapsids and to plasmodesmata, where the protein assembles into tubular structures (Storms *et al.*, 1995, and Fig. 5A,B). These structures (estimated 40–50 nm in width) extend from the cell wall into the cytoplasm at one side. These findings provide firm indications that these structures, and thereby the NSm protein, are involved in cell-to-cell translocation of TSWV. Morphologically similar tubules have been reported in cells infected with some positive-stranded RNA viruses, e.g., como- nepo-, and caulimoviruses, where it has been demonstrated that intact mature particles are transported through the tubes (Van Lent *et al.*, 1990, 1991). It is obvious that this cannot hold for TSWV since the diameter of the mature virus particle (80–110 nm) greatly exceeds not only the functional diameter of plasmodesmata (5 nm), but also that of the tubules (diameter 40–50 nm) which it induces. Thus, it is tempting to assume that TSWV moves from cell to cell along the tubule as free, nonenveloped nucleocapsids. This idea is supported by the observation that env$^-$ mutants of TSWV, though not transmissible by thrips, still have the capability to infect systemically plants with wild-type kinetics (see also Section III.A). In Fig. 6 a speculative model is presented for the cell-to-cell movement of tospoviral nucleocapsids which accommodates all observations made so far.

Interestingly, protoplasts isolated from plants at day 8 or 9 postinfection also carry the tubelike structures, which protrude from their outer surface (Fig. 5C). Similar structures are also formed on the surface of plant protoplasts transfected with a transient expression vector carrying only the NSm gene (Storms *et al.*, 1995).

V. VIRUS–VECTOR RELATIONSHIPS

A. Transmission by Thrips

Like most other plant viruses, tospoviruses have to be transmitted by arthropod vectors in order to be spread in host plant populations. Tospovirus

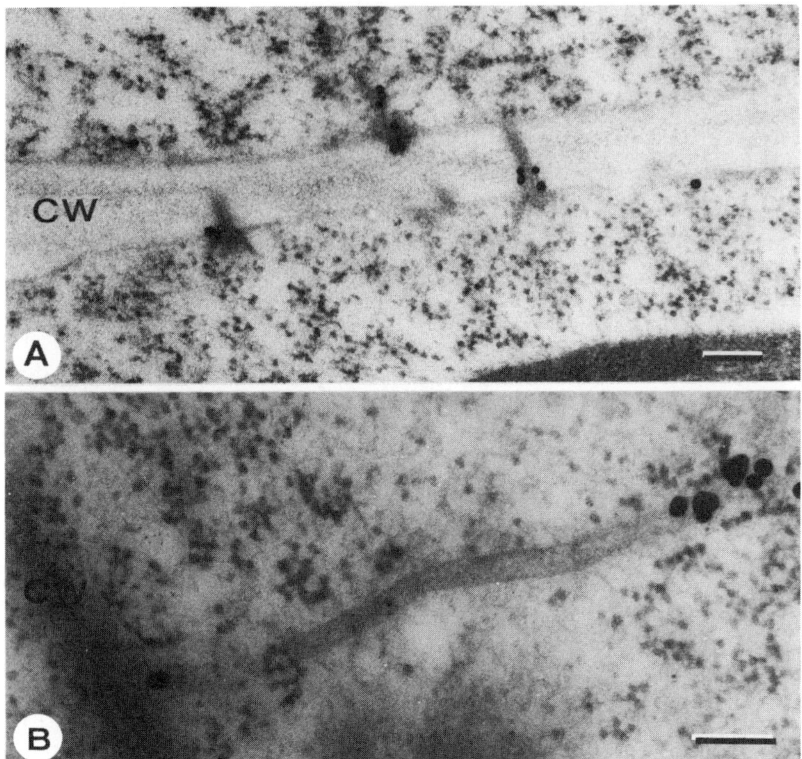

FIGURE 5.

transmission is accomplished by a limited number of thrips species, which are minute insects belonging to the order *Thripidae*. *Frankliniella occidentalis* (the Western flower thrips), *F. schultzei* (the cotton bud thrips), and *Thrips tabaci* (the onion thrips) are the most frequently reported vector species (Table III). Other species recorded as vectors are *F. fusca, T. setosus, T. palmi,* and *Scirtothrips dorsalis,* and *F. intonsa* has recently been added to this list (Wijkamp *et al.,* 1994). Those species transmitting tospoviruses feed exclusively on plants and most of them are polyphagous. Most virus transmission studies have so far been made with *T. tabaci* or *F. occidentalis* as vectors. Transmission of tospoviruses is closely linked to the life history and development of the thrips on plants. A peculiar characteristic of this transmission is that the virus can only be acquired by larvae while feeding on infected plants, while the adults cannot acquire the virus (Sakimura, 1963; Ullman *et al.,* 1992). Larvae may acquire the virus within a period of 10 min or even less, though the chance that they will become infected increases with the length of the acquisition period. Acquisition may last as long as the

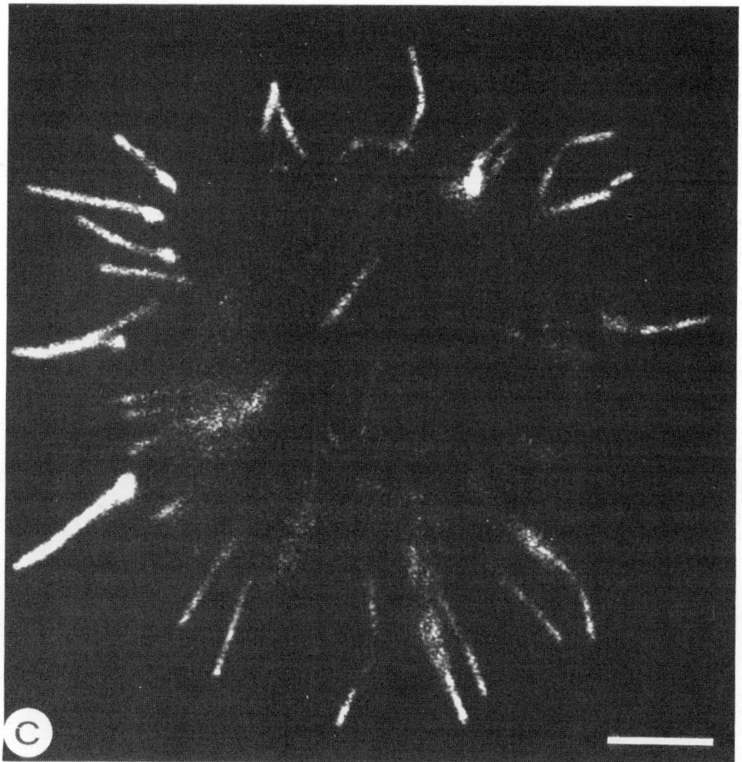

FIGURE 5. *In situ* localization of protein NSm of TSWV, in virus-infected *N. rustica* plants. Immunogold analysis reveals the presence of NSm in or near plasmodesmata (A), where it frequently forms tubular structures (B). Similar tubular structures are formed on single protoplasts infected with TSWV (C). Bars = 100 nm (A, B), 5 μm (C).

larvae remain on the infected plant. In infected thrips the virus is transstadially passaged through molting, pupation, and emergence to the adult stage. The adult may remain viruliferous for the remainder of its life, which may last 20 to 40 days depending on the environmental conditions. Transmission has been mainly ascribed to adults, although some observations were made in the past that larvae could occasionally transmit (Sakimura, 1963). Recently, it was convincingly demonstrated that second larval instars of *F. occidentalis* are excellent transmitters when newborn larvae are given a 24-hr-long acquisition period on infected plants (Wijkamp and Peters, 1993). Approximately 80% of the thrips that transmitted the virus did so when they were still larvae (Wijkamp and Peters, 1993). Most of these insects resumed transmission after emergence as adults. The median latent period for the infecting larvae ranged from 80 to 170 hr when they were kept at either 27 or 20°C, respectively. This period was similar in length for TSWV and INSV

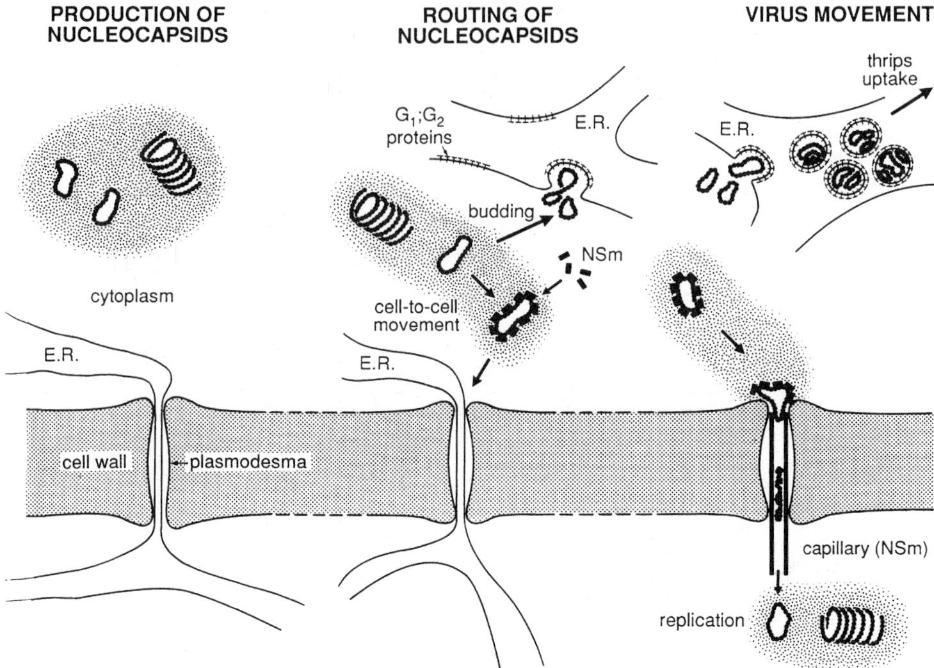

FIGURE 6. Model for the cell-to-cell movement of tospoviral nucleocapsids. On replication of the virus, newly synthesized nucleocapsids are tagged by NSm proteins and targeted to the plasmodesmata, where the NSm proteins polymerize into a rigid, tubular structure. These tubular structures (diameter 40–50 nm) penetrate through the plasmodesmata, allowing infectious nucleocapsids to pass. (Figure kindly provided by Marc Storms.)

TABLE III. Reported Vectors of Tospoviruses

Vector species		
Latin name	Common name	Tospoviruses
Frankliniella fusca	Tobacco thrips	TSWV, INSV
Frankliniella intonsa		TSWV
Frankliniella occidentalis	Western flower thrips	TSWV, TCSV, GRSV, INSV
Frankliniella schultzei	Cotton bud thrips	TSWV, TCSV, GRSV
Scirtothrips dorsalis	Chili thrips	TSWV
Thrips palmi	Melon thrips	TSWV, GBNV, WSMV
Thrips setosus		TSWV
Thrips tabaci	Onion thrips	TSWV

(Wijkamp and Peters, 1993). These results show that all cellular events determining the latent period (= "external incubation period" in arbovirological terminology), and which lead to infectious thrips, can be completed prior to pupation. The latent period of those thrips that transmit only as adults suggests they became infected while they were in their pupal stage or early days of the adult stage. However, the possibility cannot be ruled out that they were infected in their second larval stage but did not transmit the virus.

Acquisition and transmission by larvae is not a general phenomenon in the dispersion of *Bunyaviridae*. Besides the tospoviruses, only those viruses that are dispersed by ticks are transmitted by larvae. The other arthropod-borne viruses are exclusively acquired and transmitted by adults as the larval instars of these vectors do not feed on animals. The access periods required to acquire an infectious virus dose from an infected plant and to infect plants by viruliferous thrips can be as short as 5 to 15 min (Razvyazkina, 1953; Sakimura, 1963). Measurements of the acquisition and inoculation access periods in terms of the median periods have not been made. Using published data (Sakimura, 1963) a length of 36 hr could be calculated for *T. tabaci*. Observations made by Wijkamp and Peters (1993) suggest that this period can be as short as a few hours.

The inoculation activity of each individual thrips differs considerably. The number of local lesions, each of which will be the result of an infecting puncture, on petunia leaf disks by either an infectious larvae or adult varied from 1 to more than 30 in 24 hr (Wijkamp and Peters, 1993). The cause of this great variation in the number of local lesions produced by a single thrips specimen has not been analyzed thus far, but may be explained by individual differences in feeding behavior and transmission capacity. The latter may reflect the virus load of each thrips. The large number of infectious punctures made in 24 hr also shows that each individual thrips can potentially infect many plants after becoming infectious.

Evidence exists that the competence of each thrips species to transmit different tospovirus species differs considerably (Paliwal, 1974, 1976). A comparative study (Wijkamp *et al.*, 1994) showed that *F. occidentalis* transmits the tospoviruses TSWV, INSV, GRSV, and TCSV with variable efficiency. INSV was transmitted most efficiently, whereas GRSV and TCSV were poorly transmitted. In the literature, transmission of TSWV has repeatedly been ascribed to *T. tabaci* (Pittman, 1927; Sakimura, 1963). However, several other authors failed to demonstrate that this species could transmit TSWV (Paliwal, 1974; Reddy *et al.*, 1983), although a recent study in Finland showed that this thrips species transmitted TSWV very inefficiently (Lemmetty and Lindquist, 1993). It has been claimed that the failure to spread TSWV is correlated with the absence of males in *T. tabaci* populations (Zawirska, 1976). Indeed, in a recent study with four populations in which males were also absent, transmission could not be observed (Wijkamp *et al.*, 1995). Hence, a hitherto poorly investigated genetic variation within thrips populations may have an important impact on tospovirus transmission. Furthermore, the competence by which thrips species are able to transmit will certainly also depend on the plant species to be infected.

Transovarial or vertical transmission has not been demonstrated for the tospoviruses. This type of transmission can ensure the maintenance of some bunyaviruses in a reservoir in the absence of active biological transmission. For tospoviruses the large number of susceptible host plants, the ubiquitous presence of thrips, and their year-round reproduction in most climates suggest that vertical transmission is not required.

B. Epidemiology

The arthropod-borne, animal-infecting members of the *Bunyaviridae* are mainly maintained in distinct cycles comprising a limited number of preferred vectors and vertebrate hosts. In contrast, many of the plant species colonized by a single thrips species are also susceptible for tospoviruses. As a consequence, tospoviruses potentially can occur in large niches consisting of uncountable interwoven cycles. Spread will occur from these natural reservoirs to different susceptible crops, which are often also accepted as host by the thrips or are probed by them for suitability as host. As a consequence, many plants can be infected, which may lead to devastating levels of damage.

The spread of tospoviruses is thought to be primarily a function of adults. Infected adults will transmit when they move from one plant to another. They can fly over small distances at low wind speed conditions but are dispersed over large distances at higher wind speed. However, in view of the fact that the majority of the thrips that becomes viruliferous transmit already as second-instar larvae, when born on infected plants (Wijkamp and Peters, 1993), viruliferous larvae will potentially be more effective in spread of tospoviruses than anticipated, especially when a crop canopy closes and individual plants touch each other. Spread may then occur faster from primary infected plants by larvae than by adults as the latter have to pass the pupal stages before they transmit.

C. Replication in Thrips

For several animal-infecting members of the *Bunyaviridae* it is well documented that they replicate in their insect vector, and the question whether tospoviruses replicate in their vectors has been raised on many occasions. Convincing evidence now exists that the tospoviruses actually do replicate in their vectors (Wijkamp *et al.*, 1993; Ullman *et al.*, 1993). Following a decrease in the amount of the N protein seen in the first hours after ingesting virus, the amount of this protein increases steadily during the larval period. This viral antigen could be detected by electron microscopy in cells of the intestinal tract and salivary glands. The presence of this protein and also of virus particles in the salivary glands and its ducts implies that the virus replicates in the vector. This conclusion is further strengthened by the observed *de novo* synthesis of the NSs protein during the larval development (Wijkamp *et al.*, 1993) and the accumulation of this protein in midgut epithelium and salivary gland cells as shown in immunocytochemical studies (Wijkamp *et al.*, 1993; Ullman *et al.*, 1993). As this (nonstructural) protein does not occur in mature virus particles, its abundant presence in the salivary glands after the latent period conclusively demonstrates that tospoviruses replicate in their vectors.

Replication of *Bunyaviridae* in invertebrate cells is not typically associated with cytopathic effects or pathological disorders in their vectors (Beaty and Calisher, 1991). However, a high mortality has been reported for the thrips species *F. occidentalis* when infected with TSWV (Robb, 1989), and some cytopathological effects have been observed by Ullman *et al.* (1993). In these experiments the thrips were confined for long periods on infected plants. Hence, it cannot be excluded that the infected plants had a detrimental effect on the physiology of the thrips as a result of a lower food quality of these plants compared to healthy plants.

VI. NATURAL AND ENGINEERED FORMS OF RESISTANCE AGAINST TOSPOVIRUSES

A. Disease Management

As described in Section I.A, more than 650 different plant species are known to be susceptible for TSWV and some of these plants are also infected by the other hitherto identified tospoviruses. Yield losses for some of these crops can be considerable. In the field it is very difficult to limit the incidence of tospovirus infections, since it is almost impossible to properly control thrips and, moreover, weed host plants serve as virus reservoirs and are abundantly present in the neighborhood of the threatened crops. In the greenhouse the situation is slightly less complicated. Several sanitary measures can be taken to minimize losses related to tospovirus infections, such as early destruction of infected plants and biological or chemical control of thrips. In case of vegetative crop propagation, it is important to maintain virus-free stock material as well as to detect tospovirus infections early by sensitive assays. Currently various specific antisera directed against TSWV, TCSV, GRSV, and INSV are commercially available.

Another promising means of controlling tospovirus diseases could be breeding resistance to these viruses. Sections VI.B and VI.C discuss the possibilities of using natural resistance sources and of designing engineered forms of resistance, respectively.

B. Natural Forms of Host Resistance to Tospoviruses

Over the past 30 to 40 years, considerable effort has been spent on gaining crops with increased resistance or tolerance to tospovirus infections. However, the results so far obtained in numerous breeding projects are disappointingly poor. A single TSWV-tolerance gene from *Capsicum chinense* has been successfully introduced into pepper. The pepper plants, however, are only tolerant to TSWV and not to other tospoviruses. Moreover, the tolerance is linked to a number of agronomically undesired traits. Despite numerous attempts in tobacco and lettuce, no clear source of resistance to TSWV could be identified. The natural resistance in *Nicotiana alata* (A. Atanassov, personal communication) is not available for breeding programs since this species is sexually incompatible with *N. tabacum*. Despite intensive efforts over a number of years to detect resistance, none of the 7000 groundnut (*Arachis hypogaea*) genotypes tested by Indian researchers has proved to be resistant. Tolerance to the virus has been detected but this has not yet been introduced into commercially attractive ground nut cultivars (Reddy *et al.*, 1991).

Genetic resistance to tospoviruses has been found in *Lycopersicon* spp. which might possibly be introduced into commercial tomato (*Lycopersicon esculenta*) cultivars. Finlay (1953) described five different genes (two dominant, three recessive) for TSWV resistance in tomato, which were denoted Sw^a_1, Sw^b_1, sw_2, sw_3, and sw_4. All of these genes are isolate specific, however, and are broken by various different TSWV isolates and other tospoviruses (Stevens *et al.*, 1992; Boiteux and Giordano, 1993). A single dominant resistance gene in *L. peruvianum*, tentatively identified as Sw_5, provides broad resistance to TSWV isolates from many geographic areas (Stevens *et al.*, 1992) and even to two other tomato-infecting tospoviruses, TCSV and GRSV (Boiteux and Giordano, 1993). Unfortunately, the resistance is based on a hypersensitivity response (local necrosis at primary infection sites) and may coincide with severe cosmetic damage in the fruits (P. de Haan and J. Gielen, personal communication).

In summary, limited progress has been made in breeding crops for increased resistance to tospoviruses, which is partly the result of the lack of suitable forms of resistance. Furthermore, in most cases naturally occurring resistance or tolerance is polygenic, based on complex interactions between virus, vector, and plant and as a consequence difficult to use in breeding programs.

C. Engineered Forms of Host Plant Resistance

In view of the worldwide spread of TSWV and other tospoviruses and the difficulty in controlling these viruses or their vectors by conventional measures (either biologically or chemically), there is an urgent need for engineered forms of host plant resistance. For positive-strand RNA viruses of plants, several strategies have been developed, with varying success, and these have been described in several recent reviews (Beachy *et al.*, 1990; Hull and Davies, 1992; Beachy, 1993). Most of these strategies are based on the concept of "parasite-derived resistance" (Sandford and Johnson, 1985), which proposes that pathogen resistance genes may be developed from the pathogen's own genetic material. Selected genes and sequences from a virus, when inserted into the host plant genome, may render that host resistant to virus infection, based on interference of one of the steps in the infection process. Indeed, for the coat protein gene, the replicase gene, and other genomic sequences from positive-strand RNA viruses, it has been shown that their transformation into the hosts' genome leads to significant levels of resistance, although the underlying mechanism is often not fully understood. Even strategies that were originally intended to block a certain process (e.g., coat protein-mediated resistance to block uncoating, and replicase-mediated resistance to block RNA replication) appear to operate in many cases at the level of RNA. The resistance observed has recently been suggested to be linked with "cosuppression" or "gene silencing" of the virus-derived transgene rather than to blocking the step intended (Lindbo *et al.*, 1993).

Several forms of pathogen-derived resistance may be envisaged for tospoviruses, e.g., nucleoprotein-mediated resistance, antisense resistance, and DI-RNA-mediated resistance. Gielen *et al.* (1991) and later MacKenzie and Ellis (1992) showed that transformation of tobacco (*Nicotiana tabacum*) with the N gene of TSWV confers resistance to this host plant. It was envisaged that accumulation of N protein in transgenic hosts expressing the viral N gene would lead, on virus inoculation, to blocking of transcription and hence to an abortive replication process, as a result of a premature switch from transcription to replication (Gielen *et al.*, 1991; Goldbach and de Haan, 1993). Transgenic lines were obtained expressing the N protein at levels up to 1.5% of the total soluble protein and 6 out of 15 of these expressor lines showed significant to very high levels of resistance. In comparison to control plants, these transgenic plants escaped infection, and those few plants that became infected showed a significant delay (3–7 days) in symptom expression. ELISA data confirmed that symptomless plants were completely virus-free and thus exhibited full immunity to the virus. No correlation was found between the level of resistance and the amount of transgenically expressed N protein, however. Indeed, later studies (de Haan *et al.*, 1992a), using a nontranslatable N gene, revealed that the resistance observed is based on transcriptional expression, rather than translational expression. This finding implies that the role of the N protein itself is limited or even nil and that the protection observed is mainly RNA-mediated.

There are some indications, however, that high levels of TSWV N protein expression can provide a very limited protection (in fact a delay in disease development) against INSV, but interestingly not against GRSV which is even more closely related to TSWV (Pang *et al.*, 1994; Vaira *et al.*,

1994). The extreme resistance obtained by the RNA-mediated form of N transgene protection, on the other hand, is virus-specific, holding only against isolates of the same virus but being broken by other tospoviruses (de Haan *et al.*, 1992b). To obtain broad and effective tospovirus resistance, Prins *et al.* (1994) constructed a gene cassette containing the N genes of three different vegetable-crop-infecting tospoviruses, i.e., TSWV, GRSV, and TCSV. Transformation of tobacco plants with this construct rendered transgenic lines completely resistant to all three tospoviruses.

An alternative approach for engineered host resistance could be protection by transgenically expressed DI RNAs (see also Section III.B.). It should be noted that, in the case of successful DI-mediated protection, the plants are tolerant rather than immune.

Possibilities for further approaches toward genetically engineered host resistance will depend on further understanding of the molecular pathology of tospoviruses. It can be anticipated, however, that two crucial processes during tospovirus infection can be targets for antiviral approaches, i.e., cell-to-cell movement (by influencing NSm expression) and thrips transmission (where the glycoproteins are promising target proteins).

VII. CONCLUDING REMARKS

From the previous discussion it has become clear that the genus *Tospovirus* represents a distinct group within the *Bunyaviridae* which clearly differs from the other members of the family in both molecular and biological aspects. Their host range is restricted to plants instead of animals, and they are transmitted, in a propagative way, by thrips. On a molecular level, tospoviruses differ from all other genera within the family in having two ambisense genome segments. The ambisense nature of the M RNA segment is caused by the acquisition of an extra gene, the NSm gene (Fig. 2). As this gene is lacking in the animal-infecting *Bunyaviridae* (though some viruses encode a protein called "NSm" in the M segment polyprotein) and, moreover, encodes a protein that most likely is involved in the passage of tospoviruses through the rigid cell wall of plant cells (Section IV.C), it is tempting to speculate that tospoviruses have relatively recently evolved from an animal-infecting virus. This idea is further supported by the observation that, based on alignments of a number of L proteins, the genus *Tospovirus* is more closely related to the genus *Bunyavirus* than the genus *Bunyavirus* is to the genus *Hantavirus* (de Haan *et al.*, 1991). Also the relatively low number of tospoviruses, compared to the number of animal-infecting *Bunyaviridae*, and the virtual absence of effective genetic resistance in host plants are in support of the view that bunyaviruses have only recently been able to invade the plant kingdom.

ACKNOWLEDGMENTS.   The authors wish to thank Richard Kormelink, Marcel Prins, Marc Storms, and Peter de Haan for helpful discussions and comments on the manuscript, Piet Kostense for making some of the illustrations, and Rina Hartman for preparing the manuscript.

VIII. REFERENCES

Barker, I., 1989, Beware of tomato spotted wilt, *Grower* (March 2nd):17.

Beachy, R. N., (ed), 1993, Transgenic resistance to plant viruses, *Semin. Virol.* **4**:327.

Beachy, R. N., Loesch-Fries, S., and Tumer, N. E., 1990, Coat protein-mediated resistance against virus infection, *Annu. Rev. Phytopathol.* **28**:451.

Beaty, B. J., and Calisher, C. H., 1991, Bunyaviridae—Natural history, *Curr. Top. Microbiol. Immunol.* **169**:27.

Best, R. J., 1968, Tomato spotted wilt virus, *Adv. Virus Res.* **13**:65.

Best, R. J., and Gallus, H. P. C., 1953, Strains of tomato spotted wilt virus, *Aust. J. Sci.* **15**:212.

Best, R. J., and Gallus, H. P. C., 1955, Further evidence for the transfer of character determinants (recombination between strains of tomato spotted wilt virus), *Enzymologia* **17**:207.

Boiteux, L. S., and Giordano, L. de B., 1993, Genetic basis of resistance against two *Tospovirus* species in tomato (*Lycopersicon esculentum*), *Euphytica* **71**:151.

Brittlebank, C. C., 1919, Tomato diseases, *J. Agric. Victoria* **7**:231.

Brødsgaard, H. F., 1989, *Frankliniella occidentalis* (Thysanoptera; Thripidae)—a new pest in Danish glasshouses: A review, *Tidsskr. Planteavl* **93**:83.

de Àvila, A. C., Huguenot, C., Resende, R. de O., Kitajima, E. W., Goldbach, R. W., and Peters, D., 1990, Serological differentiation of 20 isolates of tomato spotted wilt virus, *J. Gen. Virol.* **71**:2801.

de Àvila, A. C., de Haan, P., Kitajima, E. W., Resende, R. de O., Goldbach, R. W., and Peters, D., 1992, Characterization of a distinct isolate of tomato spotted wilt virus (TSWV) from Impatiens sp. in the Netherlands, *J. Phytopathol.* **134**:133.

de Àvila, A. C., de Haan, P., Kormelink, R., Resende, R. de O., Goldbach, R. W., and Peters, D., 1993a, Classification of tospoviruses based on phylogeny of nucleoprotein gene sequences, *J. Gen. Virol.* **74**:153.

de Àvila, A. C., de Haan, P., Smeets, M. L. L., Resende, R. de O., Kormelink, R., Kitajima, E. W., Goldbach, R. W., and Peters, D., 1993b, Distinct levels of relationships between tospoviruses isolates, *Arch. Virol.* **128**:211.

de Haan, P., Wagemakers, L., Peters, D., and Goldbach, R., 1990, The S RNA segment of tomato spotted wilt virus has an ambisense character, *J. Gen. Virol.* **71**:1001.

de Haan, P., Kormelink, R., Resende, R. de O., Poelwijk, F. van, Peters, D., and Goldbach, R., 1991, Tomato spotted wilt virus L RNA encodes a putative RNA polymerase, *J. Gen. Virol.* **72**:2207.

de Haan, P., de Àvila, A. C., Kormelink, R., Westerbroek, A., Gielen, J. J. L., Peters, D., and Goldbach, R., 1992a, The nucleotide sequence of the S RNA of impatiens necrotic spot virus, a novel tospovirus, *FEBS Lett.* **306**:27.

de Haan, P., Gielen, J. J. L., Prins, M., Wijkamp, M. G., Van Schepen, A., Peters, D., Van Grinsven, M. Q. J. M., and Goldbach, R., 1992b, Characterization of RNA-mediated resistance to tomato spotted wilt virus in transgenic tobacco plants, *Biotechnology* **10**:1133.

D'Souza, S. E., Ginsberg, M. H., Burke, T. A., Lam, S. C.-T., and Plow, E. F., 1988, Localization of an Arg-Gly-Asp recognition site within an integrin adhesion receptor, *Science* **242**:91.

Finlay, K. W., 1953, Inheritance of spotted wilt resistance in the tomato. II. Five genes controlling spotted wilt resistance in four tomato types, *Aust. J. Biol. Sci.* **6**:153.

Francki, R. I. B., and Grivell, C. J., 1970, An electron microscopy study of the distribution of tomato spotted wilt virus in systemically infected *Datura stramonium* leaves, *Virology* **42**:969.

Francki, R. I. B., Fauquet, C. M., Knudson, D. L., and Brown, F., 1991, Classification and nomenclature of viruses: Fifth report of the International Committee on Taxonomy of Viruses, *Arch. Virol. Suppl.* **2**:1.

German, T. L., Ullman, D. E., and Moyer, J. W., 1992, Tospoviruses: Diagnosis, molecular biology, phylogeny and vector relationships, *Annu. Rev. Phytopathol.* **30**:315.

Gielen, J. J. L., de Haan, P., Kool, A. J., Peters, D., Van Grinsven, M. Q. J. M., and Goldbach, R. W., 1991, Engineered resistance to tomato spotted wilt virus, a negative-strand virus, *Biotechnology* **9**:1363.

Goldbach, R., and de Haan, P., 1993, Prospects of engineered forms of resistance against tomato spotted wilt virus, *Semin. Virol.* **4**:381.

Goldbach, R., and Peters, D., 1994, Possible causes of the emergence of tospoviruses diseases, *Semin. Virol.* **5**:113.

Hull, R., and Davies, J. W., 1992, Approaches for nonconventional control of plant virus diseases, *Crit. Rev. Plant Sci.* **11**:17.

Ie, T. S., 1971, An electron microscopic study of developmental stages of tomato spotted wilt virus in plant cells, *Virology* **43**:468.

Ie, T. S., 1982, A sap-transmissible defective form of tomato spotted wilt virus, *J. Gen. Virol.* **59**:387.

Kitajima, E. W., 1965, Electron microscopy of vira-cabeça virus (Brazilian tomato spotted wilt virus) within the host cell, *Virology* **26**:89.

Kitajima, E. W., de Àvila, A. C., Resende, R. de O., Goldbach, R. W., and Peters, D., 1992, Comparative cytological and immunogold labelling studies on different isolates of tomato spotted wilt virus, *J. Submicrosc. Cytol. Pathol.* **24**:1.

Kormelink, R. J. M., 1994, Structure and expression of the tomato spotted wilt virus genome, a plant-infecting bunyavirus, Thesis, Agricultural University, Wageningen.

Kormelink, R., Kitajima, E. W., Haan, P. de, Zuidema, D., Peters, D., and Goldbach, R., 1991, The nonstructural protein (NSs) encoded by the ambisense S RNA segment of tomato spotted wilt virus is associated with fibrous structures in infected plant cells, *Virology* **181**:459–468.

Kormelink, R., de Haan, P., Meurs, C., Peters, D., and Goldbach, R., 1992a, The nucleotide sequence of the M RNA segment of tomato spotted wilt virus, a bunyavirus with two ambisense RNA segments, *J. Gen. Virol.* **73**:2795.

Kormelink, R., de Haan, P., Peters, D., and Goldbach, R., 1992b, Viral RNA synthesis in tomato spotted wilt virus-infected *Nicotiana rustica* plants, *J. Gen. Virol.* **73**:687.

Kormelink, R., Van Poelwijk, F., Peters, D., and Goldbach, R., 1992c, Non-viral heterogeneous sequences at the 5' ends of tomato spotted wilt virus mRNAs, *J. Gen. Virol.* **73**:2125.

Kormelink, R., Storms, M., Van Lent, J., Peters, D., and Goldbach, R., 1994, Expression and subcellular location of the NSm protein of tomato spotted wilt virus (TSWV), a putative viral movement protein, *Virology* **200**:56.

Law, M. D., and Moyer, J. W., 1990, A tomato spotted wilt virus with a serological distinct N protein, *J. Gen. Virol.* **71**:933.

Law, M. D., Speck, J. V., and Moyer, J. W., 1992, The M RNA of Impatiens necrotic spot tospovirus (Bunyaviridae) has an ambisense genomic organisation, *Virology* **188**:732.

Lemmetty, A., and Lindquist, I., 1993, *Thrips tabaci* (Lind.) (Thysanoptera, Thrypidae), another vector for tomato spotted wilt virus in Finland, *Agric. Sci. Fin.* **2**:189.

Lindbo, J. A., Silva-Rosales, L., Proebsting, W. M., and Dougherty, W. G., 1993, Induction of a highly specific antiviral state in transgenic plants: Implications for regulation of gene expression and virus resisance, *Plant Cell* **5**:1749.

Lucas, W. J., Ding, B., and Van der Schoot, C., 1993, Plasmodesmata and the supracellular nature of plants, *New Phytol.* **125**:435.

MacKenzie, D. J., and Ellis, P. J., 1992, Resistance to tomato spotted wilt virus infection in transgenic tobacco expressing the viral nucleocapsid gene, *Mol. Plant Microbe Int.* **5**:34.

Maiss, E., Ivanova, L., Breyel, E., and Adam, G., 1991, Cloning and sequencing of the S RNA from a Bulgarian isolate of tomato spotted wilt virus, *J. Gen. Virol.* **72**:461.

Mantel, W. P., and Van de Vrie, M., 1988, De Californische thrips, *Frankliniella occidentalis*, een nieuwe schadelijke tripssoort in de tuinbouw onder glas in Nederland, *Entomol. Ber. (Amsterdam)* **48**:140.

Marchoux, G., Gebre-Selassie, K., and Villevieille, M., 1991, Detection of tomato spotted wilt virus and transmission by *Frankliniella occidentalis* in France, *Plant Pathol.* **40**:347.

Matthews, R. E. F., 1979, Classification and nomenclature of viruses, *Intervirology* **12**:129.

Milne, R. G., 1970, An electron microscope study of tomato spotted wilt virus in sections of infected plant material and in negative stain procedures, *J. Gen. Virol.* **6**:267.

Mohamed, N. A., Randles, J. W., and Francki, R. I. B., 1973, Protein composition of tomato spotted wilt virus, *Virology* **56**:12.

Paliwal, Y. C., 1974, Some properties and thrips transmission of tomato spotted wilt virus in Canada, *Can. J. Bot.* **52**:1177.

Paliwal, Y. C., 1976, Some characteristics of the thrips vector relationship of tomato spotted wilt virus in Canada, *Can. J. Bot.* **54**:402.

Pang, S.-Z., Bock, J. H., Gonsalves, C., Slightom, J. L., and Gonsalves, D., 1994, Resistance of transgenic *Nicotiana benthamiana* plants to tomato spotted wilt and impatiens necrotic spot tospovirus: Evidence of involvement of the N protein and N gene RNA in resistance, *Phytopathology* **84**:243.

Peters, D., de Àvila, A. C., Kitajima, E. W., Resende, R. de O., de Haan, P., and Goldbach, R. W., 1991, An overview of tomato spotted wilt virus, in: *Virus–Thrips–Plant Interactions of Tomato Spotted Wilt Virus. Proceedings of a USDA Workshop* (H. T. Hsu and R. H. Lawson, eds.), pp. 1–14, Beltsville, MD.

Pierschbacher, M. D., and Ruoslahti, E., 1984, Cell attachment activity of fibronectin can be duplicated by small synthetic fragments of the molecule, *Nature* **309**:30.

Pittman, H. A., 1927, Spotted wilt of tomatoes, *J. Aust. Counc. Sci. Ind. Res.* **1**:74.

Poch, O., Sauvaget, I., Delarue, M., and Tordo, N., 1990, Identification of four conserved motifs among the RNA-dependent polymerase encoding elements, *EMBO J.* **8**:3867.

Prins, M., de Haan, P., Luyten, R., Van Veller, M., Van Grinsven, M. Q. J. M., and Goldbach, R., 1994, Broad resistance to tospoviruses in transgenic tobacco plants expressing three different tospoviral nucleoprotein gene sequences, submitted for publication.

Razvyazkina, G. M., 1953, The importance of the tobacco thrips in the development of outbreaks of tip chlorosis of Makhorka [in Russian], *Dokl. Vses. Akad. Skh. Nauk V.I. Lemina* **18**:27.

Reddy, D. V. R., Amin, P. W., McDonald, D., and Ghanekar, A. M., 1983, Epidemiology and control of groundnut bud necrosis and other diseases of legume crops in India caused by TSWV, in: *Plant Virus Epidemiology* (R. T. Plumb and J. M. Tresh, eds.), pp. 93–102, Oxford University Press, London.

Reddy, D. V. R., Wightman, J. A., Beshar, R. J., Highland, B., Black, M., Sreenivasulu, P., Dwivedi, S. L., Demski, J. W., McDonald, D., Smith, J. W., and Smith, D. H., 1991, Bud necrosis: A disease of groundnut caused by tomato spotted wilt virus, in: *ICRISAT, Infect. Bull.* **31**, Patancheru, India.

Resende, R. de O., de Àvila, A. C., Goldbach, R. W., and Peters, D., 1991a, Comparison of polyclonal antisera in the detection of tomato spotted wilt virus using the double antibody sandwich and cocktail ELISA, *J. Phytopathol.* **132**:46.

Resende, R. de O., de Haan, P., de Àvila, A. C., Kitajima, E. W., Kormelink, R., Goldbach, R., and Peters, D., 1991b, Generation of envelope and defective interfering RNA mutants of tomato spotted wilt virus by mechanical passage, *J. Gen. Virol.* **72**:2375.

Resende, R. de O., Kitajima, E. W., de Àvila, A. C., Goldbach, R. W., and Peters, D., 1991c, Defective isolates of tomato spotted wilt virus, in: *Virus–Thrips–Plant Interactions of Tomato Spotted Wilt Virus. Proceedings of a USDA Workshop* (H. T. Hsu and R. H. Lawson, eds.), pp. 71–76, Beltsville, MD.

Resende, R. de O., de Haan, P., Van der Vossen, E., de Àvila, A. C., Goldbach, R., and Peters, D., 1992, Defective interfering L RNA segments of tomato spotted wilt virus retain both virus genome termini and have extensive internal deletions, *J. Gen. Virol.* **10**:2509.

Robb, K. L., 1989, Analysis of *Frankliniella occidentalis*, (Pergande) as a pest of floricultural crops in California green houses, Ph.D. thesis, University of California, Riverside.

Roberts, A. W., and Lucas, W. J., 1990, Plasmodesmata, *Annu. Rev. Plant Physiol. Plant Mol. Biol.* **41**:369.

Ruoslahti, E., and Pierschbacher, M. D., 1986, Arg-Gly-Asp: A versatile cell recognition signal, *Cell* **44**:517.

Ruoslahti, E., and Pierschbacher, M. D., 1987, New perspectives in cell adhesion: RGD and integrins, *Science* **238**:491.

Sakimura, K., 1962, The present status of thrips-borne viruses, in: *Biological Transmission of Disease Agents* (K. Maramorosch, ed.), pp. 33–40, Academic Press, New York.

Sakimura, K., 1963, *Frankliniella fusca*, an additional vector for the tomato spotted wilt virus, with notes on Thrips tabaci, another vector, *Phytopathology* **53**:412.

Samuel, G., Bald, J. G., and Pittman, H. A., 1930, Investigations on "spotted wilt" of tomatoes, *Aust. Commonw. Counc. Sci. Ind. Res. Bull.* **44**.

Sanders, L. C., Wang, C.-S., Walling, L. L., and Lord, E. M., 1991, A homology of the substrate adhesion molecule vitronectin occurs in four species of flowering plants, *Plant Cell* **3**:629.

Sandford, J. C., and Johnston, S. A., 1985, The concept of parasite-derived resistance genes from the parasite's own genome, *J. Theor. Biol.* **113**:395.

Simons, J. F., Hellman, U., and Petterson, R. F., 1990, Uukuniemi virus S RNA segment: Ambisense coding strategy, packaging of complementary strands into virions, and homology to members of the genus *Phlebovirus, J. Virol.* **64**:247.

Smith, K. M., 1972, Tomato spotted wilt virus, in: *A Text Book of Plant Viruses Diseases*, 3rd ed., pp. 545–549, Longman, London.

Stevens, M. R., Scott, S. J., and Gergerich, R. C., 1992, Inheritance of a gene for resistance to tomato spotted wilt virus (TSWV) from *Lycopersicon peruvianum* Mill., *Euphytica* **59**:9.

Storms, M., Kormelink, R., Van Lent, J., Peters, D., and Goldbach, R., 1995, The non-structural NSm protein of tomato spotted wilt virus induces tubular structures in plant and insect cultures, *Virology*, in press.

Tas, P. W. L., Boerjan, M. L., and Peters, D., 1977, The structural proteins of tomato spotted wilt virus, *J. Gen. Virol.* **36**:267.

Tordo, N., de Haan, P., Goldbach, R., and Poch, O., 1992, Evolution of negative-stranded RNA genomes, *Semin. Virol.* **3**:341.

Ullman, D. E., Cho, J. J., Mau, R. F. L., Westcot, D. M., and Custer, D. M., 1992, A midgut barrier to tomato spotted wilt virus acquisition by adult Western flower thrips, *Phytopathology* **82**:1333.

Ullman, D. E., German, T. L., Sherwood, J. L., Westcot, D. M., and Cantone, F. A., 1993, Tospovirus replication on insect vector cells: Immunocytochemical evidence that the nonstructural protein encoded by the S RNA of tomato spotted wilt virus is present in thrips vector cells, *Phytopathology* **83**:456.

Vaira, A. M., Roggero, P., Luisoni, E., Masenga, V., Milne, R. G., and Lisa, V., 1993, Characterization of two tospoviruses in Italy: Tomato spotted wilt virus and impatiens necrotic spot virus, *Plant Pathol.* **42**:530.

Vaira, A. M., Lisa, V., Semeria, L., Noris, E., Accotto, G. P., and Allavena, A., 1994, *Nicotiana benthaniana* plants transgenic for the N gene of TSWV but not detectably expressing it are immune to different TSWV isolates, *Abstracts Fourth International Congress of Plant Molecular Biology*, Abstract 1544.

Van Lent, J., Wellink, J., and Goldbach, R., 1990, Evidence for the involvement of the 58K and 48K proteins in the intercellular movement of cowpea mosaic virus, *J. Gen. Virol.* **71**:219.

Van Lent, J., Storms, M., Van der Meer, F., Wellink, J., and Goldbach, R., 1991, Tubular structures involved in movement of cowpea mosaic virus are also formed in infected cowpea protoplasts, *J. Gen. Virol.* **72**:2615.

Van Poelwijk, F., Boye, K., Oosterling, R., Peters, D., and Goldbach, R., 1993, Detection of the L protein of tomato spotted wilt virus, *J. Gen. Virol.* **197**:468.

Verkleij, F. N., and Peters, D., 1983, Characterization of a defective form of tomato spotted wilt virus, *J. Gen. Virol.* **64**:677.

Von Heijne, G., 1986, A new method for predicting signal sequence cleavage sites, *Nucleic Acids Res.* **14**:4683.

Wijkamp, I., and Peters, D., 1993, Determination of a median latent period of two tospoviruses in *Frankliniella occidentalis*, using a novel leaf disk assay, *Phytopathology* **83**:986.

Wijkamp, I., Van Lent, J., Kormelink, R., Goldbach, R., and Peters, D., 1993, Multiplication of tomato spotted wilt virus in its vector, *Frankliniella occidentalis*, *J. Gen. Virol.* **74**:341.

Wijkamp, I., Almar, N., and Peters, D., 1994, Median latent period and transmission of tospoviruses vectored by thrips, in: *Thrips Biology and Management*. Proceedings of the International Conference on Thysanoptera: Towards Understanding Thrips Management (P. L. Parker, M. Skinner, and T. Lewis, eds.), pp. 153–161, Plenum Press, New York.

Wijkamp, I., Almarza, N., Goldbach, R., and Peters, D. 1995, Distinct levels of specificity in thrips transmission of tospoviruses, *Phytopathology*, in press.

Yeh, S.-D., Lin, Y.-C., Cheng, Y.-H., Jih, C.-L., Chen, M.-J., and Chen, C.-C., 1992, Identification of tomato spotted wilt-like virus on watermelon in Taiwan, *Plant Dis.* **76**:835.

Zawirska, I., 1976, Untersuchungen über zwei biologische Typen von *Thrips tabaci* Lind. (Thysanoptera, Thripidae) in der VR Polen, *Arch. Phytopathol. Pflanzenschutz* **12**:411.

Synthesis, Assembly, and Intracellular Transport of *Bunyaviridae* Membrane Proteins

RALF F. PETTERSSON AND LARS MELIN

I. INTRODUCTION

One of the original criteria for classifying, in the early 1970s, a diverse collection of serologically related and unrelated arthropod-borne viruses into a new family, *Bunyaviridae*, was the site of maturation (budding) in a perinuclear region, later unambiguously shown to represent the Golgi complex (Murphy *et al.*, 1973; Kuismanen *et al.*, 1982). With only one documented exception (Anderson and Smith, 1987), all members of this large family studied so far have been found to mature in this intracellular organelle, raising the intriguing question of the molecular and cell biological mechanisms underlying this process.

Enveloped viruses acquire their lipoprotein coat by budding through one of the cellular membranes. Many viruses, such as alpha-, arena-, orthomyxo-, paramyxo-, rhabdo-, and retroviruses, mature at the plasma membrane. In these cases, virions are released after completion of the budding process directly into the extracellular space. In many other cases, the budding occurs

RALF F. PETTERSSON AND LARS MELIN • Ludwig Institute for Cancer Research, Stockholm Branch, S-17177 Stockholm, Sweden.

The Bunyaviridae, edited by Richard M. Elliott, Plenum Press, New York, 1996.

at intracellular membranes representing waystations on the exocytic pathway that normally transports secretory and membrane proteins out to the plasma membrane. These include the endoplasmic reticulum (ER; rota- and flaviviruses), the newly identified intermediate compartment located between the ER and the Golgi complex (Saraste and Kuismanen, 1992; Hauri and Schweizer, 1992) (corona- and poxviruses; Sodeik *et al.*, 1993; Krijnse-Locker *et al.*, 1994), and the Golgi complex (bunya-, corona-, and rubellaviruses). In addition, herpesviruses have picked the rather exotic inner nuclear membrane as the site of the initial budding event (for reviews see Stephens and Compans, 1988; Pettersson, 1991; Matsuoka *et al.*, 1991; Griffiths and Rottier, 1992; Hobman, 1993). In these latter cases, mature virions are released either by cell lysis (rotaviruses), or after transport of the particles within vesicles to the plasma membrane. Virions are then released on fusion of these transport vesicles with the plasma membrane.

The details of the budding process of any virus are still poorly understood. For the last two decades, it has generally been assumed that viral nucleocapsids (cores) recognize regions on the cytoplasmic face of the membrane in the budding compartment, which have been modified by the accumulation of viral membrane glycoproteins (spikes or peplomers). A specific interaction between a core protein and the cytoplasmic tail of one of the spike proteins (equivalent to a receptor) is thought to trigger budding (Garoff and Simons, 1974; Whitt *et al.*, 1989; Suomalainen *et al.*, 1992). As more such interactions are recruited, the budding process progresses and host proteins are concomitantly excluded from the bud. In some viruses (e.g., rhabdo-, orthomyxo-, paramyxo-, retroviruses), a membrane (M) protein located between the lipid bilayer and the nucleocapsid may contribute to this interaction and play a role in the budding process (Rhee and Hunter, 1991). Such an M protein is missing from viruses such as the alpha-, flavi-, rubella-, bunya-, corona-, and arenaviruses. However, in some of these latter cases, a bulky cytoplasmic domain of one of the spike proteins may substitute for the M protein function. Immunolocalization of viral spike proteins has indicated that accumulation of one or several of these proteins appears to determine the site of budding. Thus, the membrane glycoproteins of viruses maturing at the plasma membrane are readily transported through the Golgi complex out to the cell surface. In contrast, one or several of the spike proteins of intracellularly maturing viruses, accumulate in the budding compartment and are not in most cases transported to the plasma membrane. It is thought that such proteins are retained in the budding compartment because of a retention motif or signal located somewhere in the protein. In general, viral spike proteins have been used as excellent models to dissect the exocytic pathway, and to map structural motifs in proteins conferring compartment-specific retention (Pettersson, 1991; Doms *et al.*, 1993).

Viral membrane proteins have also been useful in elucidating the early events taking place in the ER, such as core glycosylation, protein folding, and oligomerization (Doms *et al.*, 1993). The hemagglutinin (HA) of influenza virus and the vesicular stomatitis virus (VSV) G protein have been particularly useful for studying these steps. The general picture that has emerged can be summarized as follows. The nascent polypeptide chain of spike proteins translocating through the ER membrane is cotranslationally coreglycosylated at Asn-X-Ser/Thr sites, and in some cases also proteolytically cleaved (e.g., the G1–G2 precursor of *Bunyaviridae* members). Folding of the nascent chain is also initiated cotranslationally and proceeds by disulfide-bridge formations catalyzed by protein disulfide isomerase (PDI). During folding, many proteins are found associated with the IgG heavy chain binding protein (BiP/grp78) and/or calnexin (Ou *et al.*, 1993), two chaperones that are believed to monitor and assist in the folding process, to prevent aggregation of incompletely folded proteins, and to prevent incompletely or misfolded proteins from leaving the ER. Other chaperones may also contribute to this process. Spike complexes, either homo-, or hetero-oligomers, are formed from subunit proteins in the ER. Oligomerization may start already while the subunits are still being folded, or may in other cases occur only after the individual subunits have been properly folded. Only when proteins have folded correctly, and assembled into oligomers, may they exit the ER and become transported via the intermediate compartment to the Golgi complex enclosed in transport vesicles. The process leading to mature, transport-competent complexes has been coined "quality control" (Hurtley and Helenius, 1989; Doms *et al.*, 1993). In the Golgi, N-linked glycans are further trimmed, leading to terminal glycosylation. Spike proteins destined to the plasma membrane may also undergo proteolytic cleavage at a late stage of the exocytic pathway.

The purpose of this chapter is to review the synthesis, assembly, and intracellular transport of the bunyavirus membrane glycoproteins G1 and G2 that constitute the structural unit of *Bunyaviridae* spikes. For a more detailed description of the structure and expression of the medium-sized (M) RNA segment encoding the membrane glycoproteins, the reader is refered to the chapters in this volume covering the molecular biology of the members of the individual genera.

II. MORPHOGENESIS AND ASSEMBLY

Despite considerable variation in the size of the structural proteins and the genomic RNA segments, the overall structure of the various members of the *Bunyaviridae* family is very similar. Virus particles measure about 90–100 nm in diameter (Pettersson and von Bonsdorff, 1987) and contain four proteins: two glycoproteins, G1 and G2, associated with the envelope and forming the spikes, and two internal proteins associated with the RNA genome, the major nucleocapsid protein, N, and the minor RNA-dependent RNA polymerase, L.

Early electron microscopic studies (Murphy *et al.*, 1968) showed that virus particles mature intracellularly by budding into smooth vesicles in a perinuclear region of infected mouse brain cells and tissue culture cells (for references see Murphy *et al.*, 1973; Bishop and Shope, 1979; Pettersson *et al.*, 1988). Using organelle-specific markers, this site of accumulation of viral proteins and budding was later unambiguously shown to represent the Golgi complex in UUK (Kuismanen *et al.*, 1982; Gahmberg *et al.*, 1986b; Rönnholm, 1992), Bunyamwera (BUN) (Nakitare and Elliott, 1993), and Rift Valley fever (RVF) (Wasmoen *et al.*, 1988) virus-infected cells. The only animal *Bunyaviridae* member reported to bud at a site other than the Golgi is a strain of the phlebovirus RVF virus that was found to mature both intracellularly (in the Golgi) and at the plasma membrane in primary rat hepatocytes (Anderson and Smith, 1987). Tospoviruses, with tomato spotted wilt (TSW) virus as the prototype member, appear to have a more complex and as yet ill-defined mode of maturation. Particles seem to bud into the ER lumen or at Golgi membranes. Double enveloped particles are found in the cytoplasm and particles accumulate within ER-like membranes. In infected cells, envelope proteins have so far been localized by immunogold labeling only to virus particles, but not to any particular membranes (Kitajima *et al.*, 1992).

The morphogenesis and the maturation process have been studied at the EM or light microscopic levels for UUK (von Bonsdorff *et al.*, 1970; Kuismanen *et al.*, 1982, 1984, 1985; Gahmberg *et al.*, 1986a,b), RVF (Anderson and Smith, 1987), and sandfly fever (SF) viruses (Smith and Pifat, 1982), which are all members of the *Phlebovirus* genus (Francki *et al.*, 1991), as well as Dugbe virus (a nairovirus; Booth *et al.*, 1991).

Using immunofluorescence (Kuismanen *et al.*, 1982, 1984) and immuno-EM (Kuismanen *et al.*, 1985) both the glycoproteins and the nucleocapsid protein of UUK virus were found to accumulate in the Golgi area (Fig. 1). The helical nucleocapsids were found to line up underneath the membrane of distended Golgi vesicles. As G1 and G2 accumulated in the Golgi complex, progressively more nucleocapsids also entered the Golgi region. Little if any N protein was seen associated with the ER or the plasma membrane. Thus, a specific interaction between nucleocapsids and membranes containing the viral glycoproteins seems to exist only in the Golgi complex. Why no such interactions appear to occur already in the ER, which also contains high amounts of G1 and G2, is not clear, but it may relate to incorrect conformation or organization of the spikes, or to the topology or accessibility of the cytoplasmic tail of one of the glycoproteins that is likely to interact with the nucleocapsids. However, nucleocapsids can associate with the ER membrane in UUK-infected cells treated with tunicamycin, which inhibits N-glycosylation (Kuismanen *et al.*, 1984), or in cells infected with a temperature-sensitive mutant of UUK virus at the restrictive temperature (Gahmberg, 1984; Gahmberg *et al.*, 1986b). In these cases, exit from the ER of G1 and G2 is arrested. Whether budding into the ER lumen can occur under these conditions has not been studied. However, in the presence of brefeldin A, which

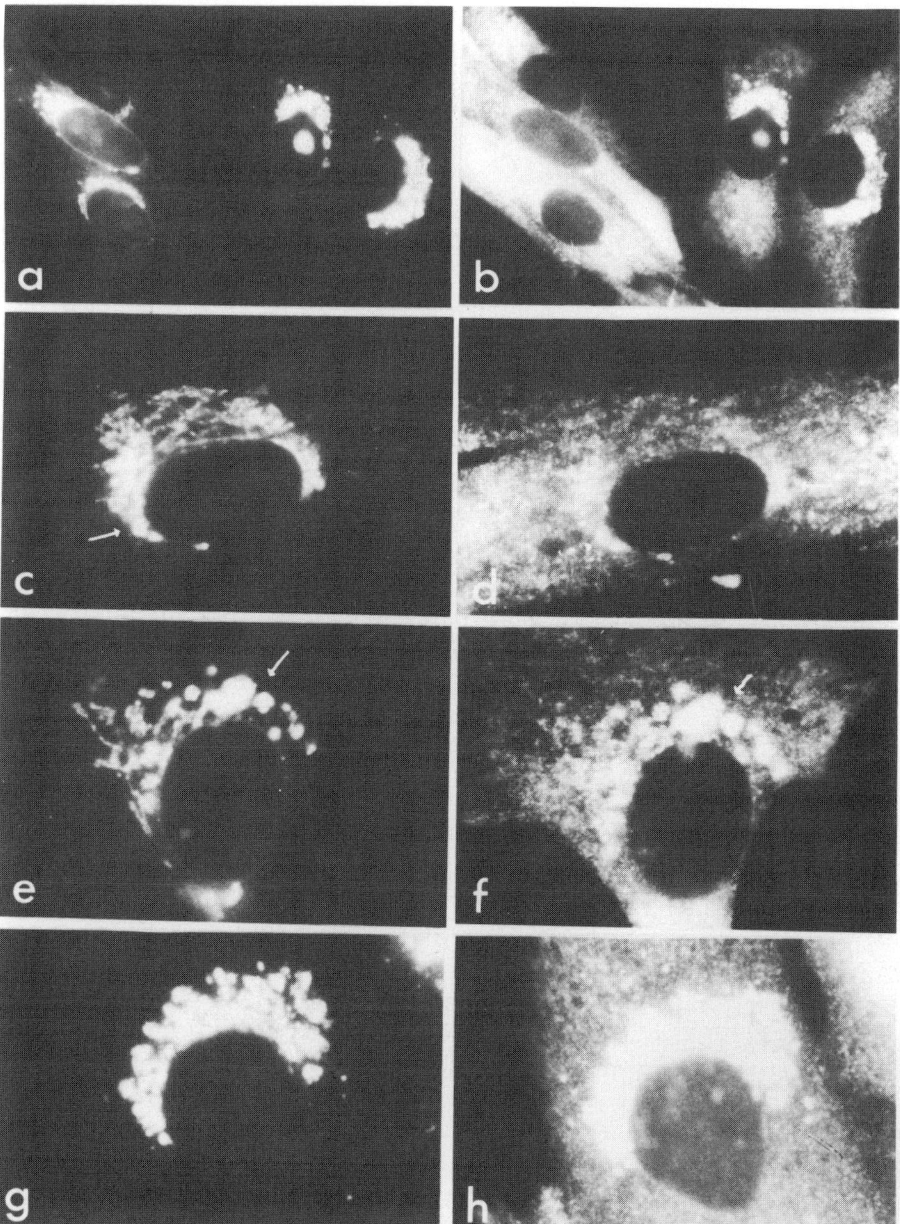

FIGURE 1. Progressive accumulation of Uukuniemi virus glycoproteins G1/G2 (a, c, e, and g) and nucleoprotein N (b, d, f, and h) in the Golgi complex of virus-infected BHK21 cells. The glycoproteins were visualized by indirect immunofluorescence using monoclonal antibodies against G2 (a) or G1 (c, e, and g). The same cells were double-stained with polyclonal antiserum against the nucleoprotein (b, d, f, and h). Note the progressive vacuolization of the Golgi stack concomitant with the accumulation of glycoproteins and nucleocapsids. The monoclonal antibodies used here preferentially recognize the Golgi conformation of G1 and G2. (From Kuismanen *et al.*, 1984).

inhibits the transport of proteins out of the ER and induces the redistribution of resident Golgi proteins to the ER, Punta Toro (PT), a phlebovirus, was found to bud into the ER. Such particles remained in dilated ER vacuoles and were not transported to the cell surface (Chen *et al.*, 1991a).

As is the case for other viruses with a segmented genome, each infectious virion has to package at least one copy of each of the three ribonucleoprotein (RNP) segments. How this is ensured and to what extent it is a regulated process is unknown. Another open intriguing question is whether there is an active transport of nucleoproteins to the site of budding in the Golgi. Because of the helical structure, the RNPs are morphologically identified only as more electron-dense thickenings underneath the Golgi membrane in budding profiles (Kuismanen *et al.*, 1982; Anderson and Smith, 1987). In thin sections of intracellular virus particles, the RNPs appear to lie in close contact with the lipid bilayer, while the interior of the particle appears empty (von Bonsdorff and Pettersson, 1975).

During UUK virus infection, the Golgi complex typically undergoes a morphological change. The stack of flat cisternae is progressively distorted, the Golgi vacuolizes, and large and small vesicles become partly dispersed in the cytoplasm (Fig. 1) (Kuismanen *et al.*, 1984; Gahmberg *et al.*, 1986b). Whether the morphological change that is observed for UUK virus is a hallmark for the *Bunyaviridae* in general has not been systematically studied. The vacuolization is probably caused by the accumulation of the glycoproteins in the Golgi complex, as shown by using a *ts* mutant defective in virus maturation at the restrictive temperature (Gahmberg *et al.*, 1986b). The morphologically altered Golgi is still functional in that it can terminally glycosylate Semliki Forest virus glycoproteins and transport them to the plasma membrane (Gahmberg *et al.*, 1986a). Budding of *Bunyaviridae* members is inhibited by the ionophore monensin (Cash, 1982; Kuismanen *et al.*, 1985; Schmaljohn *et al.*, 1986; Chen *et al.*, 1991a), whereas the association of the nucleocapsids with the Golgi-derived vesicles seems to be unaffected (Kuismanen *et al.*, 1985). Since monensin exchanges protons for sodium ions, this suggests that the pH or ionic conditions prevailing in the Golgi complex are important for virus budding. Budding may thus not only be dependent on a certain critical concentration of glycoproteins, but also on a conformational change of the glycoproteins induced by the milieu present in the Golgi complex. It is still not clear whether budding of *Bunyaviridae* members can occur throughout the Golgi stack or just in a subcompartment.

Vacuoles of varying sizes containing one or several particles are thought to be transported to the plasma membrane (Fig. 2), where virions are released on fusion of the vesicles with the plasma membrane. The nature of the exocytic transport vacuoles/vesicles and their relation to normal exocytic trafficking are unknown. Release of RVF virus (Anderson and Smith, 1987) and PT virus (Chen *et al.*, 1991a) occurs at the basolateral surface of polarized hepatocytes and Vero cells, respectively.

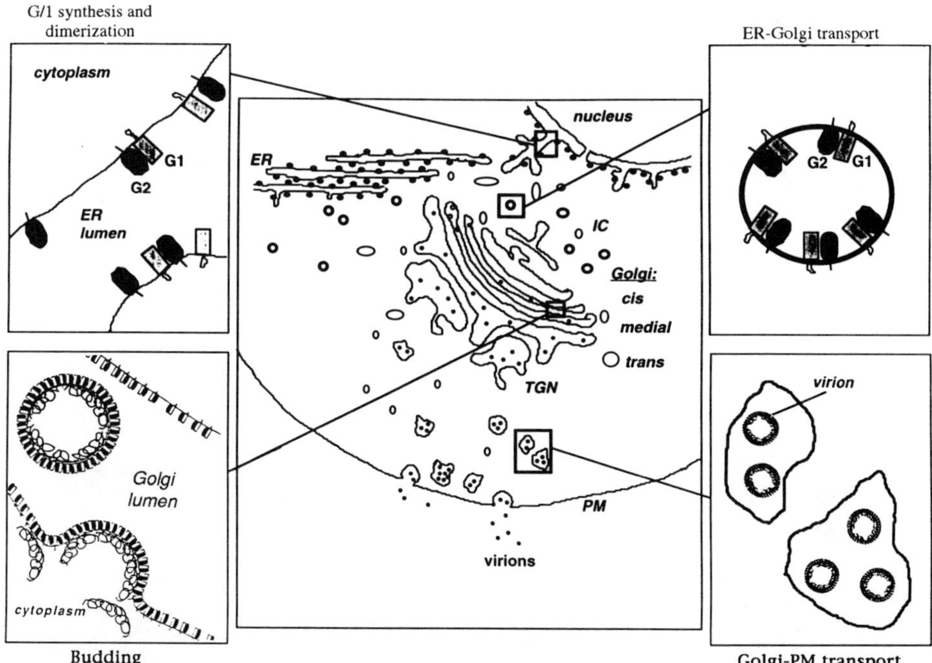

FIGURE 2. Schematic representation of *Bunyaviridae* maturation. The glycoproteins are cleaved cotranslationally in the ER from a precursor encoded by a single ORF in the M RNA segment. Following glycosylation and folding, G1 and G2 heterodimerize (upper left panel). The heterodimers are then transported to the Golgi complex (upper right panel) where further transport is arrested. G1/G2 accumulate in the Golgi and budding of virions is probably triggered by the association of helical nucleocapsids with the cytoplasmic tail of G1 (lower left panel). Virions are transported within vacuoles (lower right panel) to the cell surface where they are released on fusion of the vacuoles with the plasma membrane. IC, intermediate compartment; TGN, trans-Golgi network; PM, plasma membrane.

III. STRUCTURE AND SYNTHESIS OF THE MEMBRANE GLYCOPROTEINS

The spike proteins of all *Bunyaviridae* members are encoded by the medium-sized M RNA segment. Since the details on the structure of the M RNAs and the primary sequences of their protein products will be dealt with in the chapters reviewing the molecular biology of the individual genera, the structure of the glycoproteins will only be summarized here. The coding strategies for the various M RNA segments are depicted in Fig. 3. The spikes of all *Bunyaviridae* members are made of two glycoproteins, G1 and G2. The nomenclature is somewhat confusing in that the equivalent protein in different viruses may have different names. In each case, G1 corresponds to the

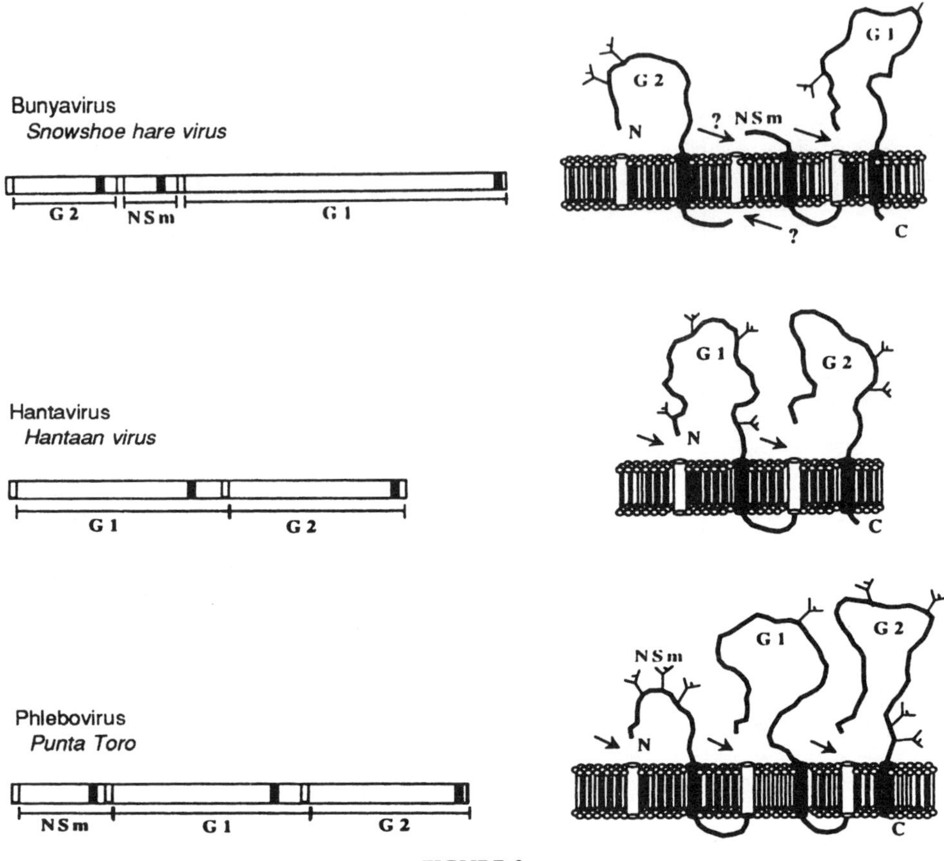

FIGURE 3.

protein with the slower mobility on an SDS–polyacrylamide gel, while G2 is the faster migrating. However, since the size of G1 and G2 in some viruses (e.g., phleboviruses) is quite similar, their mobility may depend on the conditions used during electrophoresis. Thus, under nonreduced conditions, G1 of UUK virus migrates slower than G2 (Kuismanen, 1984), while after reduction and alkylation the order is reversed, G2 now migrating slower than G1 (Persson and Pettersson, 1991). In PT virus, G1 and G2 corresponds to G2 and G1 in RVF virus (Ihara *et al.*, 1985). Recently, it has been proposed that the N-terminally located protein should be named G_N, and the C-terminal one G_C (Lappin *et al.*, 1994). It remains to be seen whether this new nomenclature will be approved by the International Committee on the Taxonomy of Viruses.

As is apparent from Fig. 3, the way the glycoproteins are synthesized, and their sizes, vary considerably between members of the different genera. Following a brief summary of the structure and synthesis of the glycopro-

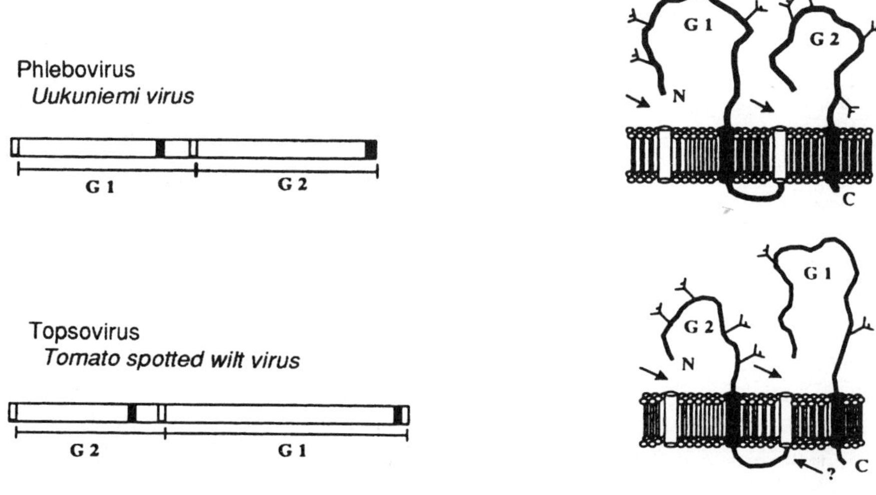

FIGURE 3. Schematic representation of the gene organization of the M RNA-encoded open reading frame of *Bunyaviridae* and the suggested processing and membrane insertion of their products. Representative examples of four genera are shown. The localization of the G1, G2, and NSm genes in the precursor are indicated. Open box depicts the N-terminal signal sequence and closed box the transmembrane anchor. Arrows on the luminal side indicate cleavage by signal peptidase. The exact site for processing the N-terminus of NSm of bunyaviruses is not known. The number of potential glycosylation sites are indicated.

teins, a more detailed description of the individual genera will be presented. In all cases studied, G1 and G2 have been found to be type I membrane-spanning glycoproteins. They are thus thought to span the lipid bilayer only once and to have their N-terminus oriented toward the ER lumen, and C-terminus facing the cytoplasm and the interior of the virus. Although this interpretation, which is deduced from the primary sequence and hydropathy plots, may be correct, limited experimental evidence in support of this topology has been published. G1 and G2 (and NSm where present) are cotranslationally cleaved from the primary translation product encompassing the single open reading frame in the M RNA. Each membrane protein is preceded by a separate signal sequence for targetting of the nascent chain to, and facilitating its translocation through, the ER membrane. The cleavages that follow to create mature G1 and G2 (and NSm) are probably all carried out by the lumenally located signal peptidase (Fig. 3). There is no evidence for additional cleavages occurring on the cytoplasmic side. Following translocation through the ER membrane, folding, disulfide-bond formation, and core-glycosylation at Asn-X-Ser/Thr sites, G1 and G2 appear to heterodimerize. After a considerable lag period, G1 and G2 are transported to the Golgi complex, where they become arrested and undergo trimming of the glycans. Although G1 and G2 accumulate in the Golgi complex, a fraction escapes this retention and is transported to the cell surface. In the Golgi, G1–G2 dimers could form higher oligomers, but such complexes have not yet been demonstrated for any *Bunyaviridae* member.

A. *Bunyavirus* Genus

The M segment of several members of this genus has been cloned and sequenced, and the deduced primary sequence of the membrane proteins compared (reviewed by Elliott, 1990; Elliott *et al.*, 1990; Bouloy, 1991). The gene order is NH_2–G2–NSm–G1–COOH, where NSm is a short 174-residue-long nonglycosylated membrane protein with unknown function (Fig. 3). The exact cleavage sites between G2 and NSm (position 299) and NSm and G1 (position 473) have been determined only for snowshoe hare (SSH) virus by sequencing the C- and N-terminal ends directly (Fazakerley *et al.*, 1988). Since the proteins within this genus are quite well conserved (Elliott, 1990), the cleavage sites for the other sequenced members [La Crosse (LAC), Bunya-mwera (BUN), and Germiston (GER) viruses] can also be deduced. G2 (M_r 29–41 kDa) is preceded by a typical signal sequence that is cleaved off in the ER, while an internal signal sequence for G1 (M_r 108–125 kDa) is apparently located at the C-terminus of NSm (Fig. 3). What proteolytic trypsin-like protease releases NSm from G2, and on which side of the ER membrane this takes place is not clear. The most likely hypothesis is that all cleavages are carried out by the ER luminal signal peptidase. None of the processing events require the concomitant expression of any other viral proteins (Nakitare and Elliott, 1993). The organization and processing of the *Bunyavirus* precursor proteins are quite similar to those of the alphavirus membrane proteins, which also contain an internal (6K) protein located between the p62 and E1 spike proteins (Garoff *et al.*, 1980). The 6K peptide has been found to be dispensible, although its presence enhances virus production (Liljeström *et al.*, 1991). The role of the NSm peptide is not known, but it may, in analogy to the alphavirus 6K peptide (Liljeström and Garoff, 1991), provide the signal sequence for the downstream G1. The fact that it localizes to the Golgi in BUN virus-infected cells suggests that it may have a role in virus maturation (Nakitare and Elliott, 1993; Lappin *et al.*, 1994). That processing of *Bunya-virus* glycoproteins occurs cotranslationally is inferred from the fact that no precursor has been found in pulse–chase experiments (Pennington *et al.*, 1977; Fazakerley *et al.*, 1988). It has not been possible to study the processing *in vitro*, since translation of the M mRNA has for some reason been unsuc-cessful.

Cysteine residues, N-glycosylation sites, and hydropathy profiles are well conserved between bunyaviruses (Elliott *et al.*, 1990), suggesting a highly conserved three-dimensional structure. Both G1 and G2 are acylated (shown only for LAC virus; Madoff and Lenard, 1982) and mature virions contain primarily N-linked complex-type glycans (Vorndam and Trent, 1979; Madoff and Lenard, 1982; Pesonen *et al.*, 1982b). Partial sequences of the glycans from Inkoo bunyavirus have been determined and found to be of three types: high-mannose, complex, and small endoglycosidase H-resistant intermediate glycans (Pesonen *et al.*, 1982b).

By analogy to the phleboviruses, one would assume that the G1 and G2 proteins of bunyaviruses would also form heterodimeric complexes. However, attempts to demonstrate such dimers have so far failed (Gerbaud *et al.*, 1992).

B. *Hantavirus* Genus

The M segment of several hantavirus strains have been cloned and sequenced (for references see Parrington *et al.*, 1991; Antic *et al.*, 1992a; Spiropoulou *et al.*, 1994). Hantaan (HTN) virus G1 and G2 have similar sizes (M_r about 65 and 55 kDa, respectively) and are synthesized in the order NH_2–G1–G2–COOH (Schmaljohn *et al.*, 1987). There is no evidence for an NSm protein. G1 and G2 are thought to be cotranslationally cleaved in the ER, since no full-length precursor has been detected in virus-infected cells (Schmaljohn *et al.*, 1986, 1990), nor in cells expressing the M segment from a cloned cDNA (Pensiero *et al.*, 1988; Ruusala *et al.*, 1992). Both proteins are preceded by a hydrophobic signal sequence (Fig. 3). That the internal signal sequence preceding G2 is functional is supported by the fact that it can target G2 correctly to the ER in the absence of G1 (Schmaljohn *et al.*, 1990; Pensiero and Hay, 1992; Ruusala *et al.*, 1992). In a recent report, Kamrud and Schmaljohn (1994) found that translation of G2 could be initiated internally, albeit inefficiently, in the full-length M mRNA from an AUG codon preceding G2. They propose that initiation occurs via a leaky scanning, rather than direct internal ribosome entry. The physiological importance, if any, of this internal initiation is unclear.

In the ectodomain of G1 there are four, and in G2 two, potential sites for N-linked glycosylation; however, one of the sites in G2 probably cannot be used (Bause, 1983). Pulse–chase experiments, as well as analyses of extracellular virions indicate that the glycans are primarily of the high-mannose, endo H-sensitive type (Schmaljohn *et al.*, 1986; Ruusala *et al.*, 1992). The fact that HTN virus G1- or G2-specific monoclonal antibodies can precipitate both proteins either without or after prior cross-linking (Arikawa *et al.*, 1989; Antic *et al.*, 1992b), and that G1 and G2 have to be coexpressed to enable exit of both proteins from the ER (Ruusala *et al.*, 1992), suggests that G1 and G2 form heterodimers already in the ER.

C. *Nairovirus* Genus

The structure and synthesis of nairovirus glycoproteins have so far been poorly studied. Only the M RNA segment of Dugbe (DUG) virus has been cloned and sequenced (Marriott *et al.*, 1992). In nairovirus-infected cells large nonstructural glycoproteins have been identified (Clerx and Bishop, 1981; Cash, 1985; Watret and Elliott, 1985). Sequence analysis of the DUG virus M segment (4888 nucleotides) and identification of its products by specific antisera prepared against fusion proteins, indicate that the processing of the primary translation product is more complex than for other *Bunyaviridae* members. The M segment harbors a single ORF corresponding to 1551 amino acids (173,300 Da). The gene order is NH_2–G2–G1–COOH. Following cleavage (probably) by signal peptidase, an 85-kDa glycosylated precursor seems to be further processed to the mature G1 (655 residues, M_r 70 kDa). A 110-kDa G1-related product has also been found in infected cells, but this is probably not a precursor of gp85 or G1. G2 (M_r 35 kDa) is probably also cleaved from a larger (M_r 70 kDa) precursor. The further fate of G1 and G2 in infected cells is not known, except that they are found in virions. A third virion glycoprotein (M_r 45 kDa) has been reported for Hazara virus, a member of the Crimean-Congo hemorrhagic fever serogroup (Foulke *et al.*, 1981).

D. *Phlebovirus* Genus

Based on sequence relationships, the previously independent *Uukuvirus* genus has been merged with the phleboviruses, forming the present *Phlebovirus* genus (Francki *et al.*, 1991). The strategy of M segment gene expression differs between UUK virus and other phleboviruses in that the mature glycoproteins are preceded by an NSm presequence only in the latter viruses (Fig. 3). The M segment of RVF (Collett *et al.*, 1985; Takehara *et al.*, 1989), PT (Ihara *et al.*, 1985), and UUK (Rönnholm and Pettersson, 1987) viruses have been cloned and sequenced.

The gene order in RVF is NH_2–NSm–G2–G1–COOH and in PTV NH_2–NSm–G1–G2–COOH, the confusing nomenclature being the result of the differences in mobility on SDS gels between G1 and G2 of the two viruses (see above). NSm, which has not been found in virions, is much larger for PT virus (30 kDa) than for RVF virus (14 kDa). Expression of wild-type and mutant forms of PT virus G1 and G2 has shown that both G1 and G2 are inserted in the membrane as type I proteins (Chen *et al.*, 1991b). Both proteins are preceded by a signal sequence that is cotranslationally processed, and they are anchored in the lipid bilayer by a hydrophobic transmembrane C-terminal domain. Whereas the transmembrane domain of G2 can be deduced to span residues 1282 to 1300, the corresponding domain has not as yet been exactly defined in G1, since this putative region encompasses some 50 amino acids (residues 684–733) (Ihara *et al.*, 1985).

The expression of the RVF virus M segment has been studied in greater detail than that of PT virus. *In vitro* translation of the M mRNA of RVF virus yields a 133-kDa uncleaved primary translation product, which in the presence of microsomal membranes is cotranslationally cleaved to yield G1 and G2, as well as 78-, 21-, and 14-kDa products (Suzich and Collett, 1988). In addition to G1 (M_r 65 kDa) and G2 (M_r 56 kDa), a glycosylated 78-kDa and a nonglycosylated 14-kDa protein (but no 21-kDa product) were found in cells infected with RVF virus, or with a recombinant vaccinia virus harboring a complete M cDNA (Kakach *et al.*, 1988; Wasmoen *et al.*, 1988). Mature G2 is preceded by 5 (in PT virus by 13) in-frame AUG initiation codons. *In vitro* mutagenesis followed by expression of the mutants from vaccinia virus recombinants and analyses of the products using peptide antibodies indicate that the first AUG is utilized for the synthesis of the 78-kDa product, whereas the second AUG is utilized for the synthesis of the 14-kDa product (Kakach *et al.*, 1988, 1989; Wasmoen *et al.*, 1988; Suzich *et al.*, 1990). The 78-kDa product, which contains the whole of G2, the 14-kDa peptide, and some upstream residues, was not found to be the precursor for mature G2. Thus, G2 and the 14-kDa presequence may be generated by cotranslational proteolytic cleavage of the protein initiated at the second AUG. The first and fourth (or fifth) AUG codons precede typical signal peptide-like sequences that could direct the 78-kDa protein and G2 through the ER membrane. The second and third AUG codons are not, however, followed by such hydrophobic sequences, making it unclear how the 14-kDa protein could be translocated through the ER membrane. It is therefore not clear whether the 14-kDa protein is indeed translocated into the ER lumen and whether it becomes an integral membrane protein. The finding that the potential N-glycosylation site at Asn_{88} is utilized only in the 78-kDa product, but not in the 14-kDa product (Kakach *et al.*, 1989), suggests that the latter protein is not translocated through the ER membrane. Mature G2 contains only one N-linked glycan, while three out of four sites in G1 seem to be utilized (Kakach *et al.*, 1989). The presequence of the 30-kDa NSm of PT virus shows no apparent sequence homology with the 14-kDa NSm of RVF virus and has not been detected in infected cells.

The role of NSm in the life cycle of phleboviruses is still obscure. Using specific antisera, Wasmoen *et al.* (1988) located G1, G2, and the 78- and 14-kDa proteins to the Golgi complex and to a reticular network, probably representing the ER, both in cells infected with RVF, and with mutant recombinant vaccinia viruses expressing the proteins. In the absence of the 78- and 14-kDa products, G1 and G2 were also localized to, and retained in, the Golgi, indicating that the information for Golgi localization resides in G1/G2 (see below). Thus, the 14-kDa product is not required for proper processing and transport of the mature G1 and G2 proteins. Possible functions for the NSm polypeptide could be that it simply provides a suitable signal sequence for translocation of the downstream protein through the ER membrane. Alternatively, it could play a role in the folding and assembly of the viral spikes in the ER, or budding in the Golgi complex. The lack of an NSm protein in UUK virus (and hantaviruses) suggests that it is dispensible for the formation of infectious virions. Whether this is true for RVF and PT viruses will, however, have to await the development of a reverse genetics system that would enable the engineering of a virus from which the NSm sequence has been deleted.

Using the T7 polymerase-driven vaccinia virus expression system, Chen and Compans (1991) found that G1 and G2 form a noncovalently linked heterodimeric complex in the ER within minutes after synthesis. Evidence for low levels of G2 disulfide-linked homodimers was also found. The complexes were identified by cross-linking, polyacrylamide gel electrophoresis, and sucrose gradient centrifugation. Both types of complexes were transported to and retained in the Golgi complex. Disulfide-linked heterodimers between a full-length G1 and a membrane anchor minus (soluble) G2 mutant were also observed and these were transported out of the ER, suggesting that the cytoplasmic tail and transmembrane domain of G2 are not required for dimerization or transport.

The primary product made *in vitro* by translating the full-length M mRNA of Uukuniemi virus in the absence of microsomal membranes is a 110-kDa precursor (p110) of G1 and G2. In the presence of dog pancreas microsomes, p110 is cotranslationally cleaved to G1 and G2 (Ulmanen , 1981). As for other *Bunyaviridae* members, the precursor has not been found in infected cells. Pulse–chase analysis (Kuismanen, 1984) and direct sequencing (Rönnholm and Pettersson, 1987) have shown that the gene order is NH_2– G1–G2–COOH. Recent indirect evidence indicates that there is only a single cleavage event between G1 and G2. This cleavage takes place between a serine and a cysteine residue and is probably carried out by the ER signal peptidase. Expression of mutant cDNAs, analyses using peptide antibodies directed against different portions of the C-terminal cytoplasmic tail of G1, and treatment of microsomes isolated from infected cells with proteases, all suggest that the internal signal sequence for G2 remains attached to the tail of G1 (Figs. 3 and 4) (unpublished results). This means that the cytoplasmic tail of G1 consists of an 81-residue largely hydrophilic region followed by a 17-residue hydrophobic region (the G2 signal sequence) (Fig. 4). That the C-terminus of G1 is exposed on the cytoplasmic face of the lipid bilayer has been shown by immunofluoresence of streptolysin O-permeabilized cells by using tail-specific peptide antibodies (unpublished results). The G1 tail is likely to interact with the nucleoproteins to facilitate budding. It also seems to contain a signal for targetting of G1 to the Golgi complex (see below). The tail of G2 is very short (5 residues) and hydrophilic, and may therefore not have any other function than to prevent G2 from slipping into the ER lumen.

Following translocation and cleavage, both G1 and G2 undergo core glycosylation at asparagine sites. There are four potential sites in both G1 and G2, and all of them are probably used (Kuismanen, 1984; Rönnholm and Pettersson, 1987). During intracellular transport, the glycans are trimmed to

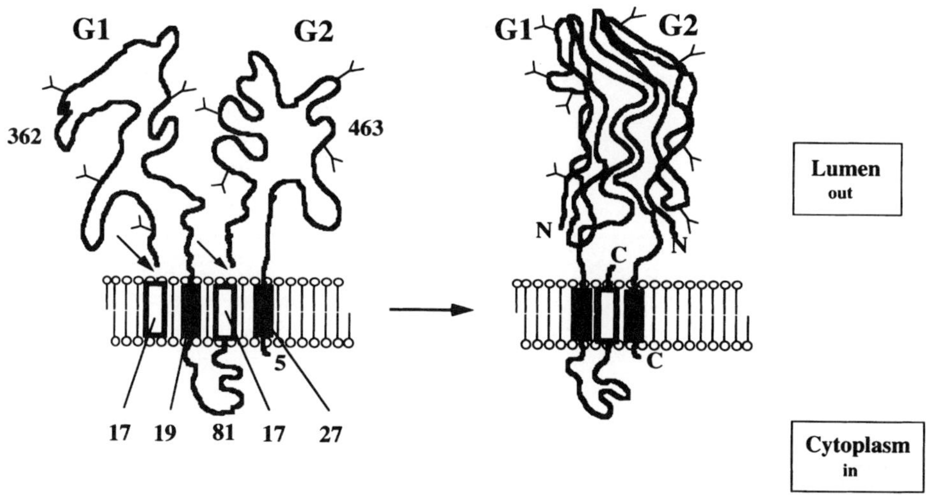

FIGURE 4. Processing and dimerization of Uukuniemi virus G1 and G2. G1 and G2 are cleaved cotranslationally in the ER from the precursor p110 and probably inserted in the ER membrane as depicted. Following glycosylation and folding, G1 and G2 heterodimerize already in the ER. Mature G1 and G2 are thought to be processed at two sites by the signal peptidase (arrows). The internal signal sequence of G2 remains covalently attached to the C-terminus of G1. However, it is not known whether this hydrophobic region remains membrane-associated as shown. The number of residues comprising the different domains as well as the glycosylation sites are indicated.

varying degrees. In purified virions, G1 contains primarily complex, sialylated, endo H-resistant glycans, while G2 contains a mixture of complex, intermediate (both endo H-resistant), and high-mannose (endo H-sensitive) chains (Pesonen *et al.*, 1982a; Kuismanen, 1984). The same pattern of glycans is also found in Inkoo virus (*Bunyavirus* genus) (Pesonen *et al.*, 1982b). Pulse–chase experiments have shown that G1 acquires endo H-resistant glycans slowly with a $t_{1/2}$ of about 45 min, while the acquisition of partial endo H resistance in G2 occurs even more slowly (Kuismanen, 1984). The kinetics of folding, as monitored by disulfide-bond formation, also differs between the two proteins. G1 was found to fold rapidly within minutes ($t_{1/2}$ about 10 min), while folding of G2 was much slower ($t_{1/2}$ about 45–60 min) (Persson and Pettersson, 1991; Pettersson *et al.*, 1993). Since both proteins have 26 cysteine residues in their respective ectodomain, and therefore the potential to form 13 disulfide bonds, this indicates that the number of disulfide bonds *per se* does not determine the folding kinetics of membrane proteins. During folding, both G1 and G2 are transiently associated with the IgG heavy chain-binding protein (BiP), an ER luminal chaperone. In conformity with the different folding kinetics, G1 is bound to BiP for a shorter time than G2. Protein disulfide isomerase (PDI) can also be coprecipitated with the glycoproteins, in agreement with the role for PDI in catalyzing disulfide-bond formation (Persson and Pettersson, 1991; Bardwell and Beckwith, 1993). Having folded correctly, G1 and G2 form heterodimers in the ER (Fig. 4). Pulse–chase experiments showed that newly made G1 rapidly dimerizes with G2. Since G2 folds much slower than G1, this means that newly synthesized G1 is unable to dimerize with its companion G2 made from the same p110 molecule. Thus, G1 and G2 cleaved from a p110 enter a glycoprotein pool in the ER from which partners are drawn into heterodimeric complexes. This conclusion is also supported by experiments in which G1 and G2 were expressed from separate cDNAs (Melin *et al.*, 1995). Following dimerization, there is a lag of some 30 to 45 min ($t_{1/2}$), before the G1–G2 heterodimers appear in the Golgi complex. It is at present unclear where the proteins reside during this long delay, as is also the mechanism for the transport delay. Both G1 and G2 become palmitylated (Pettersson, unpublished), and no evidence for O-linked glycosylation has been found (Pesonen *et al.*, 1982a). As discussed below, G1 and G2 accumulate in the Golgi complex, where G1 acquires endo H-resistant glycans, while G2 remains largely endo H-sensitive (Kuismanen, 1984). A minor fraction of the spike proteins will end up at the plasma membrane, in particular late in infection (Kuismanen *et al.*, 1985). This may represent glycoproteins left over from the budding. The fate of this surface expressed fraction is not known. It may recycle back to the Golgi complex, or be transported to the lysosomes for degradation.

Recently, it has been suggested that G1 and G2 in UUK virions are present as homodimers (E. Kuismanen, personal communication). On treatment of virions with low pH, the G2 homodimers dissociate into monomers. This suggests that the spike proteins of UUK virus undergo conformational changes and rearrangements analogous to those demonstrated for Semliki Forest virus (an alphavirus) E1 and E2 (Wahlberg and Garoff, 1992).

Whether G1/G2 forms higher-order complexes is not known. Cross-linking experiments and sucrose gradient centrifugations have failed to demonstrate such complexes (Persson and Pettersson, unpublished data). In virions G1/G2 are seen by EM as hollow cylinders organized in a T = 12 icosahedral surface lattice. This architecture is most apparent at low pH. The symmetry indicates the presence of 720 structural units (60 × T), each of which probably is made up of a G1–G2 heterodimer. This means that each particle should contain 720 molecules each of G1 and G2 (von Bonsdorff and Pettersson, 1975).

E. *Tospovirus* Genus

Although the M RNA segments of two distinct tospoviruses have been cloned and sequenced, little is known about the synthesis, transport, or intracellular localization of their products. The M segments of tomato spotted wilt (TSW) virus and impatiens necrotic spot (INS) virus are 4821 (Kormelink *et al.*, 1992) and 4972 (Law *et al.*, 1992) residues in size. In contrast to all other *Bunyaviridae*, the M segment of tospoviruses displays an ambisense coding strategy. An mRNA transcribed from the 3' part of the virion RNA encodes a precursor (TSW virus: 1136 residues, 127.4 kDa; INS virus: 1110 residues, 124.9 kDa) to the two glycoproteins G1 (M_r 78 kDa) and G2 (M_r 58 kDa) in the order NH_2–G2–G1–COOH, while a subsegmental mRNA derived from the viral-complementary strand encodes a nonstructural protein [NSm; 302 residues (33.4 kDa) and 303 (34.1 kDa) residues, respectively]. The NSm is not a membrane protein and has recently been found to associate with nucleocapsids and plant cell plasmodesmata, and to be responsible for viral nucleocapsid movement from cell to cell (Kormelink *et al.*, 1994; R. Goldbach, personal communication). Because of these properties, NSm will therefore not be discussed here further.

Based on hydropathy plots, it seems likely that tospovirus G1 and G2, similar to those of other *Bunyaviridae* members, are both type I proteins spanning the membrane once (Fig. 3). G1 of TSW virus shows clear sequence homology with G1 of BUN virus (45% similarity, 22% identity), while G2 is much less homologous (Kormelink *et al.*, 1992). G2–G1 of INS virus displays 36 and 39% similarities to PT and RVF virus glycoproteins, respectively (Law *et al.*, 1992). This indicates an evolutionary relationship between plant and animal bunyaviruses. The synthesis, processing, and transport of tospovirus glycoproteins have not yet been analyzed.

IV. G1 AND G2 ACCUMULATE IN THE GOLGI COMPLEX

As described above, members of all *Bunyaviridae* genera with the possible exception of the tospoviruses, mature by budding through membranes of the Golgi complex (Pettersson *et al.*, 1988; Matsuoka *et al.*, 1991; Pettersson, 1991; Hobman, 1993). Evidence from several bunyaviruses indicates that the site of maturation is determined by the accumulation of viral glycoproteins in the Golgi. Following synthesis, folding, glycosylation, and dimerization in the ER, G1 and G2 move to the Golgi, where further transport is arrested (Fig. 1). Accumulation of G1/G2 in the Golgi complex has been analyzed in infected cells or in cells expressing G1 and G2 from cloned cDNA by immunofluorescence, and in a few cases by immuno-EM (Kuismanen *et al.*, 1982, 1985; Anderson and Smith, 1987), and subcellular fractionation (Persson and Pettersson, 1991; Ruusala *et al.*, 1992). Colocalization of viral glycoproteins with markers for the Golgi apparatus, such as thiamine pyrophosphatase (nucleoside diphosphatase) (Kuismanen *et al.*, 1982; Pensiero *et al.*, 1988), wheat germ agglutinin (WGA) (Kuismanen *et al.*, 1982; Wasmoen *et al.*, 1988; Chen and Compans, 1991; Nakitare and Elliott, 1993; Lappin *et al.*, 1994), CTR433 (Rönnholm, 1992), Golgizone (Pensiero and Hay, 1992), or mannosidase II (Kuismanen *et al.*, 1984; Gahmberg *et al.*, 1986b; Melin *et al.*, 1995), have unambiguously shown that G1/G2 indeed accumulate in this organelle (Fig. 5). However, it has to be stressed that the exact sublocalization

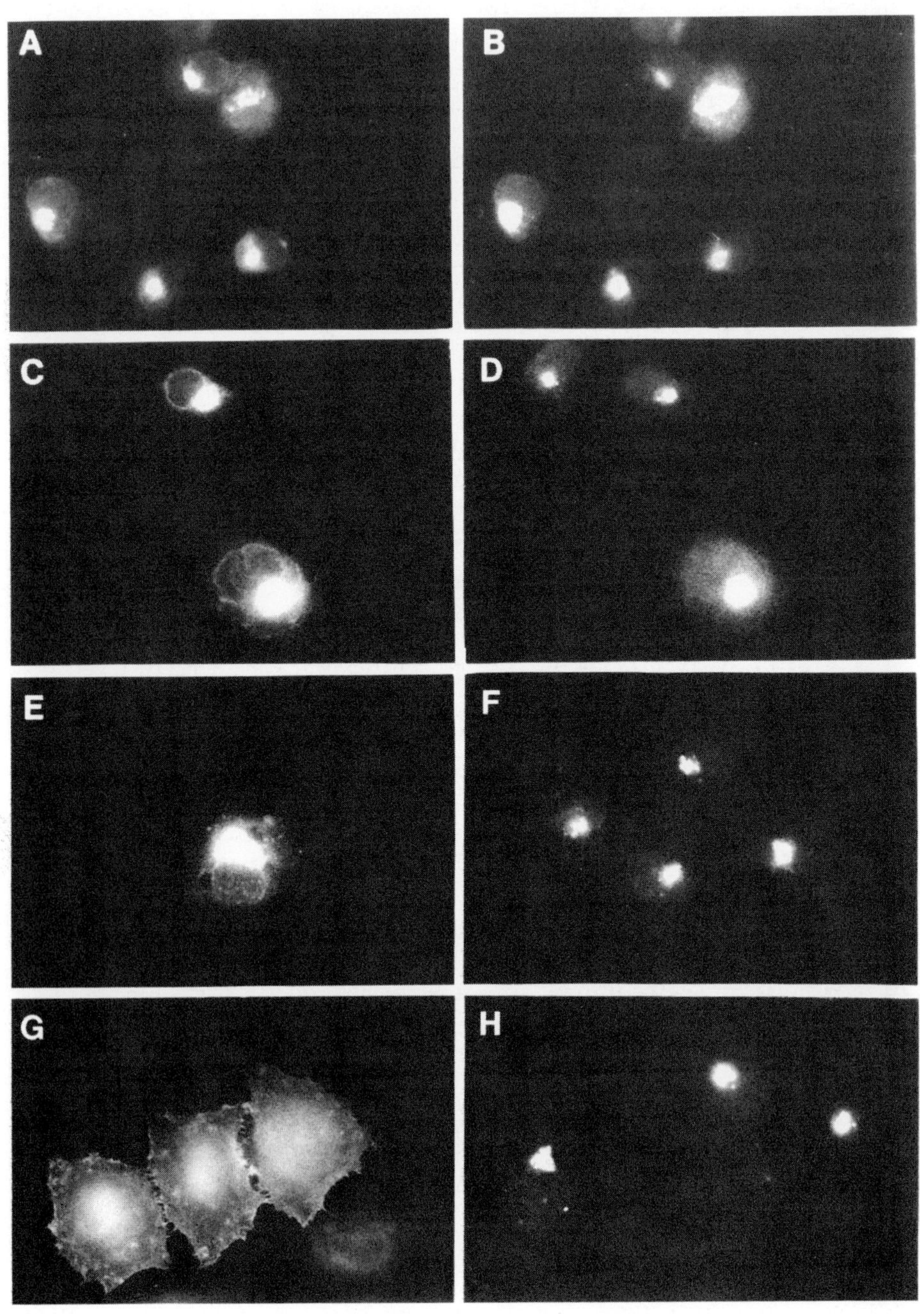

in the Golgi complex has not been determined on the EM level using appropriate markers. Because of the altered morphology induced by the viral glycoproteins (Fig. 1) (Kuismanen *et al.*, 1984; Gahmberg *et al.*, 1986b), it is very difficult to identify the *cis, medial,* and *trans* cisternae of the Golgi complex. It is therefore not known whether the glycoproteins are localized to particular cisternae of the Golgi, and whether localization is exactly the same for all members of the family. There is thus a great need for localization of G1/G2 on the EM level.

Proteins en route to the plasma membrane become transiently concentrated in the Golgi complex. This is seen by immunofluorescence as an apparent accumulation of the proteins in the Golgi. To distinguish between this type of Golgi localization and true retention and accumulation, chase in the presence of cycloheximide to stop further protein synthesis is necessary. Proteins passing through the Golgi will be chased out of the Golgi, while truely retained proteins will remain. In the case of *Bunyaviridae* glycoproteins, many investigators have shown that G1/G2 cannot be chased out of the Golgi even after 4- to 6-hr chases with cycloheximide (Gahmberg *et al.*, 1986b; Matsuoka *et al.*, 1994). In most cases no or very little glycoprotein has been localized to the plasma membrane. However, the glycoproteins of RVF virus grown in primary rat hepatocytes were expressed on the cell surface (Anderson and Smith, 1987), and UUK virus glycoproteins are present on the cell surface at later stages of infection (Kuismanen *et al.*, 1982). In the latter case, cell-associated virions may account for most of the observed patchy immunofluorescence.

V. MAPPING THE GOLGI-RETENTION SIGNAL IN G1/G2

G1 and G2 accumulate in the Golgi complex when expressed in the absence of any other viral proteins, showing that Golgi retention is an inherent property of G1/G2 (Pensiero et al., 1988; Matsuoka *et al.*, 1988, 1994;

FIGURE 5. Intracellular localization of Uukuniemi virus G1, G2, and chimeric G1 proteins in transfected cells. The cDNA constructs were expressed in HeLa (panels A–E) or BHK21 (panels F–H) cells using the T7 polymerase-driven vaccinia virus system. Proteins were localized by indirect immunofluorescence using polyclonal antiserum against G1 (A), CD4 (F–H), or mannosidase II (D), and monoclonal antibodies against G2 (B) or G1 (C, E). In panels A and B, G1 and G2 were coexpressed from two different plasmids and the cells were double-stained to detect the proteins in the same cells. Cells expressing G1 alone were double-stained with antisera against G1 (C) and mannosidase II (D), a Golgi marker. Panel E shows the localization of G1 in which its transmembrane domain and 10 flanking residues on both sides were exchanged for the corresponding domains of VSV G protein. Panels F–H show localization of chimeras between G1 and CD4: Golgi localization is apparent for chimeras in which the CD4 ectodomain and transmembrane domain were fused to the whole cytoplasmic tail of G1 (98 residues) (F), or the 81 proximal residues, i.e., with the G2 signal sequence deleted (H) (see Fig. 3). In contrast, a chimera between the CD4 ectodomain and cytoplasmic tail, and the G1 transmembrane domain resulted in surface expression (G), indicating that the G1 transmembrane domain is not necessary for Golgi localization.

Wasmoen *et al.*, 1988; Chen *et al.*, 1991a,b; Ruusala *et al.*, 1992; Rönnholm, 1992; Pensiero and Hay, 1992; Nakitare and Elliott, 1993; Lappin *et al.*, 1994; Melin *et al.*, 1995). The question thus arises as to which of the two glycoproteins determines Golgi localization, or whether both proteins are necessary.

Cellular proteins destined to different locations in the cells have been found to contain address tags (sorting, targetting, retention signals). Such signals are found on both membrane and secreted proteins, as well as on soluble cytosolic proteins. These include the classical N-terminal signal sequence for ER targetting, signals for import into mitochondria, peroxisomes, and the nucleus (von Heijne, 1990). In the exocytic pathway, signals for sorting proteins from the *trans*-Golgi network (TGN) to the lysosomes, endosomes, secretory granules, apical and basolateral surfaces of polarized cells, as well as retention and recycling signals for keeping proteins in defined compartments, such as the ER and the Golgi, have been identified and in many cases narrowed down to short amino acid stretches. The generally held view is that proteins targetted onto the exocytic pathway are transported out to the cell surface by default (bulk flow) unless they contain retention or sorting signals (Pfeffer and Rothman, 1987). There are no reasons to believe that viral membrane proteins differ from host cell proteins in this respect (Pettersson, 1991; Griffiths and Rottier, 1992).

Against this background, it has been assumed that *Bunyaviridae* glycoproteins also contain a signal for targetting to, and retention in, the Golgi complex. The identification of such a signal has been hampered by the fact that G1 and G2 form heterodimeric complexes. In many cases both proteins have to be coexpressed to become competent to exit the ER (Hurtley and Helenius, 1989). Expression of mutant proteins in these cases requires the coexpression of the other subunit for proper transport. G1 and G2 are complex glycoproteins with 5–6% cysteines in their ectodomains. This makes them vulnerable to misfolding as a consequence of mutagenesis, particularly in their ectodomains. Two approaches can be taken to map the retention signal. One involves removing the signal by mutagenesis resulting in the relief of the Golgi block and transport to the cell surface. The other possibility is to make chimeric proteins by using domains from reporter proteins normally transported to the cell surface. The identification of a Golgi-retention signal in this case relies on the ability of defined domains from either G1 or G2 to retain such reporters in the Golgi. The search for a retention signal for *Bunyaviridae* glycoproteins has just begun and no defined signal has as yet been identified. Only a few *Bunyaviridae* members, notably BUN, PT, HTN, and UUK viruses, have been used as models and the results from these examples are summarized individually below.

A. Bunyamwera Virus

BUN virus glycoproteins localize to the Golgi either in infected cells, or in HeLa cells expressing G1 and G2 from a full-length M cDNA using the T7 polymerase-driven vaccinia virus system (Nakitare and Elliott, 1993; Lappin *et al.*, 1994). In the presence of brefeldin A, the glycoproteins were redistributed to the ER. Recent results by Lappin *et al.* (1994) indicate that G1, G2, and NSm are all localized to the Golgi complex when expressed from the full-length cDNA. G2 and NSm expressed separately also localize to the Golgi, while G1 expressed alone remains in the ER. G1 can be rescued out from the ER by coexpression with G2 (but not with NSm) from a different plasmid. Using recombinant vaccinia viruses, similar results have been reported for LAC virus glycoproteins (Bupp *et al.*, 1994). These results show that G1 requires coexpression of G2 to become transport competent, suggesting heterodimeric interactions between the two proteins. They also indicate that at least G2 contains a Golgi-retention signal.

B. Hantaan Virus

HTN virus G1 and G2 accumulate in the Golgi either in infected cells (Pensiero *et al.*, 1988), or if expressed together from a full-length cDNA clone (Pensiero *et al.*, 1988; Pensiero and Hay, 1992; Ruusala *et al.*, 1992). Regarding the fate of G1 and G2 expressed separately, some inconsistencies have been reported. Pensiero and Hay (1992) found that G1 expressed in CV-1 cells on its own from a recombinant vaccinia virus was targetted to and accumulated in the Golgi complex, while G2 expressed alone remained in the ER. When the two proteins were coexpressed from different plasmids, G2 could be rescued from the ER and was transported to the Golgi. A mutant G1, truncated at its C-terminus by removing 82 residues upstream of the cleavage site of mature G2, was also localized to the Golgi. These results suggest that the first 567 residues of G1 contain a signal for Golgi retention. It should be noted that these localizations were made only on the basis of low-resolution immunofluorescence. Whether G2 accumulates in the Golgi indirectly by binding to G1, or whether it contains its own targetting signal is not clear.

Using similar recombinant viruses (Schmaljohn *et al.*, 1990), Ruusala *et al.* (1992), in contrast, found that neither protein was able to exit the ER on its own in HeLa cells. Only when coexpressed from the same full-length insert, or from separate plasmids, were both proteins localized to the Golgi. These results were confirmed by subcellular fractionation analyses on sucrose gradients. It is not apparent how to reconcile these different results, but differences in host cells, cDNA constructs, or experimental design (in particular the duration of recombinant vaccinia virus infection) are possible explanations.

C. Punta Toro Virus

G1 and G2 expressed together from a recombinant vaccinia virus containing a cDNA lacking most of NSm were found to be targetted to the Golgi complex. G1 acquired endo H resistance and very little protein was found on the cell surface (Matsuoka *et al.*, 1988). The proteins were retained in the Golgi even after a 6-hr chase in the presence of cycloheximide. In a subsequent report, Chen *et al.* (1991b) found that G2 expressed alone was transported to the cell surface, and an anchor minus, soluble, G2 was secreted into the medium. However, if this truncated G2 was coexpressed with full-length G1, then it was targetted to and retained in the Golgi. These results suggested that G2 lacks a Golgi-retention signal and that it is retained in the Golgi indirectly via its binding to G1. In this report, G1 was not expressed alone and therefore a direct proof that G1 de facto contains the Golgi-retention signal was not obtained. However, Matsuoka *et al.* (1994) have recently reported on the expression of full-length and C-terminally truncated mutants of G1. A mutant with a stop codon at the G2 cleavage site was transported to and retained in the Golgi complex. Progressive deletions of the cytoplasmic tail of G1 up to 10 residues from the G1 transmembrane domain resulted in mutants retained intracellularly, i.e., both in the ER and in the Golgi complex. The ER-retained fraction could not be chased out to the Golgi even after 4 hr of cycloheximide treatment. The inefficient exit of G1 tail deletion mutants from the ER makes it difficult to assess the role of the cytoplasmic tail in keeping G1 in the Golgi. Mutants lacking the complete cytoplasmic tail were transported to the cell surface and a soluble anchor minus mutant was secreted out of the cell. A chimeric molecule containing the ectodomain and cytoplasmic tail of the murine leukemia virus (MCF) envelope protein and the transmembrane domain of G1 was partially retained intracellularly (ER and Golgi), while a chimera in which both the transmembrane domain and cytoplasmic tail were attached to the ectodomain of MCF_{env} was more efficiently localized to the ER and Golgi. A construct containing only the cytoplasmic tail of G1 and the rest from the MCF_{env} was transported to the cell surface. These results, which are partly similar to those obtained for UUK virus G1 (see below), suggest that the Golgi retention signal of PT G1 is complex with contributions both from the transmembrane domain and from the cytoplasmic tail proximal to the membrane. Interpretation of the results is complicated by the difficulty in defining the borders of the transmembrane domain of the PT virus G1. As noted above, the hydrophobic region harboring the putative transmembrane domain is some 50 residues long (Ihara *et al.*, 1985).

D. Uukuniemi Virus

G1 and G2 of UUK virus expressed from a full-length cDNA by using an SV40 vector (Rönnholm, 1992), the T7 polymerase-driven VV expression system (Melin *et al.*, 1995), or the Semliki Forest virus (SFV) replicon (Andersson *et al.*, unpublished) are transported to and retained in the Golgi complex. Both proteins colocalize with markers for the Golgi complex. G1 acquires partial endo H resistance with kinetics resembling that of G1 expressed during virus infection (Kuismanen, 1984; Melin *et al.*, 1995). G2 expressed alone is retained in the ER and may be transported to the cell surface very slowly (Rönnholm, 1992; Melin *et al.*, 1995). Thus, there seems to be a difference between PT and UUK virus G2 in this respect (see above). G1 expressed alone is transported to and retained in the Golgi (Fig. 5, panels C and D) (Rönnholm, 1992), albeit rather inefficiently (Persson and Pettersson, 1991; Melin *et al.*, 1995). This indicates the presence of a Golgi-targetting signal in G1. When G1 and G2 are coexpressed from separate plasmids (mRNAs), the proteins colocalize to the Golgi (Fig. 5, panels A and B). Thus, G2 requires coexpression of G1 to become transport competent. In addition, the efficiency of G1 transport is increased by coexpressing G2. These findings support the notion that G1 and G2 interact with each other (Persson and Pettersson, 1991).

To map the Golgi retention signal in G1, attention has been focused on the transmembrane domain and cytoplasmic tail of the molecule. Chimeras between these domains of G1 and various domains of proteins efficiently transported to the plasma membrane have been expressed using the T7-VV and Semliki Forest virus systems. In addition, mutants with progressive deletions of the cytoplasmic tail have been analyzed. Chicken lysozyme fused to the transmembrane domain and cytoplasmic tail (including the signal sequence for G2) was targetted to and retained in the Golgi. A G1 construct in which the transmembrane domain and 10 flanking residues on each side were exchanged with those from VSV G was retained in the Golgi (Fig. 5E), while G1 with the transmembrane domain and tail from VSV G was efficiently transported to the cell surface. These results clearly suggested that the retention signal resides in the cytoplasmic tail rather than in the ectodomain or the transmembrane domain. The results were confirmed by analyzing chimeras between CD4 (a plasma membrane protein) and G1. CD4 containing the G1 transmembrane domain and the CD4 tail was efficiently transported to the cell surface (Fig. 5G), while CD4 with its own trans-membrane domain and the G1 cytoplasmic tail was retained in the Golgi complex (Fig. 5F). Deleting the G2 signal sequence in this latter hybrid had no effect on the Golgi localization (Fig. 5H). In summary, these results show that the Golgi retention signal is localized to the cytoplasmic tail between residues 10 and 50 (counting from the G1 transmembrane domain) (Pettersson *et al.*, 1996). In contrast to the PT virus G1 (see above), we have found very little effect of the transmembrane domain of UUK virus G1 on Golgi retention. Thus, for UUK virus G1 the cytoplasmic tail is both necessary and sufficient for Golgi localization.

VI. CONCLUSIONS AND PERSPECTIVES

There is one interesting property that unites the diverse members of the *Bunyaviridae* family, namely, their site of maturation in the Golgi complex. At present, we do not understand what specifies these viruses to bud into this organelle. However, it seems clear that the accumulation of the G1/G2 spike protein complex in this organelle is a major determinant. Further, it seems that only one of the two proteins contains a Golgi-targetting and retention signal, while the other becomes Golgi-localized indirectly by interacting (dimerizing) with the other. The sparse information so far obtained indicates that the Golgi retention signal resides in the tail of one of the proteins, with possible contribution from the transmembrane domain. Since there is no apparent sequence homology between the tails or transmembrane domains of different *Bunyaviridae* glycoproteins, retention may depend more on the conformation than on a primary sequence motif.

The mechanism by which G1/G2 are retained in the Golgi complex remains to be elucidated. There are at least three main possibilities. First, the cytoplasmic tail (and part of the transmembrane domain) could be interacting with cytoskeletal components in the Golgi matrix (Slusarewicz *et al.*, 1994), thereby preventing further transport. Second, the tails could interact with each other to form a lattice underneath the membrane. This could result in the formation of large G1/G2 complexes in the Golgi, which because of their size would be excluded from the transport vesicles. Such complexes have as yet not been demonstrated. *Bunyaviridae* members lack a submembranous matrix (M) protein found in, for example, rhabdo-, orthomyxo-, and paramyxoviruses. The relatively long tail (e.g., 100 residues in UUK virus G1) could serve the function provided by M proteins. The tail is likely to interact with the ribonucleoproteins, since the tail of the other protein is usually short (5 residues in the case of UUK virus G2). The fact that G1–G2 of UUK virus form an icosahedral surface lattice in virions (von Bonsdorff and Pettersson, 1975) suggests that such complexes might be formed in the Golgi complex prior to budding. Membrane proteins are also transported from the ER to the Golgi complex in small vesicles (Pryer *et al.*, 1992). This would mean that the formation of large oligomers in the ER should be prevented in order to facilitate inclusion in such transport vesicles. Third, G1/G2 could be transported to the plasma membrane and then efficiently recycled back to the Golgi complex as has been shown for the *trans*-Golgi network-specific protein TGN38/41 (Stanley and Howell, 1993).

What bearing does the retention of *Bunyaviridae* glycoproteins in the Golgi complex have on the problem of compartmentalization of cellular proteins in general, and vice versa? Few cellular compartment-specific retention signals have so far been identified. These include the KDEL sequence at the C-terminal end of resident ER luminal (soluble) proteins (e.g., BiP/grp78, protein disulfide isomerase) (Pelham, 1990), the double-lysine motif at the C-terminus of ER-retained membrane proteins (Jackson *et al.*, 1990), and the transmembrane domain (and flanking luminal stalk region) of Golgi-localized glycosyl transferases (Munro, 1991; Nilsson *et al.*, 1991; Burke *et al.*, 1992). The KDEL proteins that have escaped ER retention are recycled back to the ER with the help of a membrane receptor (Pelham, 1990). The mechanism for the retention of membrane proteins is not yet known, although binding to the microtubules in the ER (Dahllöf *et al.*, 1991) or a Golgi matrix (Slusarewicz *et al.*, 1994) have been suggested. Glycosyl transferases co-localizing to the same cisternae in the *medial*-Golgi may in addition form large aggregates by binding to each other via the transmembrane domain ("kin recognition"; Nilsson *et al.*, 1994).

In the case of other viral compartment-specific membrane proteins, equally little is known regarding the retention mechanism. The adenovirus E3/19K is retained in the ER by the double-lysine motif (Jackson *et al.*, 1990). The NS28 of rotavirus is retained in the ER by an unknown mechanism, and no retention motif has been identified (Bergmann *et al.*, 1989). The rotavirus ER luminal VP7 has been shown to be retained by an interaction of its own cleaved signal sequence with the first 31 N-terminal residues of the mature protein. Three residues were found to be critical for ER retention within this latter region (Maass and Atkinson, 1994). Finally, the first transmembrane domain of the coronavirus infectious bronchitis virus M (E1) protein specifies its localization to the *cis*-Golgi; retention may be the result of protein oligomerization (Machamer *et al.*, 1990; Weisz *et al.*, 1993). In the murine hepatitis virus (MHV A59), another coronavirus, the cytoplasmic tail also seems to play an important role (Armstrong and Patel, 1991). It is likely that different proteins are targetted to, and retained in, various membranes by different mechanisms. The elucidation of these mechanisms is a challenge both in cell biology and in virology. It is possible that viruses, including the *Bunyaviridae*, budding at various cellular membranes have during evolution adopted host membrane proteins to serve as spike proteins. In the past, enveloped viruses have been excellent models in cell biology and they will continue to play an important role in the future, e.g., in the elucidation of targetting and retention signals in membrane proteins.

Finally, a commonly asked question is whether budding at intracellular membranes offers any advantage to the virus, in particular in regard to the possibility of evading the immune response. With the detailed knowledge available today on how the humoral and cell-mediated immune responses operate on a molecular and cell biological level, we find this unlikely. A simpler explanation is that different viruses have utilized a diversity of alternative strategies offered by the complex cellular organization and metabolism. A similar broad diversity has evolved in regard to the strategies by which viruses replicate and express their genomes.

VII. REFERENCES

Anderson, G. W., Jr., and Smith, J. F., 1987, Immunoelectron microscopy of Rift Valley fever viral morphogenesis in primary rat hepatocytes, *Virology* **161**:91.
Antic, D., Kang, C. Y., Spik, K., Schmaljohn, C., Vapalahti, O., and Vaheri, A., 1992a, Comparison of the deduced gene products of the L, M and S genome segments of hantaviruses, *Virus Res.* **24**:35.

Antic, D., Wright, K. E., and Kang, C. Y., 1992b, Maturation of Hantaan virus glycoproteins G1 and G2, *Virology* **189**:324.

Arikawa, J., Schmaljohn, A. L., Dalrymple, J. M., and Schmaljohn, C. S., 1989, Characterization of Hantaan virus envelope glycoprotein antigenic determinants defined by monoclonal antibodies, *J. Gen. Virol.* **70**:615.

Armstrong, J., and Patel, S., 1991, The Golgi sorting domain of coronavirus E1 protein, *J. Cell Sci.* **98**:567.

Bardwell, J. C. A., and Beckwith, J., 1993, The bonds that tie: Catalyzed disulfide bond formation, *Cell* **74**:769.

Bause, E., 1983, Structural requirements of N-glycosylation of proteins, *Biochem. J.* **209**:331.

Bergmann, C. C., Maass, D., Poruchynsky, M. S., Atkinson, P. H., and Bellamy, A. R., 1989, The topology of the non-structural rotavirus receptor glycoprotein NS28 in the rough endoplasmic reticulum, *EMBO J.* **8**:1695.

Bishop, D. H. L., and Shope, R. E., 1979, Bunyaviridae, *Compr. Virol.* **14**:1.

Booth, T. F., Gould, E. A., and Nuttall, P. A., 1991, Structure and morphogenesis of Dugbe virus (Bunyaviridae, *Nairovirus*) studied by immunogold electron microscopy of ultrathin cryosections, *Virus Res.* **21**:199.

Bouloy, M., 1991, Bunyaviridae: Genome organization and replication strategies, *Adv. Virus Res.* **40**:235.

Bupp, K., Prabakaran, I., Nathanson, N., and Gonzalez-Scarano, F., 1994, Expression of La Crosse virus glycocprotein G1 alone is insufficient for its proper Golgi targeting, Scientific Program and Abstracts book of the 13th Annual Meeting of the American Society for Virology.

Burke, J., Pettitt, J. M., Schachter, H., Sarkar, M., and Gleeson, P. A., 1992, The transmembrane and flanking sequences of β–1,2-N-acetylglucosaminyltransferase I specify *medial*-Golgi localization, *J. Biol. Chem.* **267**:24433.

Cash, P., 1982, Inhibition of LaCrosse virus replication by monensin, a monovalent ionophore, *J. Gen. Virol.* **59**:193.

Cash, P., 1985, Polypeptide synthesis of Dugbe virus, a member of the *Nairovirus* genus of the Bunyaviridae, *J. Gen. Virol.* **66**:141.

Chen, S.-Y., and Compans, R. W., 1991, Oligomerization, transport, and Golgi-retention of Punta Toro virus glycoproteins, *J. Virol.* **65**:5902.

Chen, S.-Y., Matsuoka, Y., and Compans, R. W., 1991a, Assembly and polarized release of Punta Toro virus and effects of brefeldin A, *J. Virol.* **65**:1427.

Chen, S.-Y., Matsuoka, Y., and Compans, R. W., 1991b, Golgi complex localization of the Punta Toro virus G2 protein requires its association with G1 protein, *Virology* **183**:351.

Clerx, J. P. M., and Bishop, D. H. L., 1981, Qalyub virus, a member of the newly proposed *Nairovirus* genus (Bunyaviridae), *Virology* **108**:361.

Collett, M. S., Purchio, A. F., Keegan, K., Frazier, S., Hays, W., Anderson, D. K., Parker, M. D., Schmaljohn, C., Schmidt, J., and Dalrymple, J., 1985, Complete nucleotide sequence of the M RNA segment of Rift Valley fever virus, *Virology* **144**:228.

Dahllöf, B., Wallin, M., and Kvist, S., 1991, The endoplasmic reticulum retention signal of the E3/19K protein of adenovirus-2 is microtubule binding, *J. Biol. Chem.* **266**:1804.

Doms, R. W., Lamb, R. A., Rose, J. K., and Helenius, A., 1993, Folding and assembly of viral membrane proteins, *Virology* **193**:545.

Elliott, R. M., 1990, Molecular biology of the Bunyaviridae, *J. Gen. Virol.* **71**:501.

Elliott, R. M., Schmaljohn, C. S., and Collett, M. S., 1991, Bunyaviridae genome structure and gene expression, *Curr. Top. Microbiol. Immunol.* **169**:91.

Fazakerley, J. K., Gonzalez-Scarano, F., Strickler, J., Dietzschold, B., Karush, F., and Nathanson, N., 1988, Organization of the middle RNA segment of snowshoe hare bunyavirus, *Virology* **167**:422.

Foulke, R. S., Rosato, P. R., and Feench, G. R., 1981, Structural polypeptides of Hazara virus, *J. Gen. Virol.* **53**:169.

Francki, R. I. B., Fauquet, C. M., Knudson, D. L., and Brown, F., 1991, Classification and nomenclature of viruses. Bunyaviridae, *Arch. Virol.* Suppl. **2**:273.

Gahmberg, N., 1984, Characterization of two recombination-complementation groups of Uukuniemi virus temperature-sensitive mutants, *J. Gen. Virol.* **65:**1079.

Gahmberg, N., Pettersson, R. F., and Kääriäinen, L., 1986a, Efficient transport of Semliki Forest virus glycoproteins through a Golgi complex morphologically altered by Uukuniemi virus glycoproteins, *EMBO J.* **5:**3111.

Gahmberg, N., Kuismanen, E., Keränen, S., and Pettersson, R. F., 1986b, Uukuniemi virus glycoproteins accumulate in and cause morphological changes of the Golgi complex in the absence of virus maturation, *J. Virol.* **57:**899.

Garoff, H., and Simons, K., 1974, Location of the spike glycoproteins in the Semliki Forest virus membrane, *Proc. Natl. Acad. Sci. USA* **71:**3988.

Garoff, H., Frischauf, A.-M., Simons, K., Lehrach, H., and Delius, H., 1980, Nucleotide sequence of cDNA coding for Semliki Forest virus membrane glycoproteins, *Nature* **288:**236.

Gerbaud, S., Pardigon, N., Vialat, P., and Bouloy, M., 1992, Organization of Germiston bunyavirus M open reading frame and physicochemical properties of the envelope glycoproteins, *J. Gen. Virol.* **73:**2245.

Griffiths, G., and Rottier, P., 1992, Cell biology of viruses that assemble along the biosynthetic pathway, *Semin. Cell Biol.* **3:**367.

Hauri, H. P., and Schweizer, A., 1992, The endoplasmic reticulum–Golgi intermediate compartment, *Curr. Opin. Cell Biol.* **4:**600.

Hobman, T. C., 1993, Transport of viral glycoproteins into the Golgi complex, *Trends Microbiol.* **1:**124.

Hurtley, S. M., and Helenius, A., 1989, Protein oligomerization in the endoplasmic reticulum, *Annu. Rev. Cell Biol.* **5:**277.

Ihara, T., Smith, J., Dalrymple, J. M., and Bishop, D. H. L., 1985, Complete sequences of the glycoproteins and M RNA of Punta Toro phlebovirus compared to those of Rift Valley fever virus, *Virology* **144:**246.

Jackson, M. R., Nilsson, T., and Peterson, P. A., 1990, Identification of a consensus motif for retention of transmembrane proteins in the endoplasmic reticulum, *EMBO J.* **9:**3153.

Kakach, L. T., Wasmoen, T. L., and Collett, M. S., 1988, Rift Valley fever virus M segment: Use of recombinant vaccinia viruses to study *Phlebovirus* gene expression, *J. Virol.* **62:**826.

Kakach, L. T., Suzich, J. A., and Collett, M. S., 1989, Rift Valley fever virus M segment: Phlebovirus expression strategy and protein glycosylation, *Virology* **170:**505.

Kamrud, K. I., and Schmaljohn, C. S., 1994, Expression strategy of the M genome segment of Hantaan virus, *Virus Res.* **31:**109.

Kitajima, E. W., de Àvila, A. C., de O. Resende, R., Goldbach, R. W., and Peters, D., 1992, Comparative cytological and immunogold labelling studies on different isolates of tomato spotted wilt virus, *J. Submicrosc. Cytol. Pathol.* **24:**1.

Kormelink, R., de Haan, P., Meurs, C., Peters, D., and Goldbach, R., 1992, The nucleotide sequence of the M RNA segment of tomato spotted wilt virus, a bunyavirus with two ambisense RNA segments, *J. Gen. Virol.* **73:**2795.

Kormelink, R., Storms, M., van Lent, J., Peters, D., and Goldbach, R., 1994, Expression and subcellular location of the NS_M protein of tomato spotted wilt virus (TSWV), a putative viral movement protein, *Virology* **200:**56.

Krijnse-Locker, J., Ericsson, M., Rottier, P. J. M., and Griffiths, G., 1994, Characterization of the budding compartment of mouse hepatitis virus: Evidence that transport from the RER to the Golgi complex requires only one vesicular transport step, *J. Cell Biol.* **124:**55.

Kuismanen, E., 1984, Posttranslational processing of Uukuniemi virus glycoproteins G1 and G2, *J. Virol.* **51:**806.

Kuismanen, E., Hedman, K., Saraste, J., and Pettersson, R. F., 1982, Uukuniemi virus maturation: Accumulation of virus particles and viral antigens in the Golgi complex, *Mol. Cell. Biol.* **2:**1444.

Kuismanen, E., Bång, B., Hurme, M., and Pettersson, R. F., 1984, Uukuniemi virus maturation: Immunofluorescence microscopy with monoclonal glycoprotein-specific antibodies, *J. Virol.* **51:**137.

Kuismanen, E., Saraste, J., and Pettersson, R. F., 1985, Effect of monensin on the assembly of Uukuniemi virus in the Golgi complex, *J. Virol.* **55**:813.

Lappin, D. F., Nakitare, G. W., Palfreyman, J. W., and Elliott, R. M., 1994, Localization of Bunyamwera bunyavirus G1 glycoprotein to the Golgi requires association with G2 but not with NSm, *J. Gen. Virol.* **75**:1.

Law, M. D., Speck, J., and Moyer, J. W., 1992, The M RNA of impatiens necrotic spot *Tospovirus* (Bunyaviridae) has an ambisense genomic organization, *Virology* **188**:732.

Liljeström, P., and Garoff, H., 1991, Internally located cleavable signal sequences direct the formation of Semliki Forest virus membrane proteins from a polyprotein precursor, *J. Virol.* **65**:147.

Liljeström, P., Lusa, S., Huylebroeck, D., and Garoff, H., 1991, In vitro mutagenesis of a full-length cDNA clone of Semliki Forest virus: The 6,000-molecular-weight membrane protein modulates virus release, *J. Virol.* **65**:4107.

Maass, D. R., and Atkinson, P. H., 1994, Retention by the endoplasmic reticulum of Rotavirus VP7 is controlled by three adjacent amino-terminal residues, *J. Virol.* **68**:366.

Machamer, C. E., Mentone, S. A., Rose, J. K., and Farquhar, M. G., 1990, the E1 glycoprotein of an avian coronavirus is targeted to the *cis* Golgi complex *Proc. Natl. Acad. Sci. USA* **87**:6944.

Madoff, D. H., and Lenard, J., 1982, A membrane glycoprotein that accumulates intracellularly: Cellular processing of the large glycoprotein of La Crosse virus, *Cell* **28**:821.

Marriott, A. C., El-Ghorr, A. A., and Nuttall, P. A., 1992, Dugbe Nairovirus M RNA: Nucleotide sequence and coding strategy, *Virology* **190**:606.

Matsuoka, Y., Ihara, T., Bishop, D. H. L., and Compans, R. W., 1988, Intracellular accumulation of the Punta Toro virus glycoproteins expressed from cloned cDNA, *Virology* **167**:251.

Matsuoka, Y., Chen, S.-Y., and Compans, R. W., 1991, Bunyavirus protein transport and assembly, *Curr. Top. Microbiol. Immunol.* **169**:161.

Matsuoka, Y., Chen, S.-Y., and Compans, R. W., 1994, A signal for Golgi retention in the bunyavirus G1 glycoprotein, *J. Biol. Chem.* **269**:22565.

Melin, L., Persson, R., Rönnholm, R., Bergström, A., and Pettersson, R. F., 1995, The membrane glycoprotein G1 of Uukuniemi virus contains a signal for localization to the Golgi complex, *Virus Res.* **36**:49–66.

Munro, S., 1991, Sequences within and adjacent to the transmembrane segment of α-2,6-sialyltransferase specify Golgi retention, *EMBO J.* **10**:3577.

Murphy, F. A., Whitfield, S. G., Coleman, P. H., Calisher, C. H., Rabin, E. R., Jenson, A. B., Melnick, J. L., Edwards, M. R., and Whitney, E., 1968, California group arboviruses: Electron microscopic studies, *Exp. Mol. Pathol.* **9**:44.

Murphy, F. A., Harrison, A. K., and Whitfield, S. G., 1973, Bunyaviridae: Morphologic and morphogenetic similarities of Bunyamwera serologic supergroup viruses and several other arthropod borne viruses, *Intervirology* **1**:297.

Nakitare, G. W., and Elliott, R. M., 1993, Expression of the Bunyamwera virus M genome segment and intracellular localization of NSm, *Virology* **195**:511.

Nilsson, T., Lucocq, J. M., Mackay, D., and Warren, G., 1991, The membrane spanning domain of β-1,4-galactosyltransferase specifies *trans* Golgi localization, *EMBO J.* **10**:3567.

Nilsson, T., Hoe, M. H., Slusarewicz, P., Rabouille, C., Watson, R., Hunte, F., Watzele, G., Berger, E. G., and Warren, G., 1994, Kin recognition between *medial* Golgi enzymes in HeLa cells, *EMBO J.* **13**:562.

Ou, W.-J., Cameron, P. H., Thomas, D. Y., and Bergeron, J. M., 1993, Association of folding intermediates of glycoproteins with calnexin during protein maturation, *Nature* **364**:771.

Parrington, M. A., Lee, P.-W., and Kang, C. Y., 1991, Molecular characterization of the Prospect Hill virus M RNA segment: A comparison with the M RNA segments of other hantaviruses, *J. Gen. Virol.* **72**:1845.

Pelham, H. R. B., 1990, The retention signal for soluble proteins of the endoplasmic reticulum, *Trends Biochem. Sci.* **15**:483.

Pennington, T. H., Pringle, C. R., and McCrae, M. A., 1977, Bunyamwera virus-induced polypeptide synthesis, *J. Virol.* **24**:397.

Pensiero, M. N., and Hay, J., 1992, The Hantaan virus M-segment glycoproteins G1 and G2 can be expressed independently, *J. Virol.* **66:**1907.

Pensiero, M. N., Jennings, G. B., Schmaljohn, C. S., and Hay, J., 1988, Expression of the Hantaan virus M genome segment by using a vaccinia virus recombinant, *J. Virol.* **62:**696.

Persson, R., and Pettersson, R. F., 1991, Formation and intracellular transport of a heterodimeric viral spike protein complex, *J. Cell Biol.* **112:**257.

Pesonen, M., Kuismanen, E., and Pettersson, R. F., 1982a, Monosaccharide sequence of protein-bound glycans of Uukuniemi virus, *J. Virol.* **41:**390.

Pesonen, M., Rönnholm, R., Kuismanen, E., and Pettersson, R. F., 1982b, Characterization of the oligosaccharides of Inkoo virus envelope glycoproteins, *J. Gen. Virol.* **63:**425.

Pettersson, R. F., 1991, Protein localization and virus assembly at intracellular membranes, *Curr. Top. Microbiol. Immunol.* **170:**67.

Pettersson, R. F., and von Bonsdorff, C.-H., 1987, Bunyaviridae, *Animal Virus Structure* (M. V. Nermut and A. Steven, eds.), pp. 147–157, Elsevier, Amsterdam.

Pettersson, R. F., Gahmberg, N., Kuismanen, E., Kääriäinen, L., Rönnholm, R., and Saraste, J., 1988, Bunyavirus membrane glycoproteins as models for Golgi-specific proteins, *Modern Cell Biol.* **6:**65.

Pettersson, R. F., Simons, J. F., Melin, L., Persson, R., and Rönnholm, R., 1993, Uukuniemi virus as a model for elucidating the molecular biology of bunyaviruses, in: *Concepts in Virology from Ivanovsky to the Present* (B. W. J. Mahy and D. K. Lvov, eds.), pp. 309–319, Harwood Academic.

Pettersson, R. F., Andersson, A., and Melin, L., 1996, Mapping a retention signal for Golgi-localization of a viral spike protein complex, in: *Protein Kinesis: The Dynamics of Protein Trafficking and Stability*, Cold Spring Harbor Symp. Quant. Biol, Vol. 60, in press.

Pfeffer, S. R., and Rothman, J. R., 1987, Biosynthetic protein transport and sorting by the endoplasmic reticulum and Golgi, *Annu. Rev. Biochem.* **56:**829.

Pryer, N. K., Wuestehube, L. J., and Schekman, R., 1992, Vesicle-mediated protein sorting, *Annu. Rev. Biochem.* **61:**471.

Rhee, S. S., and Hunter, E., 1991, Amino acid substitutions within the matrix protein of type D retroviruses affect assembly, transport and membrane association of a capsid, *EMBO J.* **10:**535.

Rönnholm, R., 1992, Localization to the Golgi complex of Uukuniemi virus glycoproteins G1 and G2 expressed from cloned cDNAs, *J. Virol.* **66:**4525.

Rönnholm, R., and Pettersson, R. F., 1987, Complete nucleotide sequence of the M RNA segment of Uukuniemi virus encoding the membrane glycoproteins G1 and G2, *Virology* **160:**191.

Ruusala, A., Persson, R., Schmaljohn, C. S., and Pettersson, R. F., 1992, Coexpression of the membrane glycoproteins G1 and G2 of Hantaan virus is required for targeting to the Golgi complex, *Virology* **186:**53.

Saraste, J., and Kuismanen, E., 1992, Pathways of protein sorting and membrane traffic between the rough endoplasmic reticulum and the Golgi complex, *Semin. Cell Biol.* **3:**343.

Schmaljohn, C. S., Hasty, S. E., Rasmussen, L., and Dalrymple, J. M., 1986, Hantaan virus replication: Effects of monensin, tunicamycin and endoglycosidase on the structural glycoproteins, *J. Gen. Virol.* **67:**707.

Schmaljohn, C. S., Schmaljohn, A. L., and Dalrymple, J. M., 1987, Hantaan virus M RNA: Coding strategy, nucleotide sequence, and gene order, *Virology* **157:**31.

Schmaljohn, C. S., Chu, Y.-K., Schmaljohn, A. L., and Dalrymple, J. M., 1990, Antigenic subunits of Hantaan virus expressed by baculovirus and vaccinia virus recombinants, *J. Virol.* **64:**3162.

Slusarewicz, P., Nilsson, T., Hui, N., Watson, R., and Warren, G., 1994, Isolation of a matrix that binds *medial* Golgi enzymes, *J. Cell Biol.* **124:**405.

Smith, J. F., and Pifat, D., 1982, Morphogenesis of sandfly viruses (Bunyaviridae family), *Virology* **121:**61.

Sodeik, B., Doms, R. W., Ericsson, M., Hiller, G., Machamer, C. E., van't Hof, W., van Meer, G.,

Moss, B., and Griffiths, G., 1993, Assembly of vaccinia virus: Role of the intermediate compartment between the endoplasmic reticulum and the Golgi stacks, *J. Cell Biol.* **121:**521.

Spiropoulou, C. F., Morzunov, S., Feldmann, H., Sanchez, A., Peters, C. J., and Nichol, S. T., 1994, Genome structure and variability of a virus causing hantavirus pulmonary syndrome, *Virology* **200:**715.

Stanley, K. K., and Howell, K. E., 1993, TGN38/41: A molecule on the move, *Trends Cell Biol.* **3:**252.

Stephens, E. B., and Compans, R. W., 1988, Assembly of animal viruses at cellular membranes, *Annu. Rev. Microbiol.* **42:**489.

Suomalainen, M., Liljeström, P., and Garoff, H., 1992, Spike protein–nucleocapsid interactions drive the budding of alphaviruses, *J. Virol.* **66:**4737.

Suzich, J. A., and Collett, M. S., 1988, Rift Valley fever virus M segment: Cell-free transcription and translation of virus complementary RNA, *Virology* **164:**478.

Suzich, J. A., Kakach, L. T., and Collett, M. S., 1990, Expression strategy of a phlebovirus: Biogenesis of proteins from the Rift Valley fever M segment, *J. Virol.* **64:**1549.

Takehara, K., Min, M.-K., Battles, J. K., Sugiyama, K., Emery, V. C., Dalrymple, J. M., and Bishop, D. H. L., 1989, Identification of mutations in the M RNA of a candidate vaccine strain of Rift Valley fever virus, *Virology* **169:**452.

Ulmanen, I., Seppälä, P., and Pettersson, R. F., 1981, *In vitro* translation of Uukuniemi virus-specific RNAs: Identification of a nonstructural protein and a precursor to the membrane glycoproteins, *J. Virol.* **37:**72.

von Bonsdorff, C.-H., and Pettersson, R., 1975, Surface structure of Uukuniemi virus, *J. Virol.* **16:**1296.

von Bonsdorff, C.-H., Saikku, P., and Oker-Blom, N., 1970, Electron microscope study on the development of Uukuniemi virus, *Acta Virol.* **14:**109.

von Heijne, G., 1990, Protein targeting signals, *Curr. Opin. Cell Biol.* **2:**604.

Vorndam, A. V., and Trent, D. W., 1979, Oligosaccharides of the California encephalitis viruses, *Virology* **95:**1.

Wahlberg, J. M., and Garoff, H., 1992, Membrane fusion process of Semliki Forest virus I: Low pH-induced rearrangement in spike protein quaternary structure precedes virus penetration into cells, *J. Cell Biol.* **116:**339.

Wasmoen, T. L., Kakach, L. T., and Collett, M. C., 1988, Rift Valley fever M segment: Cellular localization of M segment-encoded proteins, *Virology* **166:**275.

Watret, G. E., and Elliott, R. M., 1985, The proteins and RNAs specified by Clo Mor virus, a Scottish nairovirus, *J. Gen. Virol.* **66:**2513.

Weisz, O. A., Swift, A. M., and Machamer, C. E., 1993, Oligomerization of a membrane protein correlates with its retention in the Golgi complex, *J. Cell Biol.* **122:**1185.

Whitt, M. A., Chong, L., and Rose, J. K., 1989, Glycoprotein cytoplasmic domain sequences required for rescue of a vesicular stomatitis virus glycoprotein mutant, *J. Virol.* **63:**3569.

CHAPTER 8

Genetics and Genome Segment Reassortment

C. R. PRINGLE

I. SEGMENTED GENOME VIRUSES

Nineteen of the ninety-five taxonomic groups of viruses (families and floating genera) listed in the Sixth Report of the International Committee for Taxonomy of Viruses (Murphy *et al.*, 1995) possess segmented genomes. All except the geminiviruses of plants are RNA viruses. These viruses are listed in Table I according to nucleic acid type, segment number, and virion composition. Segmentation of the genome is most commonly observed among the negative-stranded and double-stranded RNA genome viruses, and these viruses are single-component viruses, where the complete complement of genome segments is contained within one particle and the infectious unit is a single virion. Segmentation among the positive-stranded RNA genome viruses, on the other hand, is a phenomenon restricted to multicomponent viruses, where the individual genome segments are contained within different particles and the infectious unit is a full complement of particles. With the single exception of the *Nodaviridae*, these viruses are all viruses infecting plants.

A feature of some segmented negative-stranded RNA viruses is the ambisense encoding of genetic information, i.e., the nonoverlapping encoding of genes in the 5' halves of both the viral RNA (negative sense) and the viral-complementary RNA (positive sense). Both segments of the bipartite

C. R. PRINGLE • Biological Sciences Department, University of Warwick, Coventry CV4 7AL, United Kingdom.

The Bunyaviridae, edited by Richard M. Elliott, Plenum Press, New York, 1996.

arenaviruses exhibit ambisense encoding of information (Bishop, 1986; Bishop and Auperin, 1987), as do at least three of the four/five segments of the genomes of the filamentous tenuiviruses of plants (Ramirez and Haenni, 1994). The family *Bunyaviridae* is more heterogeneous in this respect since ambisense encoding of information is observed in the smaller RNA subunit only of viruses belonging to two of the five genera (Table I).

The three families of viruses with unsegmented negative-stranded RNA genomes are closely related in terms of genome structure and function, and accordingly they have been grouped together as the order *Mononegavirales*. Although the three families of viruses with segmented negative-stranded RNA genomes, the bipartite *Arenaviridae*, the tripartite *Bunyaviridae*, and the multipartite *Orthomyxoviridae*, are more diverse in terms of genome structure and function, they also form a coherent (super)group when phylogenetic trees for RNA-dependent RNA polymerases are constructed (Goldbach and de Haan, 1994). It has been argued that these three families of segmented genome negative-stranded RNA viruses should be combined into a second order, designated the *Multinegavirales* for consistency (Pringle, 1991a; Ward, 1993), both subsumed within the class *Negavirata* (Tordo *et al.*, 1992; Ward, 1993). The principal argument against the acceptance of this proposal is the potential for reassortment of genome subunits which defines the group. Reassortment of genome subunits can obscure evolutionary lineages and may result in the establishment of spurious phylogenetic associations. The absence of reassortment (and intermolecular recombination) in the case of the unsegmented negative-stranded genome RNA viruses (Pringle, 1987, 1990, 1991b; Chao, 1994), on the other hand, guarantees that phylogenetic relationships can be inferred with reasonable confidence, thereby allowing the creation of higher taxonomic categories.

Segmentation of the viral genome has several potential advantages. It maximizes the effects of positive Darwinian selection and promotes greater evolutionary plasticity; the epidemic and pandemic behavior of the human influenza A viruses are the best example of this in nature. It may also facilitate the elimination of deleterious mutations by purifying selection and a more rapid exit of defective interfering genomes. It has been suggested also that segmentation may be a device to increase the genetic content of RNA-containing viruses where the high mutation rate may impose a limit on the absolute size of informational RNA molecules (Reanney, 1984). However, this argument is difficult to sustain since the total genome sizes of segmented genome viruses are in the same range as those of nonsegmented viruses, and the maximum genome size of any segmented genome virus (the 23 kb of the nairoviruses or the 27.5 kbp of the double-stranded rice ragged stunt virus) does not reach the 31 kb of some of the nonsegmented coronaviruses. In the multicomponent plant viruses, segmentation of the genome is probably an adaptation to facilitate movement of virus within the infected plant, and a consequence of the absence of specific receptors on plant cells and the nonspecific mode of entry of the virus into the host plant. Segmenta-

TABLE I. The Segmented Genome Viruses

Nucleic acid type	Sense	No. of segments	Particle type	Family or genus[a]	Principal host	Evidence of reassortment
ssDNA		1	—	4 families	—	n.a.[b]
		1 or 2	Multicomponent	*Geminiviridae*	Plant	yes
dsDNA		1	—	18 families or genera	—	n.a.
		>1	—	none	—	n.a.
ssRNA	+	1	—	26 families or genera	—	n.a.
	+	2	Multicomponent	G. *Bymovirus*	Plant	?[c]
	+	2	Multicomponent	*Comoviridae*	Plant	yes
	+	2	Multicomponent	G. *Furovirus*	Plant	yes
	+	2	Multicomponent	G. *Dianthovirus*	Plant	yes
	+	2	Multicomponent	G. *Idaeovirus*	Plant	?
	+	2	Multicomponent	*Nodaviridae*	Insect	?
	+	2	Multicomponent	G. *Enamovirus*	Plant	yes
	+	2	Multicomponent	G. *Tobravirus*	Plant	yes
	+	3	Multicomponent	*Bromoviridae*	Plant	yes
	+	3	Multicomponent	G. *Hordeivirus*	Plant	yes
	−	1	—	3 families	—	n.a.
	ambisense	2	Single-component	*Arenaviridae*	Animal	yes
	−/ambisense	3	Single-component	*Bunyaviridae*	Animal/plant	yes
	−/ambisense	4 or 5	Multicomponent	G. *Tenuivirus*	Plant	?
	−	7 or 8	Single-component	*Orthomyxoviridae*	Animal	yes
dsRNA	+/−	1	—	2 families	—	n.a.
	+/−	2	Single-component	*Birnaviridae*	Animal	yes
	+/−	2	Multicomponent	*Partitiviridae*	Plant	?
	+/−	3	Single-component	*Cystoviridae*	Bacteria	yes
	+/−	10–12	Single-component	*Reoviridae*	Animal/plant	yes
DNA/RNA	+/− or +	1	Single-component	4 families or genera	Animal/plant	n.a.

[a]The genus *Bymovirus* is classified in the family *Potyviridae*. The genera *Furovirus*, *Dianthovirus*, *Enamovirus*, *Tobravirus*, *Hordeivirus*, and *Tenuivirus* are floating genera.
[b]n.a., not applicable.
[c]?, not reported.

tion may more often be a stratagem to provide control of biosynthesis at the transcriptional level; the temporal control of gene product synthesis in the bunyaviruses is an example of this (Pennington *et al.*, 1977). The evolution of segmentation in RNA viruses has been equated with the evolution of sex by Chao (1994), allowing the participation of two, three, or more parents in the production of offspring.

A consequence of segmentation of the genome is the possibility of increasing genetic heterogeneity by reassortment of subunits during replication and morphogenesis. Plant virologists use the term *pseudorecombination* rather than *reassortment* to describe the exchange of genome subunits that can accompany the maturation of multicomponent viruses (Table I). Pseudorecombination has been observed between different viruses belonging to the family *Bromoviridae* and the floating genera *Tobravirus* and *Dianthovirus*, and between different strains of the same virus in all three genera (*Nepovirus*, *Comovirus*, and *Fabavirus*) of the family *Comoviridae*, and in the genera *Furovirus*, *Hordeivirus*, *Enamovirus*, and *Geminivirus*. It is likely that all segmented genome viruses have the ability to interchange genome subunits between related strains.

II. THE *BUNYAVIRIDAE* GENOME

The viruses belonging to the family *Bunyaviridae* are well suited to genetic study because of their antigenic and ecological diversity. This provides a wealth of phenotypic variation as a substrate for genetic analysis. The tripartite nature of the genome favors the analysis of the phenomenon of reassortment, since the number of reassortant genotypes is small enough to allow complete analysis of the progeny from individual crosses. Table II shows the number of possible genotypes generated as a consequence of reassortment for the different groups of single-component segmented genome viruses. The number of reassortant genotypes from crosses of the bipartite arenaviruses is too small to address critically questions such as the randomness of the reassortment process; in addition, the ambisense encod-

TABLE II. The Number of Reassortant Genomes
in Relation to Segment Number

Virus family or group	Number of genome segments	Number of reassortant genotypes in progeny (less parental types)
Arenaviridae	2	$2^2 - 2 = \mathbf{2}$
Bunyaviridae	3	$2^3 - 2 = \mathbf{6}$
Tenuiviruses	4 or 5	$2^4/2^5 - 2 = \mathbf{14}$ or $\mathbf{30}$
Orthomyxoviridae	7 or 8	$2^7/2^8 - 2 = \mathbf{126}$ or $\mathbf{254}$
Reoviridae	10, 11, or 12	$2^{10}/2^{11}/2^{12} - 2 = \mathbf{1022}, \mathbf{2046},$ or $\mathbf{4094}$

ing of information in both segments of the genome may place constraints on mutation type and frequency. The number of progeny genotypes from crosses of tenuiviruses (Table II) might be manageable, but their multicomponent nature and the vagaries of assay of plant viruses *in vitro* render these viruses unsuitable. The number of progeny genotypes from crosses of the orthomyxoviruses or the reoviruses is too large for comprehensive analysis and genetic analysis must rely on statistical estimation. The importance of complete progeny analysis will become clear in the following sections. The genetic consequences of ambisense encoding can be investigated best in the *Bunyaviridae* because this family comprises conventional negative-stranded viruses, and several where one of the three segments exhibits ambisense encoding of information.

It is a characteristic of negative-stranded RNA viruses, whether segmented or not, that almost all of the nucleotide sequence is protein encoding. The nucleotide sequence of the genome of several viruses in the family *Bunyaviridae* (Bunyamwera virus of the genus *Bunyavirus*, Hantaan, Puumala, and Seoul viruses of the genus *Hantavirus*, and Rift Valley fever virus and Uukuniemi virus of the genus *Phlebovirus*) has been completely determined and in each case more than 95% of the sequence consists of open reading frames. Calculation of the coding potential of the bunyavirus genome has to take into account the encoding of the NSs nonstructural protein in an overlaping reading frame in the S RNA of viruses belonging to the genus *Bunyavirus*. The NSs protein gene of viruses of the *Phlebovirus* genus, however, does not overlap the N protein gene as the two genes are encoded in an ambisense fashion in opposite halves of the S RNA (see Chapter 5). Likewise the ambisense encoding of genetic information in the M RNA and S RNA of members of the genus *Tospovirus* does not involve any extension of coding capacity. The NSs protein gene is absent in viruses belonging to the *Hantavirus* and *Nairovirus* genera. The NSm nonstructural protein gene located in the M RNA is likewise not a universal feature of viruses of the family *Bunyaviridae* and does not extend the coding potential since it is derived by nascent cleavage of a polyprotein; it is absent in hantaviruses and variably present in the phleboviruses.

III. STRATEGY OF REPLICATION OF THE *BUNYAVIRIDAE*

Only those features that distinguish the viruses of the family *Bunyaviridae* from other negative-stranded RNA viruses and are relevant to the genetic properties of the *Bunyaviridae* are considered here. Replication occurs in the cytoplasm and is not inhibited by preexposure of cells to actinomycin D or α-amanitin. Nonetheless, infectious Bunyamwera virus is not released from enucleated susceptible cells, although viral protein synthesis is initiated. Furthermore, infectious virus was not released from infected BS-C-1 cells enucleated at times up to 6 hr after infection, suggesting that the presence of the nucleus may be required for normal maturation (Pringle, 1977). Membranes of the Golgi apparatus are the site of maturation of bunyaviruses, and the Golgi apparatus, although present after enucleation of BS-C-1 cells, is rapidly degraded. The presence of a functional nucleus may be necessary to maintain the function of the Golgi apparatus. On the other hand, bunyavirus-specific intranuclear inclusions have been reported in the nuclei of cells infected with the phlebovirus Rift Valley fever virus (Swanepoel and Blackburn, 1977; Struthers and Swanepoel, 1982) and the bunyavirus Ilesha virus (Campbell, 1992; and Fig. 1). In the case of Rift Valley fever virus a nonstructural protein appears to be associated with these inclusion bodies (Struthers and Swanepoel, 1982).

Anomalous nuclear inclusions have also been observed in paramyxovirus-infected cells at both early and late times after infection. Peeples (1991) has suggested that the role of the paramyxovirus protein located in the nucleus may be either to modify ribosome assembly in order to favor translation of viral rather than host mRNA, or to function as a transport protein for transfer of a nuclear factor promoting replication or assembly. The protein located in the nucleus of Newcastle disease virus-infected cells, however, has been identified using monoclonal antibodies as the viral matrix protein, a protein that viruses of the family *Bunyaviridae* lack. In the case of the *Bunyaviridae* the function of such a factor might be involved in cap transfer. Primary

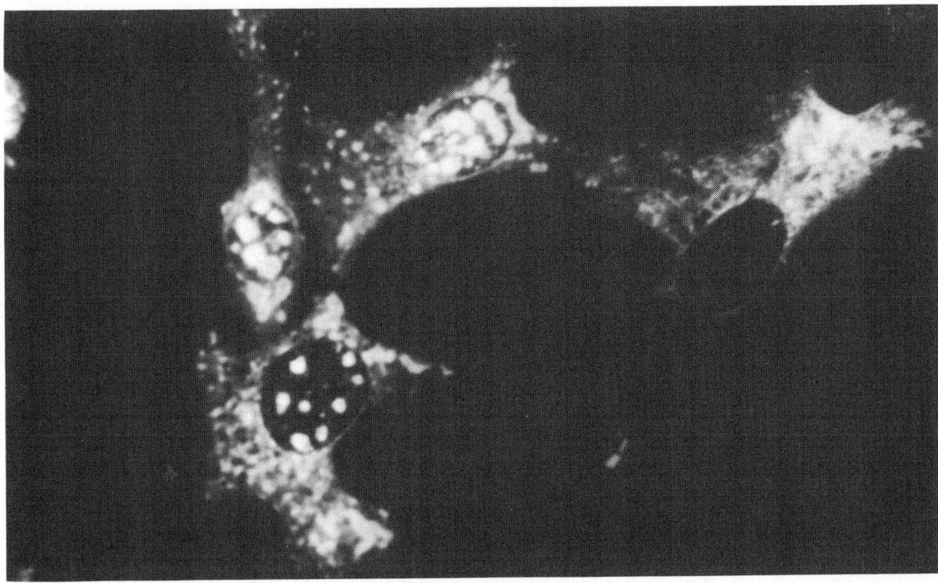

FIGURE 1. Nuclear inclusions in Ilesha virus (Bunyamwera serogroup)-infected Vero cells (photograph provided by the late P. V. Shirodaria). Ilesha virus-infected acetone-fixed Vero cells were stained at 18 hr postinfection with monoclonal antibody 6B10 and FITC-conjugated goat anti-mouse IgG. Monoclonal antibody 6B10 specifically immunoprecipitates a 30-kDa polypeptide from Ilesha virus-infected cells (Campbell, 1992).

transcription is mediated by a virion-associated RNA-dependent RNA polymerase. Synthesis of mRNA is primed by a cap-transfer mechanism analogous to that of the orthomyxoviruses, except that in the case of the *Bunyaviridae* it occurs in the cytoplasm, and the mRNAs do not appear to be polyadenylated.

The L RNA subunit has a single major open reading frame (ORF) encoding the virion polymerase; there are several additional short ORFs which do not appear to be expressed. The M RNA subunit encodes a precursor polyprotein that yields by cotranslational proteolytic cleavage the G1 and G2 envelope proteins and in some viruses a nonstructural protein NSm. The S RNA subunit encodes the nucleoprotein (N) and in some viruses a nonstructural protein (NSs). The precise functions of the NSm and NSs proteins are not known. However, the NSm protein of the tospovirus tomato spotted wilt virus, encoded in an ambisense mode in the M RNA, appears to function as a viral movement protein mediating the movement of ribonucleocapsid structures between cells and the spread of infection within plant tissue (Kormelink *et al.*, 1994).

Morphogenesis of viruses of the family *Bunyaviridae* occurs by accumulation of the G1 and G2 glycoproteins in the Golgi apparatus where terminal glycosylation takes place. The circularized nucleocapsids are enveloped by modified host cell membrane, and virions are formed by budding into the cisternae of the Golgi vesicles. The virions are released by fusion of cytoplasmic vesicles with the plasma membrane, or by direct release from ruptured cells. The absence of a matrix protein in these viruses may be related to the internal site of morphogenesis of the virions.

All viruses of the family *Bunyaviridae*, other than those classified in the genus *Hantavirus*, replicate in invertebrate and vertebrate hosts. In common with most arboviruses, little tissue damage accompanies replication in the invertebrate host. Many viruses have very narrow host ranges in nature and their biological properties are determined by vector behavior and distribution. Venereal and transovarial transmission are common and the virus can overwinter in the dormant egg. Infection of a vertebrate host is not an obligatory part of their life cycle, and there is generally less host restriction in the vertebrate host. Human and domestic animals seldom play the role of amplifying hosts and generally represent dead-end infections. The vector and ecological specificity of these viruses limits opportunities for genetic interaction and is probably responsible for the graded spectrum of restrictions on reassortment that is characteristic of viruses of the family *Bunyaviridae*.

IV. GENETICS OF THE *BUNYAVIRIDAE*

A. The Diversity of the *Bunyaviridae*

The family *Bunyaviridae* is perhaps the most diverse of virus groups, at least in terms of named viruses. The five genera represent distinct groups of viruses with distinctive modes of transmission (Beaty and Calisher, 1991). Viruses of the genera *Bunyavirus*, *Nairovirus*, and *Phlebovirus* are maintained by vertical transmission in arthropods and by a dynamic interaction with specific vertebrate hosts, usually small rodents. The bunyaviruses are associated with mosquitoes, the nairoviruses and uukuviruses with ticks, and the phleboviruses with sand flies. The plant tospoviruses are transmitted by thrips and can multiply in a number of plant hosts. The hantaviruses are maintained in rodents and have no known vectors. Identification of the determinants of these complex relationships is one of the challenges facing genetics.

The taxonomic relationships of viruses in the family *Bunyaviridae* are reflected in their molecular properties. In general, viruses classified in different genera show little or no sequence homology. Therefore, it is not surprising that genetic interactions have not been recorded between viruses belonging to different genera, although nongenetic interactions in the form of phenotypic mixing and pseudotype formation may be possible. The intracellular site of maturation, however, limits opportunities for pseudotype formation with heterologous enveloped viruses. Viruses classified within the same genus generally exhibit some sequence similarity, the extent of homology increasing in viruses that show serological relationship. The degree of sequence divergence among viruses within the same serogroup is sufficient, however, to authenticate as distinct biological entities the often bizarrely and arbitrarily named viruses.

B. Genetic Interactions

1. Mutants

Temperature-sensitive (ts) mutants of several representatives of the *Bunyaviridae* have been isolated from both untreated and mutagen-treated wild- type stocks to provide defined genetic material for study of gene function, elucidation of the process of reassortment, and identification of the determinants of host range and virulence. The mutants described so far have been derived from various viruses belonging to the genera *Bunyavirus*, *Phlebovirus*, and *Uukuvirus*. Gentsch *et al.* (1977) observed that spontaneous ts mutants were present in stocks of La Crosse virus and snowshoe hare virus (California serogroup bunyaviruses) at frequencies of 1.0 and 1.7%, respectively, whereas the frequency of ts mutants in a stock of Maguari virus (a Bunyamwera serogroup bunyavirus) was reported to be higher and estimated at 2.7% (Iroegbu, 1981). The majority of the ts mutants described have been isolated following replication in the presence of noninhibitory concentrations of 5-fluorouracil, 5-azacytidine, or N-methyl-N'-nitrosoguanidine (NTG). The data obtained by Ozden and Hannoun (1978) suggested that 5-fluorouracil was more effective as a mutagen than NTG.

2. Recombination of Genetic Markers by Subunit Reassortment

A total of 210 ts mutants isolated from different viruses have been described in some detail. Ten of these eleven viruses belong to two serogroups in the genus *Bunyavirus* and the remaining one to the genus *Uukuvirus* (Table III). These mutants have been employed principally to demonstrate that recombination of genetic markers in crosses of bunyaviruses is mediated by reassortment of genome subunits (Gentsch *et al.*, 1977, 1979; Iroegbu and Pringle, 1981a). The majority of these mutants have been assigned to reassortant groups, such that non-ts virus is only present in the progeny from mixed infections with parental ts mutants classified in different reassortant groups. An unexpected finding, however, was the predominant recovery of mutants assignable to only two reassortant groups rather than the three predicted by the tripartite nature of the viral genome.

These reassortant groups were designated groups I and II and their equivalence in the different viruses included in the same serogroup was established by analysis of the progeny from mixed infections with heterologous parental viruses. Reassortment was restricted to viruses within the same serogroup. Attempts to obtain reassortment in crosses of viruses belonging to the California and Bunyamwera serogroups have not succeeded. However, the combinations of viruses examined have been rather limited. The homologies of the reassortant groups listed in Table III do not extend beyond the boundaries of the serogroups.

Maguari virus of the Bunyamwera serogroup is anomalous in that 1 of the 46 ts mutants isolated from mutagenized wild-type virus, designated tsMAG23(III), generated non-ts reassortants in mixed infections with all mutants belonging to reassortant groups I and II tested and also with two putative double I + II mutants, suggesting that the third reassortant group predicted by the tripartite structure of the genome had been identified (Pringle and Iroegbu, 1982). Mutant tsMAG23(III) exhibited a pronounced host restriction and a small plaque phenotype which may account for the failure to isolate similar mutants in other viruses. The plaque-forming ability of tsMAG23(III) on BS-C-1 cell monolayers was 100-fold lower, and yields from mixed infections 1000-fold less than in BHK-21 cells. On the other hand, mutants of Maguari virus and other bunyaviruses classified in groups I and II did not exhibit such marked host-dependent differences. However, conclusive evidence that the three reassortment groups correspond to the three genome segments has been difficult to obtain because of conflicting evidence in the characterization of representative mutants of the three groups at the phenotypic level (see below).

3. Complementation

Intergenic complementation is difficult to discriminate from reassortment in bunyaviruses, because reassortment is an early event in the multi-

TABLE III. Classification of Temperature-Sensitive Mutants into Reassortant Groups

Genus	Serogroup	Virus	No. of ts mutants	Reassortment groups			Unclassified	References
				I	II	III		
Bunyavirus	Bunyamwera	Batai	5	1	4	0	0	Iroegbu (1981); Iroegbu and Pringle (1981a,b); Pringle and Iroegbu (1982); Elliott *et al.* (1984)
		Bunyamwera	8	5	3	0	0	
		Maguari	46	12	31	1	2	
		Germiston	8	2	6	0	0	Ozden and Hannoun (1980)
		Guaroa	4	2	2	0	0	Pringle (unpublished data)
			9	3	5	1	0	Bishop (1979)
	California	La Crosse	20	6	14	0	0	Bishop (1979); Gentsch and Bishop (1976); Gentsch *et al.* (1977, 1979, 1980)
		Snowshoe hare	48	26	20	0	2	
		Tahyna	20	1	17	0	2	
		Trivittatus	12	4	6	0	2	
		Lumbo	9	2	7	0	0	
			8	1	7	0	0	Ozden and Hannoun
Uukuvirus		Uukuniemi	13	3	2	0	8	Gahmberg (1984)
		Total	210	68	124	2	16	

plication cycle and it can occur at high frequency in mixed infections of genetically compatible viruses. Nonetheless, Gentsch and Bishop (1976) and Ozden and Hannoun (1978) were able to demonstrate intergenic complementation between mutants belonging to different reassortant groups in the case of snowshoe hare virus and Lumbo virus, respectively, both of which belong to the California serogroup of bunyaviruses.

Intragenic complementation involving ts mutants of the same complementation group has been described by Iroegbu and Pringle (1981a,b). Individual pairs of mutants classified in reassortant group I of Maguari virus (Bunyamwera serogroup) were able to complement each other, suggesting that the protein encoded in the genome subunit corresponding to group I has a multifunctional role in viral biosynthesis. Complementation was not observed in pairwise mixed infections with any of the group II mutants of Maguari virus.

4. Absence of Intrasubunit Recombination

Although reassortment of genome subunits occurs universally in negative-stranded RNA viruses with segmented genomes, intramolecular recombination (i.e., the generation of covalently joined RNAs from heterologous molecules) has not been demonstrated unequivocally in any negative-stranded RNA virus irrespective of whether the genome is segmented or not. In eukaryotic viruses the sensitivity of detection is limited by the reversion frequency of the markers employed, usually ts mutants, which is often high. However, recombination has not been observed in prokaryotic RNA viruses where nonreverting markers were used.

It is likely that recombination does occur under rare circumstances, because defective interfering viruses have been generated during high-multiplicity passage of vesicular stomatitis virus and influenza A virus which are mosaics of sequences from different regions of the genome (O'Hara *et al.*, 1984; Nayak *et al.*, 1985). Other evidence suggestive of rare recombination events is the inversion of a region of the genome of the pneumovirus turkey rhinotracheitis virus (Ling *et al.*, 1992), and an insertion of 54 nucleotides derived from the 28 S ribosomal RNA of the host cell in the hemagglutinin of influenza A virus (Khatchikian *et al.*, 1989).

Bergmann *et al.* (1992) recovered several viruses after transfection of *in vitro* reconstituted ribonucleoprotein into influenza A virus-infected cells which could only have originated by a recombinational event. Five different recombinant viruses were obtained. Two contained a neuraminidase (NA) gene whose defective polyadenylation signal had been repaired by an intergenic recombination. Surprisingly the additional sequences in the rescued gene appeared to be derived from RNA contaminating the polymerase preparation in the transfection mix and not from the helper virus in the transfected cell. Although these two recombinants originated from separate transfection experiments, the overall frequency of success was low, indicating that even in this system recombination is a rare event. Two other recombinants were

NA gene mosaics, and one a mosaic containing variable amounts of sequence derived by multiple recombinational events from matrix, PB1, and NA genes derived also from the virus used to prepare the polymerase proteins for the tranfection mix. In some experiments DNA was excluded from the transfection protocol thereby confirming that recombination involved RNA molecules only. It was not possible, however, to discriminate between a copy-choice model or a breakage-joining mechanism, nor whether the recombination occurred during *in vitro* incubation or during amplification in the helper-infected cells.

5. Homologous Reassortment and Gene Assignment

Crosses of heterologous viruses belonging to the same serogroup have been employed to correlate reassortment groups with genome subunits. Genotyping of clones of progeny virus derived from mixed infections with heterologous parental viruses has been carried out by RNA fingerprinting and dot hybridization with subunit specfic probes. These techniques were used in conjunction with phenotypic analysis using specific antibodies and by determination of the mobilities of radiolabeled polypeptides by SDS/ PAGE (Gentsch and Bishop, 1976; Gentsch *et al.*, 1977, 1979, 1980; Rozhon *et al.*, 1981; Bishop *et al.*, 1984; Pringle and Iroegbu, 1982; Pringle *et al.*, 1984a,b; Iroegbu and Pringle, 1981a,b; Endres *et al.*, 1989). The development of the polymerase chain reaction (PCR) has greatly facilitated the genotyping of progeny clones and increased the accuracy and resolving power of such experiments (Urquidi and Bishop, 1992). To enrich the isolation of reassortant clones it was usual in the early work to employ ts mutants as the parental viruses and to characterize the non-ts virus in the progeny. It soon became clear that the frequency of reassortment was sufficiently high to dispense with the need for enrichment and to make use of the ts character as an additional phenotypic marker. Overall the data obtained by the different methodologies were usually consistent although some anomalies have remained unresolved. It was observed generally that heterologous reassortment was restricted and not always reciprocal, the extent of the restriction decreasing as the serological distinctiveness of the parental viruses decreased.

It has become conventional in reassortment experiments to express the genotype of a virus as the three-letter abbreviation for each virus listed in the International Catalogue of Arboviruses (Karabatsos, 1985) to represent the parental origin of the L/M/S subunits. Thus, the genotype of wild-type snowshoe hare virus becomes SSH/SSH/SSH and that of La Crosse virus LAC/ LAC/LAC. In the California serogroup, crosses of the closely related snowshoe hare virus and La Cross virus yielded non-ts reassortants designated SSH/LAC/LAC and SSH/LAC/SSH when the snowshoe hare parental virus was a ts mutant classified in group I, and the La Crosse virus parent was a ts mutant classified in group II. This result is consistent with expectation if the group I ts mutation is located in the M RNA subunit and the group II mutant in the L RNA subunit. However, Gentsch *et al.* (1979) observed that the same two reassortants were obtained when non-ts wild-type parental viruses were employed and the remaining four expected reassortants (LAC/LAC/SSH, LAC/SSH/SSH/, SSH/SSH/LAC, LAC/SSH/LAC) were absent. No viable reassortants were isolated from the reciprocal cross of a group I La Crosse virus ts mutant and a group II snowshoe hare virus ts mutant, leaving the group assignment unconfirmed. However, by a series of ingenious backcrosses, Gentsch *et al.* (1979) were able to obtain the missing reassortants. First, ts mutants were induced in the SSH/LAC/SSH and SSH/LAC/LAC reassortants by random mutagenesis and identified as group I or group II mutants by genetic analysis. Each of these mutants was backcrossed in turn to the appropriate ts parental virus to yield further non-ts reassortants. The SSH/LAC/SSH group II reassortant backcrossed to the SSH/SSH/SSH group I grandparent yielded a new reassortant having the gene combination LAC/LAC/SSH. This new reassortant in its turn was mutagenized to obtain group I and group II ts mutants. Backcrossing of these viruses to the original SSH/SSH/SSH and LAC/LAC/LAC parental viruses yielded a non-ts reassortant virus with the fourth gene combination SSH/SSH/LAC in the progeny of the cross of a group II SSH/SSH/LAC virus and the original SSH/SSH/SSH group I mutant. The data generated in this complicated chain of experiments are consistent with the interpretation that the mutations classified in reassortment group I are located in the M RNA subunit, and those classfied in group II are located in the L RNA.

Several crosses yielded no reassortants, however, and two (LAC/SSH/SSH and LAC/SSH/LAC) of the six possible subunit combinations expected in these experiments were not recovered. These reassortants were isolated only later in experiments by Rozhon *et al.* (1981) in which the input virus was concentrated by polyethylene glycol precipitation to increase the input multiplicity. The interaction between parental virus appears to be an early event in the multiplication cycle (Iroegbu and Pringle, 1981a) and superinfection exclusion developed fairly soon after initial infection. Consequently, the relative multiplicities of infection of the parental viruses could have an important bearing on the outcome of any particular mixed infection.

It could be concluded from these experiments, therefore, that the asymmetric and restricted patterns of gene segregation observed in these experiments were not related to any inherent incompatibility of the genome subunits of these two members of the California serogroup. The restrictions are more likely a consequence of other factors such as compartmentalization of replication within the cell, aberrant morphogenesis, variable cytopathogenicity, etc. A more systematic restriction was observed when more distantly related members of the California serogroup were crossed. Crosses of group II mutants of Tahyna virus with group I mutants of snowshoe hare virus or trivitattus virus, group I mutants of Tahyna virus with group II mutants of La Crosse virus, and group I mutants of trivitattus virus with group II mutants of La Crosse virus, yielded reassortants whose genotypes were again consistent with assignment of the group I mutations to the M RNA and the group II mutations to the L RNA. There was a marked bias, however, for the L and S RNA segments to cosegregate; 58 of the 65 non-ts reassortants characterized derived their L and S RNA subunits from the same parent. The same tendency was observed in the cross of a group I La Crosse virus and a group II Tahyna virus, where 9 of 12 non-ts reassortants derived their L and S RNA from the same parent. However, two reassortant genotypes conflicting with the group I and II assignments were recovered in the progeny of this cross. This discrepancy was attributed to experimental error, but again certain combinations of parental viruses (Tahyna virus group I × trivitattus or snowshoe hare virus group II, and La Crosse virus group I × trivitattus virus group II) failed to yield the expected non-ts reassortant progeny virus.

In corresponding experiments involving characterization of non-ts progeny virus emanating from crosses of ts mutants of three Bunyamwera serogroup viruses, all six possible reassortant genotypes were recovered (Iroegbu and Pringle, 1981a,b). In contradistinction to the California serogroup data, however, analysis of the genotypes of the reassortants suggested that the group I and group II mutations of Batai, Bunyamwera, and Maguari viruses were located in the S RNA and M RNA subunits, respectively (Iroegbu and Pringle, 1981a,b; Pringle *et al.*, 1984a,b). As in the case of the California serogroup viruses, it was apparent that segregation did not occur at random except that in this case the M subunit appeared to segregate unrestricted and the L RNA and S RNA subunits showed linkage. This phenomenon was observed both when non-ts reassortants were selected from the progeny of crosses of ts mutants, and when the frequency of reassortment was sufficiently high to allow direct analysis of unselected progeny. Curiously in the former case, invariably only one of the two possible classes of non-ts progeny was obtained. The same bias was observed in unselected progeny also, and a total of 279 reassortants pooled from several experiments appeared to have derived their L and S RNA subunits from one parent only. A different situation pertained, however, when these reassortant viruses were in their turn used as the parental viruses. For example, all six possible reassortants were obtained at comparable frequencies in unselected progeny from a mixed infection of viruses of genotype BUN/MAG/BUN and BAT/BUN/BAT, i.e., containing subunits originating from three parents. This confirms that as in the case of the California serogroup viruses the restricted pattern of subunit segregation in some crosses is not the result of any inherent genetic incompatibility.

The molecular basis of these restricted segregations has not been elucidated, but the apparent linkage of the L and S RNA subunits which encode core and replicative functions suggests that the restriction acts at the gene product level. The Bunyamwera serogroup data suggest that once a homologous gene combination has been disrupted by reassortment, there is no preferred association of genes or gene products that has any selective advantage. This suggests that after the disruption of adapted gene combinations, restrictions are relaxed and the bunyaviruses may exhibit the same propensity for rapid and abrupt changes in properties as the influenza A viruses.

A major discrepancy exists between the results obtained with the California serogroup viruses, where the larger of the two groups of ts mutants was assigned to the L RNA, and the Bunyamwera serogroup viruses, where the largest group was assigned to the S RNA. Analysis of the assignment of the single ts mutant of Maguari virus should have resolved the issue, but because the phenomenon of restricted segregation was again encountered in crossing ts mutants of heterologous viruses, it was not possible to assign the group III lesion to either the L or the S RNA subunit. Non-ts reassortants of genotype BUN/MAG/BUN were obtained from the cross of MAGts23(III) with a group II Bunyamwera virus parent, which excluded location of the group III lesion in the M RNA subunit. However, the non-ts reassortant required to unequivocally assign the group III mutation to the L or S RNA subunits was not obtained, because no non-ts progeny virus was obtained from the critical cross of the MAGts23(III) and the group II Bunyamwera virus parent (Pringle and Iroegbu, 1982). Further attempts were made to resolve this problem by analysis of reassortants in the unselected progeny of crosses of various ts mutants with a heterologous wild-type virus so that the ts mutation could be employed as a fourth marker (Murphy and Pringle, 1987; Hampson, 1987). Reassortants of genotype MAG/BUN/BUN/ts$^+$ and MAG/BUN/MAG/ts were obtained from crosses of several Maguari virus group I mutants with wild-type Bunyamwera virus, and reassortants BUN/MAG/MAG/ts and MAG/MAG/BUN/ts from crosses of several Maguari virus group II mutants with wild-type Bunyamwera virus. These results are consistent with the previous tentative assignment of the group I mutants to the S RNA and the group II mutants to the M RNA. The cross of MAGts23(III) with Bunyamwera virus wild type yielded several reassortants (MAG/BUN/BUN/ts, BUN/MAG/MAG/ts$^+$, MAG/MAG/BUN/ts, BUN/MAG/BUN/ts$^+$) that were consistent with the assignment of the group III lesion to the L RNA subunit. These later experiments with Bunyamwera serogroup viruses were complicated by the frequent occurrence of heterozygous and quasidiploid particles, which were sometimes difficult to detect since in mixed infections Maguari virus appeared to have a replicative advantage and incorrect assignments may have been made.

The physical nature of these heterozygous particles is unknown, but Talmon *et al.* (1987) have demonstrated by electron microscopy of vitrified specimens that the number of subunits in particles is variable. Furthermore, in the genetic analysis of progeny virus from crosses of ts mutants of rotaviruses, it has been established that the genetic background of the recipient can affect the phenotype conferred by the gene introduced by reassortment (Chen *et al.*, 1989), which introduces an additional complication in progeny analysis. The non-ts phenotype of some bunyavirus reassortants may be determined by interactions between heterologous subunits and not solely by the absence of a ts lesion, and thereby may be responsible for the failure of gene assignment by this approach.

Nucleotide sequence analysis of the S RNA subunit of mutant MAGts23-(III) by a coupled reverse transcription/PCR method (Dunn *et al.*, 1994) revealed a transition at position 327 conferring a valine-to-alanine substitution in the N protein product and a phenylalanine-to-leucine change in the NSs protein (D. C. Pritlove and R. M. Elliott, unpublished data). This mutation at residue 327 is not present in the S RNA of mutant MAGts6(I) and mutant MAGts17(II), representing groups I and II, or in the S RNA of another 12 bunyaviruses (Dunn *et al.*, 1994). The alanine introduced into the N protein of MAGts23(III) is not present at this position in any other N protein sequence. The valine in wild-type Maguari virus is conserved in all of the other bunyavirus sequences, excepting for a substitution of isoleucine in three California serogroup viruses. The NSs protein sequence is less well conserved and the leucine substituted in MAGts23(III) also occurs in three California group viruses. These observations indicate with reasonable certainty that the group III mutation is located in the S RNA, and bring the Bunyamwera serogroup assignments into line with the California serogroup assignments. However, revertants of mutant MAGts23(III) have not been isolated and it could not be confirmed directly that the mutation at position 327 conferred temperature-sensitivity.

Additional evidence in support of this assignment comes from two other directions. By Northern blotting it has been shown that mutant ts23(III) is replication-defective but transcription-competent, which suggests that the defect lies in S RNA gene products rather than the L RNA-derived polymerase (Pritlove and Elliott, unpublished data). Dunn *et al.* (1995) have developed an *in vitro* chloramphenicol acetyl transferase (CAT) activity rescue system that is dependent on added L and S RNA gene products. In this system they have shown that the S RNA of mutants MAGts6(I) or MAGts17(II) will act in conjunction with Bunyamwera virus L RNA to rescue CAT activity at both 33 and 38°C, whereas MAGts23(III) S RNA will only rescue at 33°C (Elliott *et al.*, unpublished data). This is further strong evidence in favor of the location of the group III lesion in the S RNA.

Thus, the group I mutants represent mutations in the L RNA and group II mutants represent mutations in the M RNA. This assignment is more in accord with the frequency of isolation of the mutants and the genome target size. It is also the expectation from studies of mutants of other negative-strand RNA viruses that L protein mutants should be in excess (Pringle, 1987, 1991a,b). However, until the L and S subunits of each mutant have been sequenced, these revised assignments are not entirely conclusive. It remains to be determined why the Bunyamwera serogroup progeny analyses were so misleading, and why ts mutations were so rarely recovered in the S RNA. The overlapping encoding of information in the S RNA may account for the latter phenomenon.

C. Heterologous Reassortment and the Definition of Gene Pools

Reassortment of subunits occurs with varying facility between viruses that belong to the same serogroup. Heterologous crosses of viruses belonging to the Bunyamwera, California, and Simbu serogroups have been carried out in the laboratory to explore the extent of the ability to reassort genome subunits. The recovery of reassortants was forced by employing pairs of ts parental viruses and selecting non-ts progeny virus. The phenotype and/or genotype of progeny clones were determined by several procedures. Since the specificity of neutralization of bunyaviruses is determined by the M RNA (Gentsch *et al.*, 1980; Iroegbu and Pringle, 1981a,b; Millican and Porterfield, 1982; Murphy and Pringle, 1987), it is possible to screen for reassortment rapidly by crossing any virus for which a specific neutralizing antiserum is available with a standard group I (or group III) ts mutant which will donate a non-ts M RNA subunit. Reassortants can be detected by screening the progeny virus at the restrictive temperature in the presence of the neutralizing antibody, a procedure that should eliminate all parental virus and allow growth of two of the six possible reassortants (Table IV).

The results of such experiments are summarized graphically in Fig. 2. In the absence of an appropriate ts parental virus, the following procedure can be used to detect reassortants provided that a neutralizing antiserum against the non-ts virus is available. All six viruses of the California serogroup that have been tested were able to exchange genome subunits. On the other hand, only five of the eight Bunyamwera serogroup viruses studied were able to exchange subunits. Therefore, in addition to the restricted segregations ob-

TABLE IV. An Example of Rapid Screening
for Reassortment between Heterologous Bunyaviruses

	Genotype[a]	Growth at restrictive temperature	Growth at restrictive temperature in presence of anti-NOR serum
Parental	BUN*/BUN/BUN	−	−
	NOR/NOR/NOR	+	−
Progeny	BUN*/BUN/BUN	−	−
	NOR/NOR/NOR	+	−
	BUN*/NOR/NOR	−	−
	NOR/BUN/BUN	+	+
	BUN*/BUN/NOR	−	−
	NOR/NOR/BUN	+	−
	BUN*/NOR/BUN	−	−
	NOR/BUN/NOR	+	+

[a]BUN = Bunyamwera virus L, M, and S RNA subunits.
BUN* = A Bunyamwera virus L RNA subunit carrying a temperature-sensitive mutation.
NOR = Northway virus L, M., and S RNA subunits.

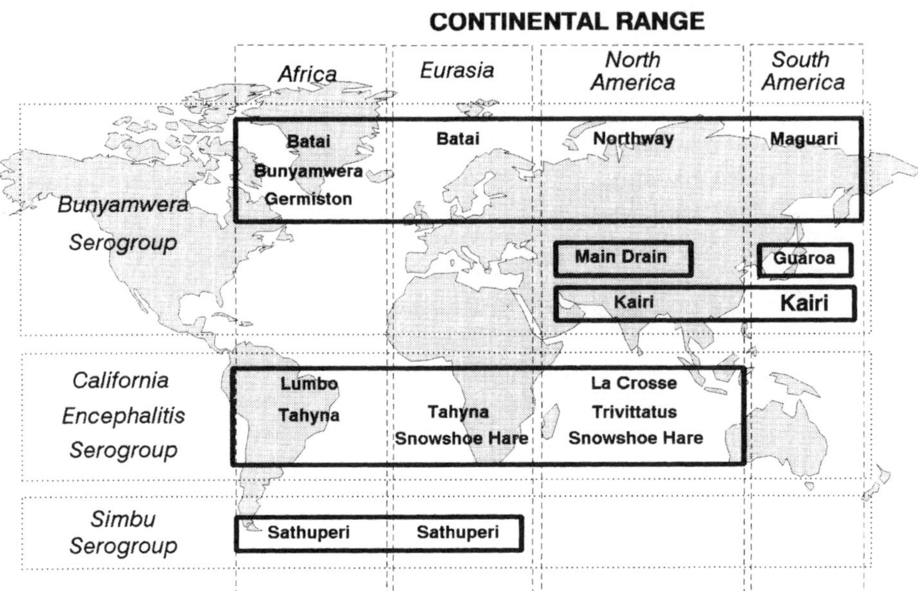

FIGURE 2. Patterns of genome subunit exchange among viruses of the Bunyamwera, California, and Simbu serogroups of the genus *Bunyavirus*. The heavy lines enclose viruses that are able to exchange genome segments by reassortment in cultured cells. Viruses in different boxes are unable to exchange genome subunits and appear to be genetically isolated. Serogroups are indicated horizontally and known geographical distribution vertically.

served in the repirocal ts mutant crosses described previously, some viruses within the same serogroup are less able to exchange genome subunits and are in effect genetically isolated. Some viruses, such as Main Drain and Kairi viruses, seem to be totally isolated from the remainder. The pattern of restriction within the Bunyamwera serogroup corresponds to the serological relationships of the viruses (Hunt and Calisher, 1979); the greater the serological divergence, the greater the restriction in subunit exchange. Guaroa virus, although included in the Bunyamwera serogroup in Fig. 2, has been ranked as a bridging virus between the Bunyamwera and California serogroups (Beaty and Calisher, 1991). In complement fixation tests, Guaroa virus is closer to the Bunyamwera serogroup viruses, whereas it reacts similarly to the California serogroup viruses in neutralization tests (Whitman and Shope, 1962). ts mutants of Guaroa virus have been isolated independently by Gentsch *et al.* (1980) and Pringle (1991c) and employed to detect heterologous reassortment. However, there was no evidence of subunit exchange between Guaroa virus and members of either the Bunyamwera serogroup or the California group, and Guaroa virus appears to be genetically isolated from both serogroups. Dunn *et al.* (1994) concluded from a comparison of the S RNA sequences of seven viruses of the Bunyamwera serogroup that Main Drain virus, Kairi virus, and Guaroa virus were less closely related than the other four (Cache Valley, Northway, Maguari, and Batai viruses), reflecting their compatibility in reassortment experiments. Germiston virus, however, does not conform to this generalization. They concluded in addition that Guaroa virus represented a reassortant virus derived from two ancestral parental viruses now included in separate serogroups.

Extensive heterologous crosses of viruses belonging to the Bunyamwera, California, and Simbu serogroups have failed to find any evidence of exchange of genome subunits between serogroups. These experiments were carried out under a variety of conditions and in different host cells, including both vertebrate and *Aedes albopictus* C6/36 cells (Elliott *et al.*, 1984). Consequently, it is likely that the different serogroups of viruses in the genus *Bunyavirus* represent discrete gene pools. Genetic variation in the family *Bunyaviridae* is the result of both genetic drift and ecological isolation, and by gene capture from heterologous viruses. The opportunity for gene exchange in viruses other than the hantaviruses is probably greatest during the arthropod-borne phase of the multiplication cycle, since more than one blood meal may be taken by an individual female mosquito. It is unlikely that immune selection is a driving force in the evolution of these viruses, given the predominant role of arthropods in the life cycles of most members of the *Bunyaviridae*. Geographical range does not seem to be a major factor in determining the genetic divergence of bunyaviruses. Figure 2 shows that viruses such as Maguari virus, which is restricted to South America, can exchange subunits with viruses originating from different continents. Conversely, viruses occupying the same geographical range (e.g., Main Drain, Kairi, and Northway) seem unable to interact. Our knowledge of the natural distribution of bunyaviruses is rudimentary, however, and the geographical isolation of different bunyaviruses may be more apparent than real. [Figure 2 has been modified from an earlier version (Pringle, 1991c) to take account of more recent information.] The pattern of genetic interaction in the *Bunyaviridae* is very different from that of the orthomyxoviruses, where it is likely that reassortment is possible between all influenza A virus, although this has not been rigorously tested.

D. Nonrandomness of Reassortment

Nonrandom reassortment of genome subunits has been described for influenza A virus (Lubeck *et al.*, 1979), bluetongue virus (Stott *et al.*, 1987), and rotavirus (Gombold and Ramig, 1986; Graham *et al.*, 1987; Ward *et al.*, 1988; Ramig and Ward, 1991), as well as for viruses of the Bunyamwera serogroup (Pringle *et al.*, 1984a,b). Urquidi and Bishop (1992) have now carried out a more critical analysis of the process of reassortment of subunits using two closely related members of the California serogroup. Monolayers of BHK-21 cells were infected with an inoculum of 5 pfu per cell of wild-type

La Crosse and snowshoe hare viruses to ensure a high probability of mixed infection. Progeny virus harvested after 72 hr incubation at 36°C was analyzed employing the polymerase chain reaction (PCR) with a set of nine primers designed to identify the parental origin of the six subunits segregating in this cross by their relative electrophoretic mobility in 2% agarose gel. This approach enabled a larger sample of progeny virus to be characterized with greater accuracy than in any previous experiment. A total of 350 plaques were isolated from monolayers incubated for a period of 7 days to ensure that small plaque-forming virus was not missed, and the parental origin of the subunits determined. The results of the analysis of the 255 clones with the proper complement of S, M, and L RNA species are given in Table V. This accounts for 73% of the progeny; the remainder not included in the analysis consisted of 52 clones (15%) where there was no amplification of one (or rarely two) of the three RNA species, 31 isolates (9%) with PCR products derived from both parents representing either diploid virus or uncloned virus, and a residue of 3% which were PCR-negative. All eight possible genotypes were identified in the progeny with the frequency distribution given in Table V. The difficulty experienced previously by Gentsch *et al.* (1979) in identifying reassortants of genotype LAC/SSH/SSH and LAC/SSH/ LAC may relate to the fact that in the earlier experiments ts mutants were used as the parental viruses. These observations reemphasize the caveat that the phenotypes of parental ts mutant viruses may influence the outcome of reassortment (Chen *et al.*, 1989).

The expected frequencies in Table V are calculated from the relative frequencies of the individual L, M, and S RNA subunits on the assumption that there are no constraints on subunit segregation. There is a highly significant difference between the observed and expected frequencies ($\chi^2 = 32.41$; $p<0.001$), and an overrepresentation (43%) of the two parental genotypes. Since the progeny virus titer exceeded the input virus titer 100-fold, this

TABLE V. Nonrandom Reassortment in a Cross
of La Crosse and Snowshoe Hare Viruses[a]

Genotype	Observed (%)	Expected[b] (%)	χ^2	Significance
LAC/LAC/LAC	81 (32%)	60 (24%)	9.32	<0.005
SSH/SSH/SSH	29 (11%)	14 (6%)	18.13	<0.001
LAC/SSH/SSH	28 (11%)	29 (11%)	0.05	>0.8
SSH/LAC/LAC	22 (9%)	28 (11%)	1.61	>0.2
LAC/LAC/SSH	30 (12%)	39 (15%)	2.23	>0.1
SSH/SSH/LAC	17 (7%)	21 (8%)	0.99	>0.3
LAC/SSH/LAC	35 (14%)	46 (18%)	2.94	>0.05
SSH/LAC/SSH	13 (5%)	18 (7%)	1.57	>0.2
Total	255	255	32.41	<0.001

[a]Data from Urquidi and Bishop (1992).
[b]Calculated from the observed relative frequencies of the genome segments: LAC L = 174/255 = 0.68; LAC M = 146/255 = 0.57; LAC S = 155/255 = 0.61; SSH L = 81/255 = 0.32; SSH M = 109/255 = 0.43; SSH S = 100/255 = 0.39.

excess of parental virus must represent a preferential association of homologous subunits in virus replication and not the reisolation of residual inoculum. The interaction between segment pairs was examined using a log-linear statistical model (Sokal and Rohlf, 1981), and the results of this analysis confirmed the existence of positive associations between certain pairs of subunits. Homologous L–M and M–S associations were preferred in this particular cross. A positive association between homologous subunits has also been described in a less rigorous analysis of progeny from crosses of viruses of the Bunyamwera serogroup (Pringle *et al.*, 1984a,b). In this case the preferential association was between the L and S RNA subunits.

Iroegbu and Pringle (1981) observed that reassortment of subunits could occur in cultured invertebrate cells multiply infected with Bunyamwera serogroup viruses perhaps more readily than in vertebrate cells, and Beaty *et al.* (1981a,b 1985) demonstrated that reassortment occurred *in vivo* when mosquitoes were infected experimentally with more than one virus of the California serogroup. All six reassortants were also found in the progeny from a mixed infection of vertebrate cells with unmutagenized stocks of La Crosse and Tahyna viruses (Janssen *et al.*, 1986; Endres *et al.*, 1991), and also *in vivo* in mosquitoes (Chandler *et al.*, 1991). The genotypes of 708 isolates obtained from *Aedes triseriatus* mosquitoes 23 days after inoculation were analyzed by Chandler *et al.* (1991) by molecular hybridization using three gene probes specific for the LAC L, M, and S RNA. High-frequency reassortment was observed and all six reassortants were identified. Analysis of these data by the method of Urquidi and Bishop (1992), however, indicates that the observed frequencies of the eight genotypes diverge significantly from the expected frequencies. This may be related to the less definitive method used in establishing genotype, but probably reflects selective effects occurring during the prolonged period of multiplication in the insect host. These experiments show, however, that reassortment can occur *in vivo* at high frequency in an arthropod vector. Preferential association of subunits during reassortment appears to be the rule in bunyaviruses, and perhaps in all other single-component segmented genome viruses. Nonrandomness is presumed to represent constraints on packaging and morphogenetic events, and the different associations observed with different viruses (or even the same viruses under different conditions) may reflect diverse replicative abilities, nonequivalence of the components of an inoculum, and deviation from equimolarity of RNA subunits at different stages of morphogenesis. It is well known that the subunit content of most virus preparations departs somewhat from equimolarity (Elliott, 1990). Interference may be another factor. Heterologous interference was observed between different members of the Bunyamwera serogroup in dual infection experiments, such that Maguari virus appeared to suppress replication of Bunyamwera virus or Batai virus (Murphy and Pringle, 1987). Urquidi and Bishop (1992) attempted to compensate for a perceived replicative advantage of La Crosse virus by delaying superinfection of snowshoe hare virus-infected cells by 4 hr. Under these conditions they observed complete interference with La Crosse virus replication and only viruses of genotype SSH/SSH/SSH were recovered in the progeny. Interference to reassortment has also been described in the case of oral infection of mosquitoes with two ts mutants of La Crosse virus (Beaty *et al.*, 1985), in mixed infections with two variants of Rift Valley fever virus (Turell *et al.*, 1990), and also in the midge *Culicoides variipennis* asynchronously infected with different serotypes of bluetongue virus of the *Reoviridae* (El-Hussein *et al.*, 1989).

E. The Role of Reassortment in Arthropods in the Evolution of the *Bunyaviridae*

Evolutionary trends within the genus *Bunyavirus* and within the *Bunyaviridae* can be recognized by their relationships to their arthropod vectors. Viruses of each genus, with the exception of the vectorless hantaviruses, have preferential relationships with one or two arthropod families only, and within genera each serogroup is preferentially associated with arthropod species belonging to one or two genera only (Labuda, 1991). Based on isolations from nature, 8 of the 16 bunyavirus serogroups are associated with mosquitoes of the genus *Culex*, two with mosquitoes of the genus *Aedes*, three with mosquitoes of the genus *Anopheles*, and one serogroup each with *Aedeomyia* mosquitoes, *Culicoides* midges, and *Hyalomma ticks*.

Chandler *et al.* (1990) have reported that replication and reassortment were enhanced in the ovaries of female mosquitoes and that the newly generated reassortants were transmitted transovarially to about 10% of their progeny. These new reassortants in their turn could be transmitted to a susceptible vertebrate host. These observations indicate that reassortment in the arthropod vector may be the driving force in the evolution of bunyaviruses in nature. *Aedes triseriatus* mosquitoes, for example, feed on a variety of hosts which increases the chance of multiple infection with different viruses since many California serogroup viruses co-circulate in the same area sharing the same vector.

Presumptive reassortant bunyaviruses have been isolated from trapped mosquitoes. Ushijima *et al.* (1981) reported that oligonucleotide fingerprints of the L and S RNA of Shark River virus and Pahayokee virus, two members of the Patois serogroup isolated from *Culex melanoconium*, were virtually identical. Their M RNA subunits on the other hand were quite distinct. These observations were interpreted to mean that these isolates of Shark River virus and Pahayokee virus were natural reassortants which had acquired their M RNA subunit from different unknown donor viruses. This interpretation is reinforced by the observations of El-Said *et al.* (1979) and Klimas *et al.* (1981) who studied the natural variation of La Crosse viruses isolated from different ecological situations and found that the L and S RNAs of these California serogroup viruses were not more conserved than the M RNA, indicating the genetic homogeneity of these viruses. The pattern of variation for all three subunits was continuous and more analogous to the progressive mutational drift observed in isolates of influenza A virus during the interpandemic phase.

The role of genetic drift in the evolution of bunyaviruses has not been studied systematically, and its relative importance has not been assessed. Klimas *et al.* (1981) found that two isolates of La Crosse virus had virtually identical oligonucleotide fingerprints, although originating from samples of human brain tissue taken 18 years apart. Likewise, 12 cycles of transovarial transmission of a phlebovirus did not induce significant changes in the oligonucleotide fingerprints of the L, M, and S RNA subunits (Bilsel *et al.*, 1988). Hewlett *et al.* (1992) have compared in terms of oligonucleotide fingerprint profiles and plaque-reduction neutralization titers the variability of field isolates of La Crosse virus and snowshoe hare virus with laboratory-derived stocks with different *in vitro* passage histories. They observed that the field isolates differed significantly, whereas the laboratory-passaged viruses exhibited little change. Battles and Dalrymple (1988) estimated the genetic variability of Rift Valley fever virus by sequencing of a region of the M gene thought to encode antigenic determinants of importance in protection. Twenty-two isolates originating from several different host species over a period of 34 years were compared and it was considered that most isolates were very similar to a reference strain at both the nucleotide (0–4.5%) and deduced amino acid sequence (0–2.4%) levels. On this rather limited and circumstantial evidence, it would appear that subunit reassortment may be the more potent force in the evolution of the *Bunyaviridae*.

By contrast, reassortment of genome subunits does not seem to play a major role in the evolution of the hantaviruses where there is no arthropod vector in the transmission cycle. Phylogenetic analyses of the nucleotide sequences of the M and S RNA subunits of the Puumala/Prospect Hill subgroup of hantaviruses suggest that the M and S RNA subunits may have evolved in parallel (Hjelle *et al.*, 1994; Spiropoulou *et al.*, 1994). It was concluded that there was no evidence to support a hypothesis that reassortment of genome subunits was responsible for the emergence of Sin Nombre virus, a new hantavirus associated with an outbreak of a previously unrecognized fatal respiratory disease in the southwestern United States in May, 1993.

Bunyaviruses are closely adapted to particular vectors (Kramer *et al.*, 1993), and the efficiency of transmission of bunyaviruses by different susceptible vectors is variable. For example, Schopen *et al.* (1991) found that the California serogroup virus La Crosse virus was transmitted transovarially by 53% of *Aedes triseriatus*, its natural vector, and by only 22% of *Culiseta inornata* mosquitoes. By contrast, the closely related snowshoe hare virus was transmitted transovarially by 89% of *C. inornata* mosquitoes, the presumptive natural vector, and by only 29% of *A. triseriatus*. Experiments monitoring the transovarial transmission of reassortant viruses showed that the La Crosse M RNA subunit was the genetic determinant favoring efficient transovarial transmission in *A. triseriatus* mosquitoes.

V. SUBUNIT REASSORTMENT IN OTHER SEGMENTED GENOME VIRUSES

Reassortment of genome subunits has been observed in the majority of viruses possessing segmented genomes, and the phenomenon is particularly well-documented for the orthomyxoviruses and the reoviruses. It is probable that reassortment of genome subunits between viruses adapted to growth in different host organisms contributes to the evolution of theses viruses in the natural environment and that it provides a mechanism for rapid adaptation to changing circumstances. Reassortment is particularly obvious where, as in the influenza A viruses, there are several reservoir host species, and is not apparent where no reservoir species exist as in the case of the influenza B viruses. The influenza C viruses present in the human and pig populations may have diverged sufficiently to have become genetically incompatible, but this has not been verified by laboratory experiment. Intrastrain reassortment of genome subunits has been described for most of the multicomponent plant viruses.

Among arenaviruses, reassortment has been demonstrated between ts mutants of the same strain and between different strains of the two arenaviruses Pichinde virus and lymphochoriomeningitis (LCM) virus. There was evidence of restriction of reassortment, as in the bunyaviruses, but gene assignments could be made without difficulty. The L and the S RNAs of arenaviruses encode two gene products in ambisense configuration, and ts mutants mapped to the S RNA subunit, but not those mapped to the L subunit, could be classified further into two complementation groups. Pichinde virus and LCM did not exchange genome segments under *in vitro* conditions and the prevalence of reassortment of arenaviruses in the natural environment is unknown. Pichinde virus and LCM virus belong to different serogroups, however, and there is some evidence of reassortment between Lassa fever and Mopeia viruses which belong to the same serogroup.

The three serotypes of human reovirus exchange genome subunits freely during mixed infection and reassortment has been employed with great success in analysis of the genetic determination of pathogenesis and virulence in these viruses. Little is known, however, regarding the role of reassortment in the evolution of these viruses. Natural reassortants of rotaviruses may be of frequent occurrence, and reassortants have been recovered from mice infected with ts mutant parental viruses (Ramig and Gombold, 1991). Genetic reassortment between simian rotavirus SA11 and rhesus rotavirus occurred at high frequency in suckling mice and the frequency of reassortment could be manipulated by passive immunization of the host (Gombold and Ramig, 1989). In contrast to the bunyaviruses, in cultured cells at least superinfecting rotaviruses are not excluded from participating in subunit exchange for at least 24 hr after infection with the virus initiating infection (Ramig, 1990). By contrast, superinfection exclusion occurred after 4 hr in cultured Vero cells infected with different serotypes of bluetongue virus (Ramig *et al.*, 1989). Reassortment was nonrandom whether the multiplicity of infection of the parental rotaviruses was equal or not. Asynchronous infections of the midge *Culicoides variipennis* by bluetongue virus, however, indicated that *in vivo* interference to superinfection did not occur until day 5 after initial infection (El-Hussein *et al.*, 1989). The fact that these vectors frequently seek a second blood meal suggests that there is frequent opportunity for reassortment and that this may play an important role in the evolution of bluetongue virus.

Evidence for gene capture from viruses present in alternate hosts is abundant. The K8 strain of human rotavirus has been identified as a reassortant derived from parental viruses belonging to the Wa and Au-1 serogroups (Nakagomi *et al.*, 1992a; Isegawa *et al.*, 1992). The Au-1 virus has a VP4-coding segment encoding a feline rotavirus-like VP4, suggesting that genetic interaction between feline and human rotaviruses is possible. Nakagomi *et al.* (1992b) have also reported evidence for presumed reassortment of an Au-1 human rotavirus with a bovine rotavirus in nature. Three epidemiologically distinct rotaviruses of the G8 serogroup have been characterized by Browning *et al.* (1992). These viruses appear to have originated from interaction between bovine and human rotaviruses and by reassortment with a rotavirus of a third, as yet unknown, host species. Finnish isolates of the G8 serogroup closely resembled bovine isolates of the G8 serogroup and may possibly have a zoonotic origin or be derived from the live bovine rotavirus used to vaccinate Finnish children. Similarly, reassortment has been recorded in field isolates of the orbivirus, bluetongue virus (De Mattos *et al.*, 1991). Sequence analysis of the S1 segment of bluetongue virus, encoding the VP7 protein carrying the major neutralizing antigen, in conjunction with hybridization data for the M3 segment, encoding the VP5 protein, suggested that the five serotypes of bluetongue virus present in North America were derived from three separate gene pools, and bluetongue virus serotype 17 had evolved by a combination of both reassortment and genetic drift (Kowalik and Li, 1991). Reassortment in the *Reoviridae* may be restricted to serologically related viruses. Genome subunit reassortment in the Kemerovo serogroup viruses appears to be restricted to taxonomically related viruses and does not correlate with geographical origin (Moss *et al.*, 1988; Brown *et al.*, 1989a,b). Brown *et al.* (1991) failed to demonstrate genetic interactions between viruses of the Orungo and Lebombo serogroups, although gene reassortment was possible between viruses of the same serogroup. Mascarenhas *et al.* (1989) have described atypical human serotype 2 rotavirus strains that appear to be reassortants in which a subgroup II rotavirus had exchanged its serotype-specific gene for the equivalent gene of a subgroup I rotavirus.

Genetic reassortment plays a major role in the evolution of the influenza A viruses and is epidemiologically important in human disease as a consequence of the generation of new pandemic forms of the virus. Avian viruses provide a reservoir of antigenic variation which can be introduced into the human virus gene pool. Much evidence suggests that the pig provides a "mixing vessel" where avian and human viruses can interact yielding progeny with new antigens and constellations of genes facilitating spread to human and other animal hosts (Gammelin *et al.*, 1989; Castrucci *et al.*, 1993). In South China the domestic duck may serve as the intermediate host carrying viruses from feral ducks to the domestic pig (Yasuda *et al.*, 1991). In North America, on the other hand, the population of viruses in swine appeared to be homogeneous, while by contrast a high degree of reassortment was observed in the domestic turkey population (Wright *et al.*, 1992). While reassortant viruses appear to have been responsible for the major influenza pandemics in the human population since the beginning of the 20th century, a recent study of avianlike viruses associated with high mortality in horses in China suggests that on occasion avian viruses can spread directly to mammals without the intervention of reassortment (Guo *et al.*, 1992). Conversely, although there is no evidence of antigenic change in current human H1N1 influenza A viruses, genetic drift and subunit reassortment are occurring undiminished (Xu *et al.*, 1993).

Genome subunit reassortment in all segmented genome viruses appears to exhibit the same general characteristics with only minor variation between families:

1. Reassortment is restricted to taxonomically related groups. In the family *Orthomyxoviridae* the boundaries are defined by species, and by serogroups within species in the other families. Genetic compatibility as measured by ability to reassort genome subunits may be a useful approach to equating taxons in different virus families.
2. Reassortment is rarely random. The outcome of a mixed infection is influenced by the genetic compatibility of the parental viruses, input multiplicity, and timing of superinfection. In mammals and birds the outcome may be affected by the immune status of the host. Preferential association of subunits has been observed wherever comprehensive progeny analysis has been carried out.
3. Transmission by vectors and multiplication in alternate vertebrate hosts are common features of viruses with segmented genomes.
4. In all segmented genome viruses there is potential for rapid adaptation to new hosts, abrupt appearance of viruses with new antigenic characteristics, and sudden extensions of host range and virulence.

VI. HOST–VIRUS INTERACTIONS

A. Persistent Infection, Autointerference, and Deletion Mutations

Deletion mutations located predominantly in the viral polymerase genes are associated with the phenomena of autointerference and persistence in all RNA viruses. Undiluted passage of Bunyamwera virus in BHK-21 cells (Kacsak and Lyons, 1978) or BS-C-1 cells (Iroegbu and Pringle, unpublished data) was associated with progressive diminution in yield of infectious virus. Other persistent infections of vertebrate cells have been established by propagation of Dugbe virus (genus *Nairovirus*) in pig kidney cells (David-West and Porterfield, 1974), and Toscana virus (genus *Phlebovirus*) in Vero cells (Verani *et al.*, 1984). In BHK-21 cells this phenomenon was associated with the appearance of S RNA-containing particles. Cunningham and Szilagyi (1987) have reported the presence of truncated L RNA in cytoplasmic extracts of BHK-21 cells infected with Germiston virus (Bunyamwera serogroup). These molecules resemble the polymerase gene deletions characteristic of the defective interfering (DI) particles which are associated with autointerference in other negative-stranded RNA viruses. Patel and Elliott (1992) detected novel polypeptides in mouse cells infected with small plaque variants of Bunyamwera virus. Four of the five variants studied interfered with the multiplication of standard virus. Subgenomic L RNAs were detected in the infected cells suggesting that interference might be mediated by truncated L proteins. Host cell components appeared to be involved in the generation of the DI virus, since subgenomic L RNAs were abundant in mouse cell-passaged virus and not apparent in BHK-21 cell-passaged virus. Similarly, DI RNA molecules derived from genomic L RNA have been observed also during sequential passage of tomato spotted wilt virus (TSWV) by mechanical inoculation of *Nicotiana rustica* plants at high multiplicity of infection. These DI molecules were single internal deletions of 60–80% of the L RNA (Resende *et al.*, 1992). Curiously, four different DI molecules retained open reading frames encoding the amino acid sequence of the carboxy-terminal region of the L protein, suggesting that translation of these proteins might interfere with the polymerase function of the intact L protein and thereby mediate inhibition of replication.

Multiplication of bunyaviruses in cultured arthropod cells is not normally accompanied by cytopathic effect (Kacsak and Lyons, 1978; Newton *et al.*, 1981; Elliott and Wilkie, 1986). Elliott and Wilkie reported that there was an overrepresentation of the S RNA subunit of Bunyamwera virus in persistently infected C6/36 *Aedes albopictus* cells and the identification of subgenomic L RNA suggested the presence of DI particles. By contrast, Rossier *et al.* (1988) failed to detect subgenomic RNAs in La Crosse virus-infected C6/36 cells and proposed that the benign nature of the interaction was a consequence of translational control mediated by the N protein. This hypothesis was supported by Hacker *et al.* (1990) who found that encapsidation of S RNA by N protein prevented its translation. The absence of subgenomic RNA in the inoculum virus (Rossier *et al.*, 1988; Hacker *et al.*, 1990) indicates that DI particles of the classic type were not responsible for establishing these persistent infections. Scallan and Elliott (1992) reinvestigated the course of persistent infection of C6/36 cells by Bunyamwera virus and confirmed that there was an overrepresentation of S RNA in C6/36 cells compared to BHK-21 cells. They found no evidence of DI particle-like RNA in the early phase of propagation of the infected cells, but after prolonged propagation of the persistently infected cultures several subgenomic L RNAs were detected. These defective RNAs were not packaged into virions and there was no direct relationship between the presence of these particles and virus titer. Nor did the amount of defective RNA present correlate with resistance to superinfection with homologous virus. Consequently, the role of DI particles in the maintenance of persistence in arthropod cells is dubious. A preliminary analysis of the viral RNAs present in individual clones of cells from C6/36 cells by Scallan and Elliott (1992) revealed great heterogeneity and they suggest that cellular factors may play an important role in the maintenance of persistence.

Deletion mutations located in the M and S RNA subunits have not been described. Non-ts revertants of the Maguari virus group II mutant tsMAG8, however, exhibited gross shifts in the electrophoretic mobility of the G1 protein (Murphy and Pringle, 1987). These revertants grew normally and they have not been investigated to determine whether sequences have been deleted. The accelerated mobility phenotype of the G1 protein was considered tentatively to be the result of exposure of a cryptic proteolytic cleavage site (Hampson, 1987).

B. Determinants of Neurovirulence and Attenuation

The biology of the family *Bunyaviridae* is dominated by the M RNA since this subunit of the genome encodes the genes concerned in many of the most important interactions of the virus with its hosts. Virulence, host range, tissue tropism, transmissibility, neutralization, hemagglutination, and membrane fusion are the principal phenotypic properties that have been attributed to the M RNA subunit gene products. Characterization of heterologous reassortants has played a leading role in recognition of the predominance of the M RNA in determining the disease-producing potential of bunyaviruses.

1. Virus Virulence in Vertebrates

The virulence of bunyaviruses often diminishes rapidly during serial passage in cultured cells. For example, field isolates of Tahyna virus were generally virulent in suckling mice, but propagation in cultured cells resulted in the rapid appearance of strains with differing *in vivo* virulence and *in vitro* cytopathogenicity. Tahyna virus passaged serially in cultured mosquito cells at 20°C replicated less well in vertebrate cells at 37°C and exhibited reduced virulence for mice by any route of inoculation (Malkova and Marhoul, 1976). Conversely, a laboratory strain remained neurovirulent but exhibited reduced peripheral virulence as a consequence of serial passage in mouse brain (Janssen *et al.*, 1984). Clearly, bunyaviruses can adapt rapidly to changed environmental circumstances without the necessity for reassort-ment of preexisting variation.

Experiments with California serogroup viruses in laboratory animals have shown that pathogenicity is determined by the M RNA (Shope *et al.*, 1981). Viruses of differing virulence for mice were crossed and disease-producing ability segregated according to the parental origin of the M RNA subunit in the progeny virus. Tahyna virus was avirulent by intracerebral or subcutaneous inoculation into 4-week-old mice, whereas La Crosse and snowshoe hare viruses and reassortants carrying LAC or SSH M RNA were virulent. Exceptionally, LAC/LAC/LAC and LAC/LAC/SSH reassortants obtained from a cross of tsLAC16(I) and a group II ts mutant of genotype SSH/LAC/SSH were significantly less virulent following intracerebral inoculation into mice. Analysis of further reassortants revealed that the L RNA of mutant tsLAC16(I) carried an additional mutation that could modify virulence (Rozhon *et al.*, 1981).

These findings were verified independently by Janssen *et al.* (1986) who further showed that the S RNA as well as the L RNA could modify virulence. Reassortants derived from the cross of an avirulent strain of Tahyna virus and a virulent La Crosse virus confirmed that in the California serogroup viruses the M RNA is the major determinant of peripheral virulence. Peripheral virulence mimics the natural route of transmission from insect vector to vertebrate host and is defined as susceptibility to infection by subcutaneous inoculation. It was apparent also, however, that the origin of the L and S RNAs influenced the level of peripheral virulence displayed by these reassortants. This is in accord with the general expectation that a complex phenotype such as pathogenicity is likely to have polygenic determinants. In fact, Gonzalez-Scarano *et al.* (1988) were able to partition the virulence phenotype into four independently varying components (neuroinvasiveness and neurovirulence in mice, and oral transmissibility and intrathoracic susceptibility in mosquitoes), establishing the essentially polygenic nature of the virulence phenotype (see Chapter 9). Griot *et al.* (1993) examined the genetic determination of extraneural replication of California serogroup viruses using two attenuated reassortants with attenuating mutations in the L and M RNA subunits, respectively. An inoculum of 1000 pfu of these attenuated viruses failed to replicate peripherally and did not reach the brain. Consequently, no visible disease resulted. An inoculum of 10^6 pfu, on the other hand, produced a transient viremia sufficient to enable virus to reach the brain and induce fatal disease. Reassortants generated from crosses of these viruses showed that the L and M RNA determinants acted independently and additively, since reassortants carrying both attenuating determinants were more attenuated than the parental viruses. Conversely, reassortants carrying neither determinant were more invasive than either parent.

Attenuation of alphaviruses has been achieved by selection for rapid penetration of BHK-21 cells (Olmsted *et al.*, 1984). Following this precedent, Endres *et al.* (1989) succeeded in isolating a novel neuroattenuated variant by serial passage of a TAH/LAC/LAC reassortant in BHK-21 cells. The neuro-attenuated phenotype (failure of intracerebral inoculation to produce disease in adult mice) of this variant is unique among California serogroup viruses. This variant also exhibited decreased neuroinvasiveness following subcutaneous inoculation of suckling mice, replicated less well in neuroblastoma cells than neurovirulent viruses, and was temperature-sensitive in BHK-21 cells at 38.9°C. Non-ts revertants simultaneously regained neurovirulence and ability to replicate efficiently in neuroblastoma cells. Thus, neuroblastoma cells provide a promising *in vitro* system for analysis of the phenomenon of neurovirulence in mice.

In another study of mouse virulence, Bishop *et al.* (1984) isolated reassortants from a cross of two parental strains of Caraparu virus (serogroup C) with different disease-producing properties and concluded that M RNA subunit gene products determined virulence in this serogroup also.

2. Mechanisms of Transmission in Invertebrates

Homologous reassortment has been demonstrated in mosquitoes dually infected with group I and group II ts mutants of La Crosse virus, either by simultaneous or by interrupted feeding, and by intrathoracic injection (Beaty and Bishop, 1989). Likewise, heterologous reassortants have been recovered from mosquitoes infected with different viruses of the California serogroup (Beaty *et al.*, 1981). Putative natural reassortants have been recovered also from mosquitoes in the wild. Transmission of bunyaviruses from an infected mosquito to a vertebrate host becomes possible only after virus reaches the salivary glands, a process that requires 7–14 days from ingestion of the virus. Characterization of heterologous reassortants has shown that dissemination of virus from the midgut to other tissues is under the control of the M RNA (Beaty *et al.*, 1982). Separate domains of the M RNA determine ability to overcome the midgut barrier on the one hand, and ability to multiply in thoracic organs on the other (Gonzalez-Scarano *et al.*, 1988). Since the M RNA encodes the viral membrane glycoproteins, it is likely that these tissue tropisms in the mosquito are receptor-mediated. The midgut also imposes a barrier to the dissemination of TSWV in adult thrips (Ullman *et al.*, 1992). The vector must ingest the virus in the larval stage to enable the virus to reach the salivary glands.

3. Vaccine Development

Experimental inactivated and live virus vaccines have been developed to protect humans and animals from the ravages of Rift Valley fever virus (genus *Phlebovirus*). The live virus vaccines have been developed by empirical routes (Smithburn, 1949; Moussa *et al.*, 1982) and by random mutagenization (Caplen *et al.*, 1985; Rossi and Turell, 1988). The latter involved propagation of a wild-type virus in BHK-21 cells in the continuous presence of 200 μg/ml

5-fluorouracil. Single plaques were picked after each cycle of multiplication
to initiate the next infection. Virus passaged by this regimen became rapidly
attenuated for mice, whereas virus passaged in the absence of the mutagen
did not. Retrospective reassortment experiments suggested that attenuating
mutations had accumulated in all segments. Takehara *et al.* (1989) sequenced
the M RNA subunits of wild-type parent and attenuated derivative and found
that 12 nucleotide substitutions had occurred resulting in seven amino acid
changes and creation of a new upstream AUG start site. The identity of the
critical attenuating mutations could not be determined, however, because
the virulent parental strain differed by 7 nucleotide substitutions and three
coding changes from another virulent strain isolated simultaneously during
the same epidemic, and because the L and S RNAs were not sequenced.

C. Genetic Determination of Host Resistance in Plants

Host resistance to tospovirus infection has been studied extensively in
the interests of crop protection. Classical genetic approaches have been
employed to identify sources of resistant germ plasma for use in plant breeding. The genetic determination of resistance to TSWV appears to be highly
specific and genetically complex. Boiteux *et al.* (1993) evaluated 70 species
and cultivars of *Capiscum* and observed different levels of resistance under
field conditions of infection. As a result of controlled greenhouse trials, a
complex pattern of resistance was observed and it was concluded that several
genetic mechanisms were involved in TSWV resistance. For example, two
lines of *C. chinense.* which were virtually immune to one isolate of TSWV,
were susceptible to another isolate of the same virus from a different region
of Brazil. Kumar and Irulappan (1992) compared the resistance of tomato
plants (*Lycopersicom peruvianum* var. *humifusum, L. hirsutum* f. *glabra-
tum, L. hirsutum*) to TSWV under field and greenhouse conditions. Fifteen
crosses involving five susceptible parents and three wild species revealed
that resistance was controlled by a small number of recessive genes. On the
other hand, Stevens *et al.* (1991) identified a single dominant gene in the
"Stevens" cultivar of *L. peruvianum* which provided solid resistance to
several isolates of TSWV. Similarly, the combined resistance to both TSWV
and *Thielaviopsis basicola* of the Polalta cultivar, derived from wild *Nicotiana alata,* was inherited as a dominant characteristic that could be introduced into several susceptible Oriental tobacco cultivars (Yancheva, 1990).

The production of transgenic plants is an alternative approach to producing host resistance which has potential for crop protection. Transgenic "Sam-
sun" cultivars of tobacco expressing the complete NP gene of TSWV have
been produced which were resistant to superinfection with TSWV (Mackenzie and Ellis, 1992). These experiments with a bunyavirus confirmed that
protection of host plants could be obtained by expression of homologous coat
protein genes of membrane-enveloped, multicomponent, negative-stranded

viruses, as well as isometric positive-stranded viruses. Pang *et al.* (1992) investigated the cross-resistance spectrum of tobacco plants transgenic for the NP gene of a strain of TSWV belonging to the L serogroup. Resistance to the homologous TSWV strain was almost universal in the transgenic plants and a fairly broad spectrum of resistance to heterologous strains of the L serogroup was observed even where NP gene expression was low. Plants that expressed high levels of NP were resistant also to an isolate belonging to serogroup I. However, all of the transgenic plants remained suceptible to a Brazilian strain of TSWV belonging to another serogroup. These results indicate that a fairly broad spectrum of resistance can be obtained by expression of the coat protein gene of TSWV in a susceptible host plant, but there are limits to the extent of cross-protection obtainable by this means. Subsequently, it was shown that protection against the homologous isolate and closely related isolates in plants expressing low levels of the N gene was related to the presence of N gene RNA, whereas protection against the homologous isolate and isolates of the distantly related impatiens necrotic spot virus in plants expressing high levels of the N gene was related to accumulation of N protein (Pang *et al.*, 1994). Thus, different mechanisms appear to be involved in protection against infection by tospoviruses that share different levels of N gene sequence identity.

VII. PROSPECTS

Future progress in the study of the genetics of bunyaviruses will depend greatly on exploitation of the reverse genetic approach whereby genes are manipulated in the form of cDNA and subsequently reintroduced into infectious viral genomes. The problems inherent in the application of reverse genetics to viruses with negative-stranded RNA genomes have been solved for the orthomyxoviruses (Enami *et al.*, 1990). Although this methodology is not directly transferable from one negative-strand virus group to another, the feasibility of the reverse genetic approach with negative-stranded RNA viruses has been established. The technical problems specific to the bunyaviruses are close to solution (Dunn *et al.*, 1995), and we are now on the threshold of a new phase in the investigation of genetic and evolutionary processes.

VIII. REFERENCES

Battles, J. K., and Dalrymple, J. M., 1988, Genetic variation among geographic isolates of Rift Valley fever virus, *Am. J. Trop. Med. Hyg.* **39**:617.
Beaty, B. J., and Bishop, D. H. L., 1989, *Bunyavirus*–vector interaction, *Virus Res.* **10**:289.
Beaty, B. J., and Calisher, C. H., 1991, *Bunyaviridae*—Natural history, *Curr. Top. Microbiol. Immunol.* **169**:27.
Beaty, B. J., Rozhon, E. J., Gensemer, P., and Bishop, D. H. L., 1981, Formation of reassortant bunyaviruses in dually infected mosquitoes, *Virology* **111**:662.

Beaty, B. J., Miller, B. R., Shope, R. E., Rozhon, E. J., and Bishop, D. H. L., 1982, Molecular basis of bunyavirus *per os* infection of mosquitoes: Role of the middle-sized RNA segment, *Proc. Natl. Acad. Sci. USA* **79**:1295.

Beaty, B. J., Sundin, D. R., Chandler, L. J., and Bishop, D. H. L., 1985, Evolution of bunyaviruses by genome reassortment in dually infected mosquitoes (*Aedes triseriatus*), *Science* **230**:548.

Bergmann, M., Garcia-Sastre, A., and Palese, P., 1992, Transfection-mediated recombination of influenza A virus, *J. Virol.* **66**:7576.

Bilsel, P. A., Tesh, R. B., and Nichol., S. T., 1988, RNA genome stability during serial transovarial transmission in the sandfly *Phlebotomus perniciosus, Virus Res.* **11**:87.

Bishop, D. H. L., 1979, Genetic potential of bunyaviruses, *Curr. Top. Microbiol. Immunol.* **86**:1.

Bishop, D. H. L., 1986, Ambisense RNA genomes of arenaviruses and phleboviruses, *Adv. Virus Res.* **31**:1.

Bishop, D. H. L., and Auperin, D. D., 1987, Arenavirus gene structure and organisation, *Curr. Top. Microbiol. Immunol.* **133**:5.

Bishop, D. H. L., and Shope, R. E., 1979, *Bunyaviridae,* in: *Comprehensive Virology* (H. Fraenkel-Conrat and R. R. Wagner, eds.), Vol. 14, pp. 1–156, Plenum Press, New York.

Bishop, D. H. L., Fuller, F., Akashi, H., Beaty, B. J., and Shope, R. E., 1984, The use of reassortant bunyaviruses to deduce their coding and pathogenic potentials, in: *Mechanisms of Viral Pathogenesis: From Gene to Pathogen* (A. Kohn and P. Fuchs, eds.), pp. 49–60, Nijhoff, The Hague.

Boiteux, L. S., Nagata, T., Dutra, W. P., and Fonseca, M. E. N., 1993, Sources of resistance to tomato spotted wilt virus (TSWV) in cultivated and wild species of *Capsicum, Euphytica* **67(1–2)**:89.

Brown, S. E., Morrison, H. G., and Knudson, D. L., 1989a, Genetic relatedness of the Kemerova serogroup viruses: II. RNA–RNA blot hybridization and gene reassortment *in vitro* of the Great Island serocomplex, *Acta Virol.* **33(3)**:206.

Brown, S. E., Morrison, H. G., and Knudson, D. L., 1989b, Genetic relatedness of the Kemerovo serogroup viruses: III. RNA–RNA blot hybridization and gene reassortment *in vitro* of the Chenuda serocomplex, *Acta Virol.* **33(3)**:221.

Brown, S. E., Morrison, H. G., Karabatsos, N., and Knudson, D. L., 1991, Genetic relatedness of two new orbivirus serogroups: Orungo and Lebombo, *J. Gen. Virol.* **72**:1065.

Browning, G. F., Snodgrass, D. R., Nakagomi, O., Kaga, E., Sarasini, A., and Gerna, G., 1992, Human and bovine serotype G8 rotaviruses may be derived by reassortment, *Arch. Virol.* **125**:121.

Campbell, S., 1992, Preparation and characterisation of monoclonal antibodies to Ilesha virus, M. Med. Sci. thesis, Queen's University of Belfast.

Caplen, H., Peters, C. J., and Bishop, D. H. L., 1985, Mutagen-directed attenuation of Rift Valley fever virus as a method for vaccine development, *J. Gen. Virol.* **66**:2271.

Castrucci, M. R., Donatelli, I., Sidoli, L., Barigazzi, G., Kawaoka, Y., and Webster, R. G., 1993, Genetic reassortment between avian and human influenza A viruses in Italian pigs, *Virology* **193**:503.

Chandler, L. J., Beaty, B. J., Baldridge, G. D., Bishop, D. H. L., and Hewlett, M. J., 1990, Heterologous reassortment of bunyaviruses in *Aedes triseriatus* mosquitoes and transovarial and oral transmission of newly evolved genotypes, *J. Gen. Virol.* **71**:1045.

Chandler, L. J., Hodge, G., Endres, M., Jacoby, D. R., Nathanson, N., and Beaty, B. J., 1991, Reassortment of La Crosse and Tahyna bunyaviruses in *Aedes triseriatus* mosquitoes, *Virus Res.* **20**:181.

Chao, L., 1994, Evolution of genetic exchange in RNA viruses, in: *The Evolutionary Biology of Viruses* (S. S. Morse, ed.), pp. 233–250, Raven Press, New York.

Chen, D., Burns, J. W., Estes, M. K., and Ramig, R. F., 1989, Phenotypes of rotavirus reassortants depend upon the recipient genetic background, *Proc. Natl. Acad. Sci. USA* **86**:3743.

Cunningham, C., and Szilagyi, J. F., 1987, Viral RNAs synthesized in cells infected with Germiston bunyavirus, *Virology* **157**:431.

David-West, T. S., and Porterfield, J. S., 1974, Dugbe virus: A tick-borne arbovirus from Nigeria, *J. Gen. Virol.* **23**:297.

De Mattos, C. C. P., De Mattos, C. A., Osburn, B. I., Ianconescu, M., and Kaufman, R., 1991, Evidence of genome segment 5 reassortment in blue tongue virus field isolates, *Am. J. Vet. Res.* **52**:1794.

Dunn, E. F., Pritlove, D. C., and Elliott, R. M., 1994a, The S RNA genome segments of Batai, Cache Valley, Guaroa, Kairi, Lumbo, Main Drain and Northway bunyaviruses: Sequence determination and analysis, *J. Gen. Virol.* **75**:597.

Dunn, E. F., Pritlove, D. C., Jin, H., and Elliott, R. M., 1994b, Transcription of a recombinant bunyavirus RNA by transiently expressed Bunyamwera virus proteins, *Virology* **211**:135–143.

El-Hussein, A., Ramig, R. F., Holbrook, F. R., and Beaty, B. J., 1989, Asynchronous mixed infection of *Culicoides variipennis* with blue tongue virus serotypes 10 and 17, *J. Gen. Virol.* **70**:3355.

Elliott, R. M., 1990, Molecular biology of the *Bunyaviridae*, *J. Gen. Virol.* **71**:501.

Elliott, R. M., and Wilkie, M. L., 1986, Persistent infection of *Aedes albopictus* C6/36 cells by Bunyamwera virus, *Virology* **150**:21.

Elliott, R. M., Lees, J. F., Watret, G. E., Clark, W., and Pringle, C. R., 1984, Genetic diversity of bunyaviruses and mechanisms of genetic variation, in: *Mechanisms of Viral Pathogenesis: From Gene to Pathogen* (A. Kohn and P. Fuchs, eds.), pp. 61–76, Nijhoff, The Hague.

El-Said, L. H., Vorndam, V., Gentsch, J. R., Clewley, J. P., Calisher, C. H., Klimas, R. A., Thompson, W. H., Grayson, M., Trent, D. W., and Bishop, D. H. L., 1979, A comparison of La Crosse virus isolates obtained from different ecological niches and an analysis of the structural components of California encephalitis serogroup viruses and other bunyaviruses, *Am. J. Trop. Med. Hyg.* **28**:364.

Enami, M., Luytjes, W., Krystal, M., and Palese, P., 1990, Introduction of site-specific mutations into the genome of influenza virus, *Proc. Natl. Acad. Sci. USA* **87**:3902.

Endres, M. J., Jacoby, D. R., Janssen, R. S., Gonzalez-Scarano, F., and Nathanson, N., 1989, The large viral RNA segment of California serogroup bunyaviruses encodes the large viral protein, *J. Gen. Virol.* **70**:223.

Endres, M. J., Griot, C., Gonzalez-Scarano, F., and Nathanson, N., 1991, Neuroattenuation of an avirulent bunyavirus maps to the L RNA segment, *J. Virol.* **65**:5465.

Gahmberg, N., 1984, Characterization of two recombination-complementation groups of Uukuniemi virus temperature-sensitive mutants, *J. Gen. Virol.* **65**:1079.

Gammelin, M., Mandler, J., and Scholtissek, C., 1989, Two subtypes of nucleoproteins (NP) of influenza A viruses, *Virology* **170**:71.

Gentsch, J. R., and Bishop, D. H. L., 1976, Recombination and complementation between temperature-sensitive mutants of the bunyavirus snowshoe hare virus, *J. Virol.* **20**:351.

Gentsch, J. R., Wynne, L. R., Clewley, J. P., Shope, R. E., and Bishop, D. H. L., 1977, Formation of recombinants between snowshoe hare and La Crosse bunyaviruses, *J. Virol.* **24**:893.

Gentsch, J. R., Robeson, G., and Bishop, D. H. L., 1979, Recombination between snowshoe hare and La Crosse bunyaviruses, *J. Virol.* **31**:707.

Gentsch, J. R., Rozhon, E. J., Klimas, R. A., El-Said, L. H., Shope, R. E., and Bishop, D. H. L., 1980, Evidence from recombinant bunyavirus studies that the M RNA gene products elicit neutralising antibodies, *Virology* **102**:190.

Goldbach, R., and de Haan, P., 1994, RNA viral supergroups and the evolution of RNA viruses, in: *The Evolutionary Biology of Viruses* (S. S. Morse, ed.), pp. 105–120, Raven Press, New York.

Gombold, J. L., and Ramig, R. F., 1986, Analysis of reassortment of genome segments in mice mixedly infected with rotaviruses SA11 and RRV, *J. Virol.* **57**:110.

Gombold, J. L., and Ramig, R. F., 1989, Passive immunity modulates genetic reassortment between rotaviruses in mixedly infected mice, *J. Virol.* **63**:4525.

Gonzalez-Scarano, F., Beaty, B., Sundin, D., Janssen, R., Endres, M. J., and Nathanson, N., 1988, Genetic determinants of the virulence and infectivity of La Crosse virus, *Microbiol. Pathogen* **4**:1.

Graham, A., Kudesia, G., Allen, A. M., and Desselberger, U., 1987, Reassortment of human rotavirus possessing genome rearrangements with bovine rotavirus: Evidence for host cell selection, *J. Gen. Virol.* **68:**115.

Griot, C., Pekosz, A., Lukac, D., Scherer, S. S., Stillmock, K., Schmeidler, D., Endres, M. J., Gonzalez-Scarano, F., and Nathanson, N., 1993, Polygenic control of neuroinvasiveness in California serogroup bunyaviruses, *J. Virol.* **67:**3861.

Guo, Y., Wang, M., Kawaoka, Y., Gorman, O., Ito, T., Saito, T., and Webster, R. G., 1992, Characterization of a new avian-like influenza A virus from horses in China, *Virology* **188:**245.

Hacker, D., Rochat, S., and Kolakofsky, D., 1990, Anti mRNAs in La Crosse bunyavirus-infected cells, *J. Virol.* **64:**5051.

Hampson, J., 1987, A study of the genetics and growth of a tripartite RNA virus using ts mutants, Ph.D. thesis, University of Warwick.

Hewlett, M. J., Clerx, J. P. M., Clerx-van Haaster, C. M., Chandler, L. J., McLean, D. M., and Beaty, B. J., 1992, Genomic and biologic analyses of snowshoe hare virus field and laboratory strains, *Am. J. Trop. Med. Hyg.* **46:**524.

Hjelle, B., Jenison, S., Torrez-Martinez, N., Yamada, T., Nolte, K., Zumwalt, R., Macinnes, K., and Myers, G., 1994, A novel hantavirus associated with an outbreak of fatal respiratory disease in the southwestern United States: Evolutionary relationships to known hantaviruses, *J. Virol.* **68:**592.

Hunt, A. R., and Calisher, C. H., 1979, Relationships of Bunyamwera serogroup viruses by neutralisation, *Am. J. Trop. Med. Hyg.* **28:**740.

Iroegbu, C. U., 1981, A genetic study of the Bunyamwera complex of the genus *Bunyavirus* (family *Bunyaviridae*), Ph.D. thesis, University of Glasgow.

Iroegbu, C. U., and Pringle, C. R., 1981a, Genetic interactions among viruses of the Bunyamwera complex, *J. Virol.* **37:**383.

Iroegbu, C. U., and Pringle, C. R., 1981b, Genetics of the Bunyamwera complex, in: *The Replication of Negative Strand RNA Viruses* (D. H. L. Bishop and R. W. Compans, eds.), pp. 159–163, Elsevier, New York.

Isegawa, Y., Nakagomi, O., Nakagomi, T., and Ueda, S., 1992, A VP4 sequence highly conserved in human rotavirus strain AU-1 and feline rotavirus strain FRV-1, *J. Gen. Virol.* **73:**1939.

Janssen, R., Gonzalez-Scarano, F., and Nathanson, N., 1984, Mechanisms of bunyavirus virulence. Comparative pathogenesis of a virulent strain of La Crosse and an avirulent strain of Tahyna virus, *Lab. Invest.* **50:**447.

Janssen, R. S., Nathanson, N., Endres, M. J., and Gonzalez-Scarano, F., 1986, Virulence of La Crosse virus is under polygenic control, *J. Virol.* **59:**1.

Kacsak, R. J., and Lyons, M. J., 1978, Bunyamwera virus: II The generation and nature of defective interfering particles, *Virology* **98:**539.

Karabatsos, N., 1985, *International Catalogue of Arboviruses 1985, Including Certain Other Viruses of Vertebrates* (N. Karabatsos, ed.), The American Society of Tropical Medicine and Hygiene, San Antonio.

Khatchikian, D., Orlich, M., and Rott, R., 1989, Increased viral pathogenicity after insertion of a 28S ribosomal RNA sequence into the haemagglutinin gene of influenza virus, *Nature* **340:**156.

Klimas, R. A., Thompson, W. H., Calisher, C. H., Clark, G. G., Grimstad, P. R., and Bishop, D. H. L., 1981, Genotypic varieties of La Crosse virus isolated from different geographic regions of the continental United States and evidence for a naturally occurring intertypic recombinant La Crosse virus, *Am. J. Epidemiol.* **114:**112.

Kormelink, R., Storms, M., Van Lent, J., Peters, D., and Goldbach, R., 1994, Expression and subcellular location of the NSm protein of tomato spotted wilt virus (TSWV), a putative viral movement protein, *Virology* **200:**56.

Kowalik, T. F., and Li, J. K. K., 1991, Bluetongue virus evolution: Sequence analyses of the genomic S1 segments and major core protein VP7, *Virology* **181:**749.

Kramer, L. D., Hardy, J. L., Reeves, W. C., Presser, S. B., Bowen, M. D., and Eldridge, B. F., 1993, Vector competence of selected mosquito species (*Diptera: Culicidae*) for California strains of Northway virus (*Bunyaviridae: Bunyavirus*), *J. Med. Entomol.* **30:**607.

Kumar, N., and Irulappan, I., 1992, Inheritance of resistance to spotted wilt virus in tomato (*Lycopersicon esculentum* Mill.), *J. Genet. Breeding* **46(2):**113.

Labuda, M., 1991, Arthropod vectors in the evolution of bunyaviruses, *Acta Virol.* **35(1):**98.

Ling, R., Easton, A. J., and Pringle, C. R., 1992, Sequence analysis of the 22K, SH, and G genes of turkey rhinotracheitis virus and their intergenic regions reveals a gene order different to that of other pneumoviruses, *J. Gen. Virol.* **73:**1709–1715.

Lubeck, M. D., Palese, P., and Schulman, J. L., 1979, Nonrandom association of parental genes in influenza A virus recombinants, *Virology* **95:**269.

Mackenzie, D. J., and Ellis, P. J., 1992, Resistance to tomato spotted wilt virus infection in transgenic tobacco expressing the viral nucleocapsid gene, *Mol. Plant-Microbe Interact.* **5:**34.

Malkova, D., and Marhoul, Z., 1976, Influence of temperature corresponding to that of the vector of Tahyna virus, *Acta Virol.* **20:**494.

Mascarenhas, J. D. P., Linhares, A. C., Gabbay, Y. B., De-Freitas, R. B., Mendez, E., Lopez, S., and Arias, C. F., 1989, Naturally occurring serotype 2/subgroup II rotavirus reassortants in northern Brazil, *Virus Res.* **14:**235.

Millican, D., and Porterfield, J. S., 1982, Relationship between glycoproteins of the viral envelope of bunyaviruses and antibody-dependent plaque enhancement, *J. Gen. Virol.* **63:**233.

Moss, S. R., Ayres, C. M., and Nuttall, P. A., 1988, The Great Island serogroup of tick-borne orbiviruses represent a single gene pool, *J. Gen. Virol.* **69:**2721.

Moussa, M. I., Wood, O. L., and Abdel-Wahab, K. S. E., 1982, Reduced pathogenicity associated with a small plaque variant of the Egyptian strain of Rift Valley fever virus (ZH501), *Trans. R. Soc. Trop. Med. Hyg.* **76:**482.

Murphy, J., and Pringle, C. R., 1987, Bunyavirus mutants: Reassortment group assignment and G1 protein variants, in: *The Biology of Negative Strand Viruses* (B. W. J. Mahy and D. Kolakofsky, eds.), pp. 357–362, Elsevier, Amsterdam.

Murphy, F. A., Fauquet, C. M., Bishop, D. H. L., Ghabrial, S. A., Jarvis, A. W., Martelli, G. P., Mayo, M. A., and Summers, M. D., 1995, *Virus Taxonomy: Classification and Nomenclature of Viruses*, pp 1–586, Springer Verlag, New York.

Nakagomi, O., Kaga, E., and Nakagomi, T., 1992a, Human rotavirus strain with unique VP4 neutralisation epitopes as a result of natural reassortment between members of the AU-1 and Wa genogroups, *Arch. Virol.* **127:**365.

Nakagomi, O., Kaga, E., Gerna, G., Sarasini, A., and Nakagomi, T., 1992b, Subgroup I serogroup 3 human rotavirus strains with long RNA pattern as a result of naturally occurring reassortment between members of the bovine and Au-1 genogroups, *Arch. Virol.* **126:**337.

Nayak, D. P., Chambers, T. M., and Akkina, R. K., 1985, Defective interfering RNAs of influenza viruses: Origin, structure, expression and interference, *Curr. Top. Microbiol. Immunol.* **114:**104.

Newton, S. E., Short, N. J., and Dalgarno, L., 1981, Bunyamwera virus replication in cultured *Aedes albopictus* (mosquito) cells: Establishment of a persistent infection, *J. Virol.* **38:**1015.

O'Hara, P. J., Nichol, S. T., Horodyski, F. M., and Holland, J. J., 1984, Vesicular stomatitis virus defective interfering particles can contain extensive genomic sequence rearrangements and base substitutions, *Cell* **36:**915.

Olmsted, R. A., Baric, R. S., Sawyer, R. A., and Johnston, R. E., 1984, Sindbis virus mutants selected for rapid growth in cell culture display attenuated virulence in animals, *Science* **225:**424.

Ozden, S., and Hannoun, C., 1978, Isolation and preliminary characterization of temperature-sensitive mutants of Lumbo virus, *Virology* **84:**210.

Ozden, S., and Hannoun, C., 1980, Biochemical and genetic characterization of Germiston virus, *Virology* **103:**232.

Pang, S. Z., Nagpala, P., Wang, M., Slightom, J. L., and Gonsalves, D., 1992, Resistance to heterologous isolates of tomato spotted wilt virus in transgenic tobacco expressing its nucleocapsid gene, *Phytopathology* **82:**1223.

Pang, S. Z., Bock, J. H., Gonsalves, C., Slightom, J. L., and Gonsalves, D., 1994, Resistance of transgenic *Nicotiana benthamiana* plants to tomato spotted wilt and impatiens necrotic spot tospoviruses: Evidence of involvement of the N protein and N gene RNA in resistance, *Phytopathology* **84:**243.

Patel, A. H., and Elliott, R. M., 1992, Characterization of Bunyamwera virus defective interfering particles, *J. Gen. Virol.* **73:**389.

Peeples, M. E., 1991, Paramyxovirus M proteins: Pulling it all together and taking it on the road, in: *The Paramyxoviruses* (D. W. Kingsbury, ed.), pp. 427–456, Plenum Press, New York.

Pennington, T. H., Pringle, C. R., and McCrae, M. A., 1977, Bunyamwera virus induced polypeptide synthesis, *J. Virol.* **24:**387.

Pringle, C. R., 1977, Enucleation as a technique in the study of virus–host interactions, *Curr. Top. Microbiol. Immunol.* **76:**49.

Pringle, C. R., 1987, Rhabdovirus genetics, in: *The Rhabdoviruses* (R. R. Wagner, ed.), pp. 167–243, Plenum Press, New York.

Pringle, C. R., 1990, The genetics of viruses, in: *Topley and Wilson's Principles of Bacteriology, Virology and Immunity*, 8th ed., Vol. 4 (L. H. Collier and M. C. Timbury, eds.), pp. 69–104, Arnold, London.

Pringle, C. R., 1991a, The order *Mononegavirales, Arch. Virol.* **117:**137.

Pringle, C. R., 1991b, The genetics of Paramyxoviruses, in: *The Paramyxoviruses* (D. W. Kingsbury, ed.), pp. 1–40, Plenum Press, New York.

Pringle, C. R., 1991c, The *Bunyaviridae* and their genetics—An overview, *Curr. Top. Microbiol. Immunol.* **169:**1.

Pringle, C. R., and Iroegbu, C. U., 1982, A mutant identifying a third recombination group in a bunyavirus, *J. Virol.* **42:**873.

Pringle, C. R., Clark, W., Lees, J. F., and Elliott, R. M., 1984a, Restriction of sub-unit reassortment in the *Bunyaviridae*, in: *Segmented Negative Strand Viruses* (R. W. Compans and D. H. L. Bishop, eds.), pp. 45–50, Academic Press, New York.

Pringle, C. R., Lees, J. F., Clark, W., and Elliott, R. M., 1984b, Genome subunit reassortment among bunyaviruses analysed by dot hybridization using molecularly cloned complementary DNA probes, *Virology* **135:**244.

Ramig, R. F., 1990, Superinfecting rotaviruses are not excluded from genetic interactions during asynchronous mixed infections *in vitro, Virology* **176:**308.

Ramig, R. F., and Gombold, J. L., 1991, Rotavirus temperature-sensitive mutants are genetically stable and participate in reassortment during mixed infection of mice, *Virology* **182:**468.

Ramig, R. F., and Ward, R. L., 1991, Genomic segment reassortment in rotaviruses and other *Reoviridae, Adv. Virus Res.* **38:**163.

Ramig, R. F., Garrison C., Chen, D., and Bell-Robinson, D., 1989, Analysis of reassortment and superinfection during mixed infection of Vero cells with blue tongue virus serotypes 10 and 17, *J. Gen. Virol.* **70:**2595.

Ramirez, B.-C., and Haenni, A. L., 1994, Molecular biology of tenuiviruses, a remarkable group of plant viruses, *J. Gen. Virol.* **75:**467.

Reanney, D., 1984, The molecular evolution of viruses, in: *The Microbe 1984* (B. W. J. Mahy and J. R. Pattison, eds.), SGM Symposium 36, Part 1, pp. 175–196, Cambridge University Press, London.

Resende, R. de O., de Haan, P., Van de Vossen, E., de Àvila, A. C., Goldbach, R., and Peter, D., 1992, Defective interfering L RNA segments of tomato spotted wilt virus retain both virus genome termini and have extensive internal deletions, *J. Gen. Virol.* **73:**2509.

Rossi, C. A., and Turell, M. J., 1988, Characterization of attenuated strains of Rift Valley fever virus, *J. Gen. Virol.* **98:**817.

Rossier, C., Raju, R., and Kolakofsky, D., 1988, La Crosse virus gene expression in mammalian and mosquito cells, *Virology* **165:**539.

Rozhon, E. J., Gensemer, P., Shope, R. E., and Bishop, D. H. L., 1981, Attenuation of virulence of a bunyavirus involving an L RNA defect and isolation of LAC/SSH/LAC and LAC/SSH/SSH reassortants, *Virology* **111:**125.

Scallan, M. F., and Elliott, R. M., 1992, Defective RNAs in mosquito cells persistently infected with Bunyamwera virus, *J. Gen. Virol.* **73:**53.

Schopen, S., Labuda, M., and Beaty, B. J., 1991, Vertical and venereal transmission of California group viruses by *Aedes triseriatus* and *Culiseta inornata* mosquitoes, *Acta Virol.* **35(4):**373.

Shope, R. E., Rozhon, E. J., and Bishop, D. H. L., 1981, Role of the bunyavirus middle-sized RNA segment in mouse virulence, *Virology* **114:**273.

Smithburn, K. C., 1949, Rift Valley fever: The neurotropic adaptation of the virus and the experimental use of this modified virus as a vaccine, *Br. J. Exp. Pathol.* **30**:1.

Sokal, R. R., and Rohlf, F. J., 1981, *Biometry*, pp. 747–765, Freeman, New York.

Spiropoulou, C. F., Morzunov, S., Feldmann, H., Sanchez, A., Peters, C. J., and Nichol, S. T., 1994, Genome structure and variability of a virus causing Hantavirus pulmonary syndrome, *Virology* **200**:715.

Stevens, M. R., Scott, S. J., and Gergerich, R. C., 1991, Inheritance of a gene for resistance to tomato spotted wilt virus (TSWV) from *Lycopersicon peruvianum* Mill., *Euphytica* **59(1)**:9.

Stott, J. L., Oberst, R. D., Channel, M. B., and Osburn, B. I., 1987, Genome reassortment between two serotypes of blue tongue virus in a natural host, *J. Virol.* **61**:2670.

Struthers, J. K., and Swanepoel, R., 1982, Identification of a major non-structural protein in the nuclei of Rift Valley fever virus-infected cells, *J. Gen. Virol.* **60**:381.

Swanepoel, R., and Blackburn, N. K., 1977, Demonstration of nuclear immunofluorescence in Rift Valley fever virus infected cells, *J. Gen. Virol.* **34**:557.

Takehara, K., Min, M. K., Battles, J. K., Sugiyama, K., Emery, V. C., Dalrymple, J. M., and Bishop, D. H. L., 1989, Identification of mutations in the M RNA of a candidate vaccine strain of Rift Valley fever virus, *Virology* **169**:452.

Talmon, Y., Prased, B. V. V., Clerx, J. P. M., Wang, G. J., Chu, W., and Hewlett, M. J., 1987, Electron microscopy of vitrified-hydrated La Crosse virus, *J. Virol.* **61**:2319.

Tordo, N., de Haan, P., Goldbach, R., and Poch, O., 1992, Evolution of negative-stranded RNA genomes, *Semin. Virol.* **3**:341.

Turell, M. J., Saluzzo, J.-F., Tammariello, R. F., and Smith, J. F., 1990, Generation and transmission of Rift Valley fever viral reassortants by the mosquito *Culex pipiens*, *J. Gen. Virol.* **71**: 2307.

Ullman, D. E., Cho, J. J., Mau, R. F. L., Westcot, D. M., and Custer, D. M., 1992, A midgut barrier to tomato spotted wilt virus acquisition by adult Western flower thrips, *Phytopathology* **82**: 1333.

Urquidi, V., and Bishop, D. H. L., 1992, Non-random reassortment between the tripartite RNA genomes of La Crosse and snowshoe hare viruses, *J. Gen. Virol.* **73**:2255.

Ushijima, H., Clerx-van Haaster, C. M., and Bishop, D. H. L., 1981, Analyses of Patois group bunyaviruses: Evidence for naturally occurring recombinant bunyaviruses and existence of immune precipitable and non-precipitable nonvirion proteins induced in bunyavirus-infected cells, *Virology* **110**:318.

Verani, P., Nicoletti, L., and Marchi, A., 1984, Establishment and maintenance of persistent infection by the *Phlebovirus* Toscana in Vero cells, *J. Gen. Virol.* **65**:367.

Ward, C. M., 1993, Progress towards a higher taxonomy of viruses, *Res. Virol. (Inst. Pasteur)* **144**:419.

Ward, R. L., Knowlton, D. R., and Greenberg, H. B., 1988, Phenotypic mixing during coinfection of cells with two strains of human rotavirus, *J. Virol.* **62**:4358.

Whitman, L., and Shope, R. E., 1962, The California complex of arthropod-borne viruses and its relationship to the Bunyamwera serogroup through Guaroa virus, *Am. J. Trop. Med. Hyg.* **11**:691.

Wright, S. M., Kawaoka, Y., Sharp, G. B., Senne, D. A., and Webster, R. G., 1992, Interspecies transmission and reassortment of influenza A viruses in pigs and turkeys in the United States, *Am. J. Epidemiol.* **136**:488.

Xu, X., Rocha, E. P., Regenery, H. L., Kendal, A. P., and Cox, N. J., 1993, Genetic and antigenic analyses of influenza A (H1N1) viruses, 1986–1991, *Virus Res.* **28**:37.

Yancheva, A., 1990, A possibility for transferring combined resistance to tomato spotted wilt virus and *Thielaviopsis basicola* to intercultivar tobacco hybrids, *Genet. Sel.* **23(3)**:194.

Yasuda, J., Shortridge, K. F., Shimizu, Y., and Kida, H., 1991, Molecular evidence for a role of domestic ducks in the introduction of avian H3 influenza viruses to pigs in southern China, where the A/Hong Kong/68 (H3N2) strain emerged, *J. Gen. Virol.* **72**:2007.

CHAPTER 9

Pathogenesis of Diseases Caused by Viruses of the *Bunyavirus* Genus

FRANCISCO GONZÁLEZ-SCARANO, KEITH BUPP, AND NEAL NATHANSON

I. INTRODUCTION

The *Bunyavirus* genus of the family *Bunyaviridae* is named after Bunyam-wera virus (Smithburn *et al.*, 1946), and includes over 150 individual viruses classified in several serogroups. In addition to their serological relationship, members of the genus have been grouped together because of similar molecu-lar features (Bishop and Shope, 1979; Bishop *et al.*, 1980; Gonzalez-Scarano and Nathanson, 1995; see Chapters 1 and 2 in this volume). Almost all of the viruses in the *Bunyavirus* genus can be transmitted by mosquitoes to a large number of vertebrate hosts. From the standpoint of human infections, the viruses in the California serogroup are the most relevant, as they have been associated with influenza-like illnesses in central Europe and are an impor-tant cause of encephalitis in the midwestern United States. However, other members of the genus are of veterinary importance, and there have been scattered reports of febrile illnesses in humans associated with several other viruses in the genus (Parsonson and McPhee, 1985; Shope, 1985). Table I lists

FRANCISCO GONZÁLEZ-SCARANO, KEITH BUPP, AND NEAL NATHANSON • Depart-ments of Neurology and Microbiology, University of Pennsylvania Medical Center, Phila-delphia, Pennsylvania 19104-6146.

The Bunyaviridae, edited by Richard M. Elliott, Plenum Press, New York, 1996.

TABLE I. Principal Members of the *Bunyavirus* Genus
Associated with Diseases of Vertebrates

Virus	Vertebrate host/disease	Vector
Akabane	Ruminants—fetal abnormalities	*Culex, Aedes* species; Culicoides (midges)
Bunyamwera	Horses, deer—? disease Humans—febrile illness	*Aedes* species
Cache Valley	Sheep—fetal abnormalities	*Aedes* species, *Culiseta inornata, Culex tarsalis*
Inkoo	Humans—(rare)	*Aedes* species
Jamestown Canyon	Humans—encephalitis (rare)	*Aedes* species, *Culiseta inornata*
La Crosse	Humans—encephalitis	*Aedes triseriatus*
Oropouche	Humans—fever, arthralgias	*Culicoides paraensis* (midge)
Snowshoe hare	Humans—encephalitis (rare)	*Aedes canadensis, Culiseta inornata*
Tahyna	Humans—influenza-like illness	*Aedes vexans, Culiseta annulata*

the members of the genus that are known to cause disease in either humans
or animals.

In addition to their natural hosts, some members of the genus can infect
laboratory animals, particularly mice, which provides a useful model for
delineating the pathogenesis of disease caused by these agents.

II. CALIFORNIA SEROGROUP VIRUSES

The California serogroup includes 14 viruses and subtypes (Table II).
Although serological studies and analysis of the genomic sequences have
conclusively demonstrated their genetic relationship, each virus is only
prevalent in a specific geographic area delimited by the presence of its mosquito vector (and mammalian hosts). Among the members of the serogroup,
La Crosse (LAC), snowshoe hare (SSH), and Tahyna (TAH) viruses have been
studied most intensively. La Crosse virus is an endemic cause of encephalitis
in the midwestern United States, particularly in areas of Wisconsin from
which the virus takes its name. Tahyna virus is responsible for influenza-like
illnesses in Central Europe, particularly in areas of the former Czechoslovakia (Traavik *et al.*, 1978).

A. California Encephalitis

Studies of arbovirus encephalitides in Kern County, California, led to
the isolation of a new virus, termed California encephalitis (CE) virus, from
Aedes melanimon mosquitoes in 1943 (Hammon and Reeves, 1952; Reeves
and Hammon, 1962); serological studies subsequently confirmed its association with three clinical cases of encephalitis. At the time of the original

TABLE II. Members of the California Serogroup of Viruses

Virus or strain	Geographic location	Vertebrate host	Human infection
California encephalitis[a]	Western United States	Rodents, rabbits	Rare
La Crosse	Midwestern United States	Chipmunks, squirrels	Endemic
Snowshoe hare	Canada, northern United States	Snowshoe hare	Rare
San Angelo	Western North America	Unknown	None
Tahyna	Eastern Europe	Rabbits, domestic animals	Endemic
Lumbo	East Africa	Unknown	Unknown
Inkoo	Finland, Russia	Unknown	Rare
Melao	South America	Unknown	Unknown
Jamestown Canyon	North America	White-tailed deer	Uncommon
South River	Northeastern United States, Quebec	Unknown	None
Keystone	Eastern United States	Unknown	Rare
Trivittatus	North America	Unknown	Rare
Guaroa	South America	Unknown	Unknown

[a]Boldface indicates groupings within the serogroup.

isolations, the prevalence of seropositivity among 188 residents of the county was 11%. CE virus has been isolated from mosquitoes in later years, and subsequent surveys have confirmed the high incidence of seropositivity. Nevertheless, CE virus has not been implicated recently in any serious human diseases in California (LeDuc, 1987; Reeves *et al.*, 1983). Antigenic studies indicated a relationship between CE virus and other arboviruses, and the group was eventually designated the California serogroup (Whitman and Shope, 1962). The most important isolate in the group, La Crosse virus, was recovered from a fatal case of encephalitis in a 4-year-old girl who died in La Crosse, Wisconsin, having apparently acquired the infection while in Minnesota (Thompson *et al.*, 1965). It was quickly recognized as an important etiologic agent in human disease in many areas of the midwestern United States (Kappus *et al.*, 1983). Testing of mosquitoes, and more widespread serological diagnosis of individuals with encephalitis identified a number of other California serogroup viruses that are enzootic in the United States, some of which have been associated with occasional cases of meningoencephalitis.

B. La Crosse Encephalitis

1. Ecology

La Crosse virus is maintained in nature by infections of its two alternate hosts: mosquitoes and small mammals, particularly chipmunks and squirrels. Its main vector is *Aedes triseriatus*, a woodland mosquito that breeds in tree holes and in discarded tires (LeDuc *et al.*, 1975a; Turell and LeDuc, 1983; Yuill, 1983). In endemic areas a high proportion of chipmunks (60%) and squirrels (30%) have been found to be seropositive (Gauld *et al.*, 1974; Thompson, 1983). Experimental studies have demonstrated that these small mammals will generate a brief but sufficiently high level of viremia following inoculation to allow subsequent infection of hematophagous mosquitoes (Patrican *et al.*, 1985; Yuill, 1983).

La Crosse virus overwinters in the northern Midwest through its persistent infection of female mosquitoes, which may emerge infected following hibernation (Balfour *et al.*, 1975). Additionally, infected females will transmit virus transovarially to eggs which also overwinter in tree holes (LeDuc *et al.*, 1975b; Patrican *et al.*, 1985a). Larvae collected in the early spring, before the emergence of adults from hibernation, have been found to contain La Crosse virus (Thompson, 1983). Transmission between mosquitoes may provide an alternative cycle that does not require a vertebrate host, and venereal transmission from infected males to uninfected females has been documented (Beaty and Thompson, 1975; Thompson, 1983; Thompson and Beaty, 1978).

Aedes triseriatus is present in the United States in areas east of the Mississippi River, where the number of mosquitoes peaks from June through September (DeFoliart, 1983; Berry *et al.*, 1983). Serologic studies indicate that human infections occur over the same geographic region and cases of encephalitis are correspondingly seasonal (Calisher, 1983). *Aedes albopictus*, an aggressive mosquito introduced into the United States from Asia, apparently in shipments of used tires, can serve as a vector for La Crosse virus, at least experimentally.

2. Epidemiology

The vast majority of cases of encephalitis caused by California serogroup viruses are associated with La Crosse virus infection. Unlike other arboviral infections, which occur in epidemics, the number of documented La Crosse virus infections has remained constant at around 100 per year (range 42–174) since reporting began in 1963 (Fig. 1; Kappus *et al.*, 1983). The stable incidence may reflect little variation in the population of *A. triseriatus*. Increased awareness and better serological follow-up of encephalitis cases during epidemics of St. Louis encephalitis led to an increase in the number of documented La Crosse virus infections in 1975 (Kappus *et al.*, 1983). This suggests that encephalitis associated with La Crosse virus is underreported.

California encephalitis is concentrated geographically in Minnesota, Wisconsin, Iowa, Illinois, Indiana, and Ohio, which report over 90% of all cases. The clear age dependence that is characteristic of experimental infection of mice (see Section II.C) must also apply to humans, since there have been very few cases of California encephalitis in adults. Campers and hikers between the ages of 1 and 19 are at greatest risk of exposure to bites from the

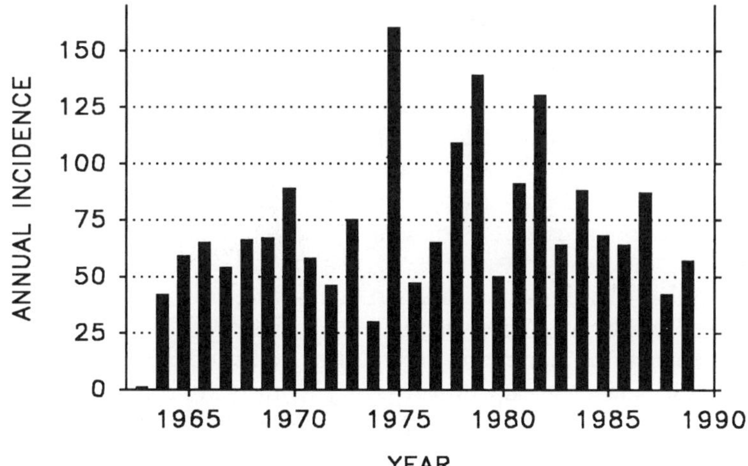

FIGURE 1. Annual incidence of cases of documented La Crosse virus encephalitis, from its description in 1963 (data from CDC).

principal vector, and tree holes near the home, or deposits of discarded tires are particularly associated with increased numbers of cases (Monath *et al.*, 1970; Woodruff *et al.*, 1992). As infected humans are unlikely to transmit the virus back to mosquitoes, humans are considered "dead-end" hosts. The mortality rate of the CNS disease is low, 0.3% among 1348 cases studied by Kappus (Kappus *et al.*, 1983), as is the incidence of chronic neurologic sequelae.

Serological surveys have demonstrated that the prevalence of antibodies in areas of high vector mosquito population can reach 20% by age 60 (Grimstad *et al.*, 1984b; Henderson and Coleman, 1971). Using these data, it was estimated that there are about 15,000 infections/year in Indiana, and perhaps 300,000 in the entire midwestern United States (Grimstad *et al.*, 1984b; Kappus *et al.*, 1983). Most of these infections are subclinical or nonspecific. For children younger than 16, it is estimated that there are more than 1000 infections, not necessarily involving the nervous system, per reported case.

3. Human Disease

The clinical disease caused by La Crosse virus is an acute encephalitis, preceded by a nonspecific febrile illness (Chun, 1983; Chun *et al.*, 1968; Gundersen and Brown, 1983; Kalfayan, 1983). The incubation period is estimated to be around 1 week, primarily on the basis of extrapolation from experimental studies. The symptoms of central nervous system disease include stiff neck, lethargy, seizures, more rarely coma, and they may last up to 10 days, during which most affected children are hospitalized. The cerebrospinal fluid demonstrates an elevated protein in about 20% of cases, and a pleocytosis of over 50 WBC/μl in 65% of the children. Unlike most viral encephalitides, many of the cells in the cerebrospinal fluid are polymorphonuclear neutrophils (Gundersen and Brown, 1983).

Seizures occur in around half of the children with La Crosse virus encephalitis, and about 10% of the total will develop epilepsy, or chronic seizures (Chun *et al.*, 1968; Chun, 1983). A small proportion of patients may develop persistent paresis (2%) or learning disabilities and other cognitive deficits (2%).

Pathological studies are based on the small number of fatalities that have occurred during La Crosse virus encephalitis (Kalfayan, 1983). There are inflammatory lesions occurring in patches, with focal necrosis and even hemorrhages in some cases. Inflammatory lesions are mainly located in the cerebral cortex, with some lesions in the brain stem and very few changes in the medulla and spinal cord.

4. Other California Serogroup Viruses

The principal vectors and vertebrate hosts of those California serogroup viruses that infect humans in North America are shown in Table II. Information available from serological surveys indicates that Jamestown Canyon virus is common in Michigan, other midwestern states, and in New York, with seroprevalence rates between 5 and 25%, depending on the locale (Deibel *et al.*, 1983; Grimstad *et al.*, 1984a; Srihongse *et al.*, 1984). A few cases of human encephalitis, mostly in adults, have been identified. Except for the age distribution of the cases, the clinical features resemble those of La Crosse virus.

Snowshoe hare virus, which is antigenically and genomically very similar to La Crosse virus, has been implicated as a cause of human encephalitis in Nova Scotia, Quebec, and Ontario (Artsob, 1983; Grimstad *et al.*, 1984b; Srihongse *et al.*, 1984) and can be isolated from mosquitoes throughout Canada.

Trivittatus virus has been isolated from mosquitoes throughout the 48 contiguous United States (Calisher, 1983; Rowley *et al.*, 1983). Although there is a considerable prevalence of antibody reactivity among residents of the Midwest, trivittatus virus has not been associated with symptomatic human illness except in a few instances (Grimstad *et al.*, 1984a; Monath *et al.*, 1970.; Rowley *et al.*, 1983).

C. Experimental Infection

1. Mosquitoes

Infection of *A. triseriatus* mosquitoes begins with infection of the epithelial cells of the midgut. The midgut may serve as a barrier that contains the spread of infection, based on the finding that under some circumstances viral antigen may accumulate in the midgut epithelium without dissemination into the rest of the organism (Tesh and Beaty, 1983). Intrathoracic injection of a virus that fails to disseminate after ingestion may lead to a more widespread infection, confirming the role of the midgut in containment (Miller, 1983). Surprisingly, some of the same viral determinants important in neuroinvasiveness (see section on infection of vertebrates) are critical in allowing penetration of the midgut barrier (Sundin *et al.*, 1987).

Viral antigen can next be detected in the hemocoel, just outside the midgut. From there, the virus spreads to heart, neural ganglia, fat body, and, critical in terms of transmission, the ovaries and the salivary glands (Beaty and Thompson, 1975; McLean *et al.*, 1975, 1979; Tesh and Beaty, 1983; Turell *et al.*, 1984). Infection of the salivary glands results in high levels of viral replication and is the main source of transmission of the virus to the vertebrate host during feeding. The interval between the initial infected blood meal and the appearance of high virus levels in mosquito saliva is around 7–14 days for La Crosse virus.

Transovarial transmission of La Crosse virus to both female and male offspring is the result of infection of ovaries, including oocytes, occurring during systemic spread of virus from the hemocoel (Tesh and Beaty, 1983; Watts *et al.*, 1973). Infection of the larvae, which does not proceed through the midgut and hemocoel, is thus widespread, includes the gonads, and serves as a powerful mechanism for further spread of the virus, both through vertebrate feeding and venereally to uninfected mosquitoes (Thompson and Beaty, 1978).

Neither mosquito cells in culture nor mosquitoes demonstrate any apparent changes as a result of La Crosse infection (DeFoliart, 1983; Turell and Hardy, 1980). Coinfection of mosquitoes with viruses of the appropriate strains (not all strains can reassort—see Chapter 8) can result in the production of progeny virions with reassorted genomes. In one series of experiments with La Crosse and Tahyna viruses, two strains that reassort readily in tissue culture, all expected genotypes were obtained from coinfected mosquitoes (Chandler *et al.*, 1991).

Reassortant viruses have also been used to map the viral determinants important for dissemination. The medium-sized segment (M RNA) appears to be the major genetic determinant of infectivity (Beaty *et al.*, 1982), as well as the major determinant of vertical and venereal transmission (Schopen *et al.*, 1991).

The Asian tiger mosquito, *A. albopictus*, was imported into the United States in two waves, apparently in association with shipments of tires. It has been spreading from initial entry points in the South to areas close to the Wisconsin border (T. Yuill, personal communication) and can serve as a vector for La Crosse virus in the laboratory. Several new strains of viruses classified in the *Bunyavirus* genus have been obtained from *A. albopictus* mosquitoes captured in the United States (Francy *et al.*, 1990). Thus, *A. albopictus* could become a clinically significant vector for La Crosse virus.

2. Vertebrate Hosts

Infection of the laboratory mouse reflects many of the features of La Crosse virus infection of humans and small mammals (Janssen *et al.*, 1984; Johnson and Johnson, 1968; Parsonson and McPhee, 1985; Tignor *et al.*, 1983). For example, (1) the route of infection, normally subcutaneous inoculation, is similar to inoculation following a mosquito bite; (2) the spectrum of disease ranges from an inapparent infection to an acute, rapidly fatal encephalitis; (3) there is a clear age-dependent susceptibility, similar to that evident in human infections; and (4) the pathological features in the central nervous system of infected mice are similar to the neuropathology of the human cases (Kalfayan, 1983). The outcome of infection can be manipulated with all of these variables. Furthermore, the experimental infection can be used to study the extraneural phase of replication—by inoculating virus peripherally—as well as replication in the nervous system—following intracerebral injection. Adult mice survive peripheral challenges with high doses of La Crosse virus, whereas an intracranial injection of 100 PFU leads to fatal encephalitis.

The sequential events in La Crosse virus replication are shown in Fig. 2. In experimentally infected newborn mice (injected subcutaneously) the virus replicates extraneurally in muscle tissue, primarily skeletal muscle, although cardiac and smooth muscles are involved as well. The robust replication in muscle leads to high levels of plasma viremia; the level of viremia is associated with penetration of the nervous system (*neuroinvasiveness*). In suckling mice the virus replicates in all CNS tissue elements; immunohisto-chemical studies demonstrate widespread production of viral antigens, with subsequent necrosis of the infected cells (Janssen *et al.*, 1984). The entire course of the encephalitic phase lasts only 3–4 days. The spread of virus to skeletal muscle may be the result of passive viremia during inoculation or via lymphatic channels. Equally obscure is the actual passage from the bloodstream (where the viremia is cell-free) to the CNS, although infection of vascular endothelium is a strong possibility supported by *in vitro* studies of murine endothelial cells (C. Griot, unpublished results).

Neural spread of viruses is an important mechanism for CNS penetration by herpesviruses, rabies virus, and some reovirus strains. Any contribution of neural spread to penetration of the CNS by California serogroup viruses was ruled out in experiments described by Griot *et al.* (1993b). In those experiments, footpad inoculation of a strain that causes a low-level viremia was performed (B1-1a, a reassortant between Tahyna and La Crosse viruses). The sciatic nerve, which supplies the footpad, was transected in some animals, and sham-transected in others, but there was no difference in mortality. In the animals with an intact sciatic nerve, sequential examination of the spinal cord, which would have been expected to demonstrate a gradient if CNS invasion were occurring via neural pathways, showed that the virus was present at all levels of the spinal cord at the same time (Griot *et al.*, 1993b; Tyler *et al.*, 1986).

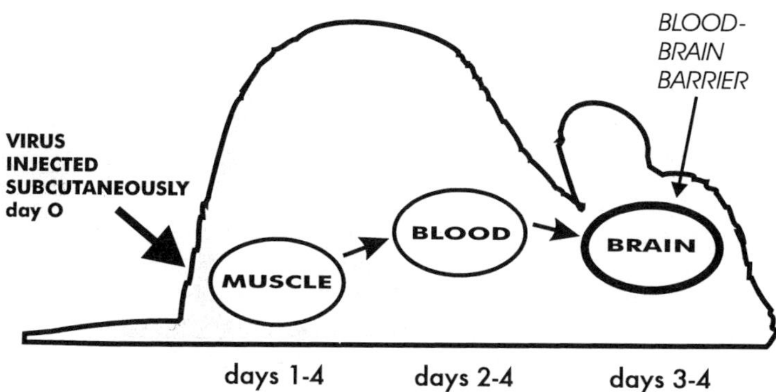

FIGURE 2. Sequential steps in the pathogenesis of La Crosse virus disease in suckling mice. Following a subcutaneous inoculation, there is robust replication in muscle, which leads to plasma viremia. Penetration of the CNS is associated with high levels of viremia, but the exact mechanism is unknown—it may involve infection of endothelial cells. Clinical disease and death are the result of widespread infection of the brain, involving all cell types.

In adult outbred mice, peripheral inoculation leads to little viral replication in extraneural tissues, and viremia is below easily detectable levels. However, adult mice can be readily infected after intracerebral infection, where the virus spreads from the ventricles to the ependyma, then replicates in neurons. Nevertheless, the level of antigen production, as estimated from immunohistochemical studies, is much lower than in newborn mice. Adult mice inoculated intracerebrally die within 1 week.

Tahyna virus, isolated in the former Czechoslovakia, is reported to be associated with influenza-like symptoms including fever and arthralgias (Bardos, 1969). Experimentally, Tahyna virus can cause encephalitis when injected peripherally at high doses (in suckling mice) or intracerebrally in adults (Bardos, 1965; Wallnerova, 1969). A neuroadapted strain of Tahyna virus has been used to map the determinants of neuroinvasiveness (see below).

Lethal infections of laboratory mice with La Crosse virus are associated with CNS pathology; no other organs appear to be affected, in spite of evidence of viral replication peripherally. In suckling mice, neuronal necrosis is the predominant finding; adult mice inoculated intracerebrally demonstrate more evidence of inflammation, including perivascular cuffs, glial nodules, and leptomeningitis (Janssen *et al.*, 1984). The few ultrastructural studies performed on infected mouse brain demonstrated that, as in electron micrographs performed in infected tissue culture cells, the viruses mature by budding into intracytoplasmic vesicles (Lyons and Heyduk, 1973; Murphy *et al.*, 1968).

La Crosse virus infection of larger mammals, including adult chipmunks, produces a viremia without overt clinical findings (Ksiazek and Yuill, 1977; Pantuwatana *et al.*, 1972; Patrican *et al.*, 1985; Seymour *et al.*, 1983). Rabbits injected with large inocula of La Crosse virus also appear to remain healthy (unpublished results). Similar results were obtained with snowshoe hare virus in adult hares (Seymour *et al.*, 1983).

3. Immune Response and Immunity

The humoral immune response to bunyaviruses, and more specifically to those viruses belonging to the California serogroup, has been measured using a variety of well-established techniques, including hemagglutination inhibition, complement fixation, neutralization, and ELISA (Gonzalez-Scarano *et al.*, 1982). The cellular immune response has not been studied to any great extent.

Antibodies against the G1 protein of California serogroup viruses are capable of neutralization, inhibition of hemagglutination, and can block cell-to-cell fusion (Gonzalez-Scarano *et al.*, 1982; Grady *et al.*, 1983; Kingsford *et al.*, 1983; Kingsford, 1991; Ludwig *et al.*, 1991). Some neutralizing anti-G1 antibodies will inhibit virus binding to susceptible cells (Pekosz *et al.*, in press), though this is not the sole mechanism for neutralization (Kingsford *et al.*, 1991). Among a panel of neutralizing monoclonal antibodies, some were specific for the immunizing strain among the California serogroup viruses, and some demonstrated wider activity (Gonzalez-Scarano *et al.*, 1982).

A single panel of monoclonal antibodies against the G2 protein has been reported (Ludwig *et al.*, 1991). Those antibodies had minimal neutralizing activity in infections of vertebrate cells, but neutralized the virus effectively in assays with mosquito cells. Ludwig *et al.* (1991) also described antibodies that they felt were directed against both G1 *and* G2, and indicated that these had neutralizing activity in assays using vertebrate cell lines. However, one of these antibodies (807-22) is only directed against G1 (Gonzalez-Scarano *et al.*, 1982).

High-titer neutralizing antibodies are probably sufficient to protect the organism against repeated challenges with the same virus strain. Administration of a neutralizing monoclonal antibody directed against the G1 protein to a weanling mouse, approximately 24 hr prior to challenge with a lethal dose of La Crosse virus, was sufficient to protect the animal from disease (Pekosz *et al.*, 1995). In view of the cross-reactivity of many monoclonal antibodies, cross protection may also occur, though this has not been tested. Similarly, immunization of suckling mice with a vaccinia virus recombinant encoding the M RNA open reading frame (expressing G1, G2, and NSm) from La Crosse virus conferred complete protection from a potentially lethal challenge (Pekosz *et al.*, 1995). Antibody against G1 is sufficient for protection. Protection from the challenge was associated with the development of neutralizing antibodies, although cellular immunity was not measured.

4. Virulence Variation

The high mutation rate characteristic of many RNA viruses can be expected to lead to some variation in the virulence properties of California serogroup viruses. Variation in mouse virulence has been studied extensively for Tahyna viruses (Bardos, 1965; Danielova, 1974; Malkova, 1971). Whereas all field isolates are virulent after intracerebral injection in mice, few will induce disease following intraperitoneal inoculation. Laboratory passage may enhance some of these biological properties. For example, a laboratory strain, TAH 181/57, was further neuroadapted by sequential passage in suckling mice. While this strain was then exquisitely able to replicate in CNS tissue, it had very low subcutaneous virulence (Janssen *et al.*, 1984), which was attributed to the inability of this strain to replicate in either muscle tissue or kidney (Janssen *et al.*, 1984).

Similarly, passage of Tahyna virus in mammalian cell culture increases replication in cultured cells, usually with a reduction in the ability of the virus to replicate in mice (Malkova and Marhoul, 1971). We have used sequential passage of a reassortant virus to derive a strain with markedly reduced neurovirulence (see next section; Endres *et al.*, 1990).

D. Genetic Determinants of Virulence

Neurovirulence versus Neuroinvasiveness

With some exceptions, the genomic segments of California serogroup viruses will reassort in infected cell cultures (Chapter 8). To perform these experiments, temperature-sensitive mutants containing lesions in appropriate segments can be used in coinfections with complementing viruses of a different strain. As described elsewhere, these reassortant viruses have provided important information about the coding relationships between genomic segments and the viral glycoproteins. The use of reassortant viruses in virulence studies depends on two requirements for meaningful results: (1) the reassortant viruses must have been derived from nonmutagenized virus stocks, to eliminate the possibility that silent mutations could contribute to the phenotypic property under study, and (2) the parent viruses used for reassortment must be sufficiently different in phenotype to avoid ambiguous results.

To identify the genetic determinants associated with nervous system disease, we have separated the ability to cause encephalitis into two properties: *neurovirulence* is the ability to infect CNS cells, and *neuroinvasiveness* refers to the ability of a virus to reach the CNS following a peripheral inoculation. For the California serogroup the latter should reflect replication in muscle, the ability to generate a high enough level of viremia, and the ability to cross the blood–brain barrier. To assay and quantify these properties, we have used the PFU/LD$_{50}$ ratio following either peripheral (subcutaneous) inoculation into suckling mice (assay for neuroinvasiveness) or the PFU/LD$_{50}$ ratio after intracerebral inoculation into adult mice (assay for neurovirulence).

Most California serogroup viruses are neurovirulent, and, when assayed in suckling mice, are also neuroinvasive. However, a strain of Tahyna virus (clone Tahyna 181/57) that had been passaged extensively in mouse brain retained its neurovirulence while having markedly decreased neuroinvasiveness. Thus, while La Crosse original virus has a neuroinvasiveness index of <1, Tahyna 181/57 requires 10^4 PFU for 1 LD$_{50}$ in suckling mice.

Reassortant viruses were then prepared from these two parents by co-infection of BHK-21 cells and the progeny tested for neuroinvasiveness (Janssen *et al.*, 1986). With one exception (see Fig;. 3), all of the reassortant viruses containing the M RNA segment from the neuroinvasive La Crosse virus were neuroinvasive, while all of the reassortants containing the M RNA segment from Tahyna 181/57 virus were not. Studies comparing the pathogenesis of Tahyna 181/57 virus with that of La Crosse virus indicated that the former replicated poorly after subcutaneous injection, with little replication in muscle and low levels of viremia.

More recently, the replication of both parent viruses and of reassortant viruses has been compared in C2C12 cells, a differentiable murine muscle line. La Crosse, and reassortant viruses with the La Crosse virus M RNA segment were able to infect these cells much more effectively than either Tahyna 181/57 virus or reassortants with the Tahyna virus M RNA segment (Griot *et al.*, 1994). Whether the more efficient infection of muscle cells is related to differences in receptor affinity or to other viral properties is not known. However, the property of neuroinvasiveness maps to the gene segment encoding the glycoproteins (and one nonstructural protein, NSm), and correlates, at least partially, with the ability of the La Crosse virus M RNA segment to mediate a more robust infection of muscle cells.

To further explore the role of the glycoproteins in neuroinvasiveness, a set of monoclonal antibody-resistant (MAR) mutants of La Crosse virus were injected subcutaneously into suckling mice. These MAR variants were selected with neutralizing monoclonal antibodies directed against the G1 protein. One group of variants (V22), those selected with monoclonal antibody 807-22, had markedly reduced neuroinvasiveness, since at a subcutaneous dose of 2200 PFU, there was no detectable replication in muscle (Fig. 4). Nevertheless, V22 was able to infect adult mouse brain when injected intracranially, with titers comparable to those of the parent virus ($\approx 10^5$ PFU/mg brain). Thus, a virus with a presumed small mutation in G1 was markedly attenuated peripherally, confirming the role of the glycoproteins in CNS invasion. Concomitantly, V22 virus was an ineffective mediator of cell-to-cell fusion and demonstrated a small plaque phenotype. These characteristics, which could represent inefficient cell-to-cell spread, or other entry deficits, appear to be important in mediating peripheral replication.

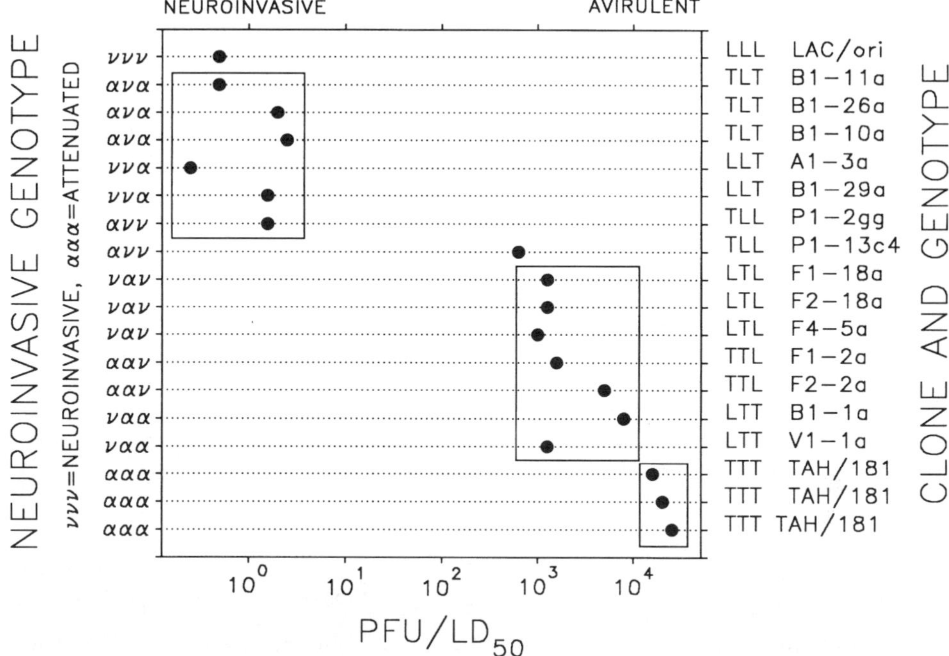

FIGURE 3. Neuroinvasiveness of reassortants derived from coinfections between La Crosse virus (genotype LLL) and Tahyna 181/57 virus (genotype TTT). The genotype indicated on the right refers to the origin of the large, middle, or small RNA segment. The neuroinvasive genotype on the left designates each genomic segment as either virulent (v) or avirulent (α). Neuroinvasiveness was quantified as the ratio PFU/LD$_{50}$ following subcutaneous inoculation of suckling mice. With one exception, reassortants with a middle RNA segment from La Crosse virus were neuroinvasive, whereas those with a middle RNA segment derived from Tahyna 181/57 virus were not. (Data from Janssen *et al.*, 1986; reprinted from Griot *et al.*, 1993a.)

Defining the genetics of neurovirulence was more complicated, since all California serogroup viruses tested were neurovirulent when assayed by intracerebral injection in adult mice. To derive a strain that was attenuated intracerebrally, a reassortant virus obtained by a cross between La Crosse and Tahyna viruses with genotype TLL (see Chapter 8; the genotype is designated by a letter indicating the source of the large, middle, or small RNA, respectively; T = Tahyna L RNA; L = La Crosse M RNA; L = La Crosse S RNA) was sequentially passaged in BHK-21 cells and plaque purified (Endres *et al.*, 1990). In comparison with either La Crosse or Tahyna viruses, this clone, termed B.5, was markedly attenuated, with a PFU/LD$_{50}$ ratio of $>10^6$ after intracerebral inoculation. This clone then served as one of the parent viruses for the preparation of a set of reassortants that were used to map neurovirulence (the other parent being a reciprocal clone with genotype LTT—see Fig. 5).

Reassortants containing the L RNA segment of B.5 (indicated as A or avirulent in Fig. 5) all replicated poorly after intracerebral injection, with

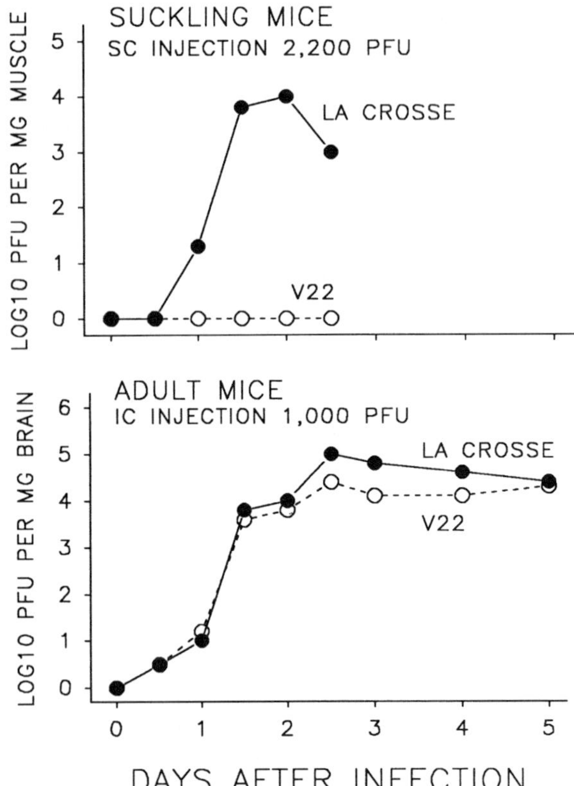

FIGURE 4. Replication of a monoclonal antibody-resistant (MAR) mutant, V22, following subcutaneous injection in suckling mice (upper panel) or intracerebral injection in adult mice (lower panel). The MAR mutant showed reduced neuroinvasiveness and high neurovirulence. (Data from Gonzalez-Scarano *et al.*, 1985; reprinted from Griot *et al.*, 1993a.)

titers below the detectable limit of the plaque assay. B.5 was also a temperature-sensitive mutant, with restricted replication at 39.8°C. The *ts* lesion also mapped to the L segment, indicating that a pleiotropic mutation is apparently responsible for both the *ts* phenotype and the decreased neurovirulence. Furthermore, *ts* revertants had returned to a neurovirulent phenotype. However, the temperature of the mouse is 37°C, so that the *ts* mutation is not *mechanistically* responsible for the attenuated phenotype (Griot *et al.*, 1993b). The exact mutation is now being determined, a preliminary effort being the sequencing of the La Crosse virus L segment, which has just been completed (Roberts *et al.*, 1995).

In summary, the neuroinvasiveness of La Crosse virus maps to its M RNA segment, and more specifically appears to map to the ectodomain of the G1 protein. Neurovirulence, at least in the one attenuated serogroup virus ever described, maps to the L segment, and is associated with a *ts* mutation. We have also prepared reassortant viruses with several attenuated segments. For example, the L segment of B.5 (originally derived from Tahyna virus) was

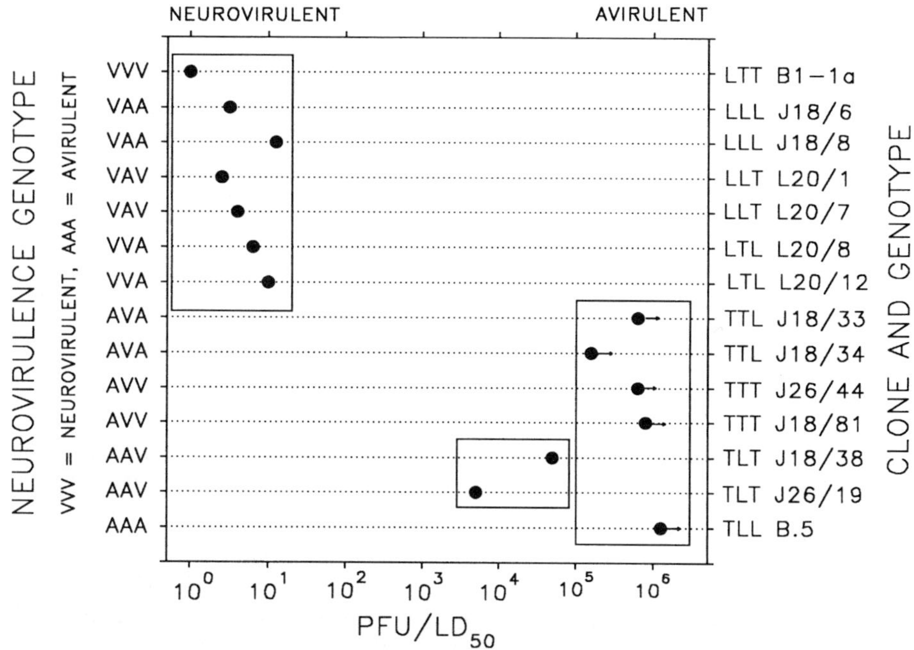

FIGURE 5. Neurovirulence of reassortants between B.5, a neuroattenuated mutant (genotype TLL, or AAA, at the start of the experiment, since the segment containing the attenuation had not been identified) and a B1-1a virulent virus with reciprocal genotype (genotype LTT or VVV). The genotype of each virus obtained is listed on the right, and the neurovirulence of each segment designated as either neurovirulent (V) or avirulent (A). The neurovirulence phenotype of each clone was obtained by determination of the PFU/LD_{50} ratio following intracerebral inoculation of adult mice. (Data from Griot *et al.*, 1993b; reprinted from Griot *et al.*, 1993a.)

combined with the M segment of Tahyna virus. Under such circumstances, the virus derived has two avirulent segments, and is more attenuated than either parent clone (Griot *et al.*, 1993b).

III. OTHER BUNYAVIRUSES

A. Bunyamwera Virus

Some other members of the *Bunyavirus* genus are also known to be pathogenic. The prototype virus of the Bunyamwera serogroup (Bunyamwera virus) was originally isolated from mosquitoes belonging to the *Aedes* genus in the Bwamba county of Uganda (Smithburn *et al.*, 1946). It was found to be pathogenic in mice inoculated by various routes. After intracerebral injection, Bunyamwera virus could be reisolated from various tissues, with the highest titers in brain and spleen. Histopathological changes were mainly seen in the kidneys and brain. Subsequent studies have indicated a possible pathological role for the virus in rare human infections (Kokernot *et al.*, 1958; Theiler and Downs, 1973).

Bunyamwera virus and other members of the serogroup can be grown in several cell types in culture including BHK, L, and Vero cells (Way *et al.*, 1976). The antiviral compound cicloxone sodium has been shown to inhibit Bunyamwera virus growth in tissue culture (Dargan *et al.*, 1992). The virus can also infect *A. albopictus* cells and persistently infected cell lines have been established. These lines are resistant to superinfection with the homologous but not heterologous viruses (Elliott and Wilkie, 1986; Newton *et al.*, 1981). Passage of Vero cell-grown Bunyamwera virus through *A. albopictus* cells results in a change in the antigenic structure of the virus (James and Millican, 1986).

Sera containing antibodies to Bunyamwera serogroup viruses have been obtained from horses or humans in Argentina (Calisher *et al.*, 1987, 1988b), Sudan (Woodruff *et al.*, 1988), Uganda (Rodhain *et al.*, 1989), and the United States (Calisher *et al.*, 1988a). Bunyamwera group viruses have been isolated from mosquitoes in Norway (Traavik *et al.*, 1985), South Africa (Worth *et al.*, 1961), Siberia (Mitchell *et al.*, 1993), and the United States (Francy *et al.*, 1990; Jakob *et al.*, 1989; Calisher *et al.*, 1986; Campbell *et al.*, 1992). They have also been isolated from midges in the United States (Calisher *et al.*, 1986; Kramer *et al.*, 1990).

B. Cache Valley Virus

Cache Valley virus is a Bunyamwera serogroup virus recently shown to be pathogenic in sheep. This virus is known to affect fetuses, causing various abnormalities including arthrogryposis, hydrancephaly, and hydrocephalus (Chung *et al.*, 1990; Edwards *et al.*, 1989; Crandell *et al.*, 1989). Such findings are similar to those seen in fetal infections of lambs by Akabane virus (a Simbu serogroup virus—see below). That Cache Valley virus is the causative agent has been demonstrated by serological studies of an outbreak of the disease in Texas (Edwards *et al.*, 1989; Crandell *et al.*, 1989) and by direct inoculation of fetuses by the virus (Chung *et al.*, 1990). The virus might also cause a febrile illness in humans, having been isolated from a U.S. soldier stationed in Panama (Mangiafico *et al.*, 1988).

Cache Valley virus has been isolated from mosquitoes in Canada (Iversen *et al.*, 1979; Burton *et al.*, 1973), the United States (Main and Crans, 1986; Clark *et al.*, 1986) and Trinidad (Jonkers *et al.*, 1968). Antibody-positive sera have been collected from ungulates in Oregon, Michigan, Minnesota, and Mexico (Aguirre *et al.*, 1992; Neitzel and Grimstad, 1991; McLean *et al.*, 1987; Eldridge *et al.*, 1987) and from Mexican jackrabbits (*Lepus californicus*) (Aguirre *et al.*, 1992).

C. Akabane Virus

Akabane virus (a member of the Simbu serogroup) was initially isolated from mosquitoes in Japan (Oya *et al.*, 1961). It was subsequently shown to be responsible for fetal abnormalities in cattle which could lead to abortions or deformities such as arthrogryposis and hydrancephaly. The virus was isolated from a fetal calf during an epizootic of abortions and arthrogryposis–hydrancephaly in Japan (Kurogi *et al.*, 1976) and could induce fetal abnormalities when injected into pregnant cows (Kurogi *et al.*, 1977). Similar observations were made concerning Akabane virus infection of sheep in Australia (Parsonson *et al.*, 1977). Effective vaccines against the virus were developed shortly afterwards (Kurogi *et al.*, 1978, 1979).

In cows, the type of pathology observed as a result of Akabane virus infection depends on the time at which the developing fetus is infected (Kirkland *et al.*, 1988)—hydrancephaly being more typical of an earlier infection and arthrogryposis of the later stages of pregnancy. The pregnant animals themselves do not appear to develop any significant disease.

Immunostaining techniques have shown that viral antigen is present in various tissues of the fetus, including brain, skeletal muscle, and liver (Parsonson *et al.*, 1988; Narita and Kawashima, 1993; Kurogi *et al.*, 1977). Detailed analysis of the spinal cord from a calf affected with Akabane virus-induced arthrogryposis revealed a reduction in several types of neurotransmitter receptors in ventral horn motor nuclei, and an increase in receptors typical of glial cells in lateral and ventral spinal columns (Gundlach *et al.*, 1990). These changes are consistent with the neurodegeneration and gliosis observed in the spinal cord. Studies of the immune response to Akabane virus in ovine fetuses showed inconsistent responses to the virus in early gestation, but significant responses later in development. However, the later response was not enough to prevent pathological changes (McClure *et al.*, 1988).

In infected mouse brain, viral antigen was found in Ammon's horn, the deep part of the cerebral cortex, and the central gray matter of the midbrain and the brain stem—corresponding to the observed encephalitic or degenerative lesions. Although severe spongy lesions were observed in the cerebral cortex, no viral antigen was detected there (Haziroglu *et al.*, 1987). Chicken embryos can also be infected with Akabane virus resulting in neuronal and muscular degeneration (Konno *et al.*, 1988). As with other bunyaviruses, Akabane virus is cytopathic for BHK cells but not for mosquito cells (Hoffmann and St. George, 1985).

In addition to being found in Japan and Australia, antibodies reacting against Akabane virus have been detected in the sera of various animals from Africa (Al-Busaidy *et al.*, 1987; Davies and Jessett, 1985; Hamblin *et al.*, 1990) and the Middle East (Sellers and Pedgley, 1985; Al-Busaidy *et al.*, 1988). The virus itself has also been isolated from midges and goats in the Sultanate of Oman (Al-Busaidy and Mellor, 1991).

D. Oropouche Virus

Oropouche virus was originally isolated in Trinidad (1955) from a patient with fever (Anderson *et al.*, 1961). It was subsequently responsible for several outbreaks of a minor febrile illness in Brazil. This syndrome was characterized by chills, headache, and soreness of the muscles and joints. Leukopenia and neutropenia were also typical of the disease (Pinheiro *et al.*, 1981b). Several studies (Pinheiro *et al.*, 1981a, 1982; Dixon *et al.*, 1981; Roberts *et al.*, 1981) have shown that the midge *Culicoides paraensis* rather than mosquitoes is responsible for the spread of the virus (Dixon *et al.*, 1981; Smith and Francy, 1991; Hoch *et al.*, 1987).

Other bunyaviruses have been rarely implicated in human febrile illnesses (Iversson *et al.*, 1987; Shope, 1985).

IV. SUMMARY

The bunyaviruses are a large group of viruses within the family *Bunyaviridae*, and include some members that are important for their disease potential as well as some that have agricultural importance. From the standpoint of human disease, the California serogroup viruses are the most relevant, and thus have been studied more intensively, with the mouse serving as an accurate and reproducible model. As in clinical disease in humans, the brain is the most important target organ, and its infection is the cause of mortality. Diseases of agricultural importance have been described with other bunyaviruses, including Akabane, Bunyamwera, Cache Valley, and Oropouche viruses. The diseases caused by Oropouche and Bunyamwera viruses are ill-defined febrile illnesses. Akabane and Cache Valley viruses are responsible for fetal wastage in sheep and cattle associated with arthrogryposis and other limb abnormalities which may also be related to CNS disease.

ACKNOWLEDGMENTS.   Supported by PHS Grants AI-24888 and NS-20904. K.B. is supported by PHS training grant NS-07180. We thank all of the members of our laboratory for their comments and for permission to quote unpublished work.

V. REFERENCES

Aguirre, A. A., McLean, R. G., Cook, R. S., and Quan, T. J., 1992, Serologic survey for selected arboviruses and other potential pathogens in wildlife from Mexico, *J. Wildl. Dis.* **28:**435.

Al-Busaidy, S. M., and Mellor, P. S., 1991, Isolation and identification of arboviruses from the Sultanate of Oman, *Epidemiol. Infect.* **106:**403.

Al-Busaidy, S., Hamblin, C., and Taylor, W. P., 1987, Neutralising antibodies to Akabane virus in free-living wild animals in Africa, *Trop. Anim. Health Prod.* **19:**197.

Al-Busaidy, S. M., Mellor, P. S., and Taylor, W. P., 1988, Prevalence of neutralising antibodies to Akabane virus in the Arabian peninsula, *Vet. Microbiol.* **17:**141.

Anderson, C. R., Spence, L., Downs, W. G., and Aitken, T. H. G., 1961, Oropouche virus: A new human disease agent from Trinidad, West Indies, *Am. J. Trop. Med. Hyg.* **10:**574.

Artsob, H., 1983, Distribution of California serogroup viruses and virus infection in Canada, in: *California Serogroup Viruses* (C. H. Calisher and W. H. Thompson, eds.), pp. 277–292, Liss, New York.

Balfour, H. H., Edelman, C. K., Cook, F. E., Barton, W. I., Buzicky, A. W., Siem, R. A., and Bauer, H., 1975, Isolates of California encephalitis (La Crosse) virus from field-collected eggs and larvae of Aedes triseriatus: Identification of the overwintering site of California encephalitis, *J. Infect. Dis.* **131:**712.

Bardos, V., 1965, Tahyna virus in experimentally infected white mice, *Acta Virol.* **9:**358.

Bardos, V., 1969, On the pathogenesis of California complex virus infectons, in: *Arboviruses of the California Complex and the Bunyamwera Group* (V. Bardos, ed.), pp. 229–236, Slovak Academy of Sciences, Bratislava.

Beaty, B. J., and Thompson, W. H., 1975, Emergence of La Crosse virus from endemic foci, *Am. J. Trop. Med. Hyg.* **24:**685.

Beaty, B. J., Miller, B. R., Shope, R. E., Rozhon, E. J., and Bishop, D. H., 1982, Molecular basis of bunyavirus per os infection of mosquitoes: Role of the middle-sized RNA segment, *Proc. Natl. Acad. Sci. USA* **79:**1295.

Berry, R. L., Parsons, M. A., Restifo, R. A., Peterson, E. O., Gordon, S. W., Reed, M. R., Calisher, C. H., Bear, G. T., and Halpin, T. J., 1983, California serogroup virus infections in Ohio: An 18-year retrospective summary, in: *California Serogroup Viruses* (C. H. Calisher and W. H. Thompson, eds.), pp. 215–223, Liss, New York.

Bishop, D. H. L., and Shope, R. E., 1979, Bunyaviridae, *Comp. Virol.* **14:**1.

Bishop, D. H. L., Calisher, C. H., Casals, J., Chumakov, M. P., Gaidamovich, S. Y., Hannoun, C., Lvov, D. K., Marshall, I. D., Oker-Blom, N., Pettersson, R. F., Porterfield, J. S., Russell, P. K., and Shope, R. E., 1980, Bunyaviridae, *Intervirology* **14:**125.

Burton, A. N., McLintock, J., and Francy, D. B., 1973, Isolation of St. Louis encephalitis and Cache Valley viruses from Saskatchewan mosquitoes, *Can. J. Public Health* **64:**368.

Calisher, C. H., 1983, Taxonomy, classification, and geographic distribution of California serogroup bunyaviruses, in: *California Serogroup Viruses* (C. H. Calisher and W. H. Thompson, eds.), pp. 1–18, Liss, New York.

Calisher, C. H., Francy, D. B., Smith, G. C., Muth, D. J., Lazuick, J. S., Karabatsos, N., Jakob, W. L., and McLean, R. G., 1986, Distribution of Bunyamwera serogroup viruses in North America, *Am. J. Trop. Med. Hyg.* **35:**429.

Calisher, C. H., Monath, T. P., Sabattini, M. S., Mitchell, C. J., Lazuick, J. S., Tesh, R. B., and Cropp, C. B., 1987, A newly recognized vesiculovirus, Calchaqui virus, and subtypes of Melao and Maguari viruses from Argentina, with serologic evidence for infections of humans and horses, *Am. J. Trop. Med. Hyg.* **36:**114.

Calisher, C. H., Lazuick, J. S., Lieb, S., Monath, T. P., and Castro, K. G., 1988a, Human infections with Tensaw virus in South Florida: Evidence that Tensaw virus subtypes stimulate the production of antibodies reactive with closely related Bunyamwera serogroup viruses, *Am. J. Trop. Med. Hyg.* **39:**117.

Calisher, C. H., Oro, J. G. B., Lord, R. D., Sabattini, M. S., and Karabatsos, N., 1988b, Kairi virus identified from a febrile horse in Argentina, *Am. J. Trop. Med. Hyg.* **39:**519.

Campbell, G. L., Reeves, W. C., Hardy, J. L., and Eldridge, B. F., 1992, Seroepidemiology of California and Bunyamwera serogroup Bunyavirus infections in humans in California, *Am. J. Epidemiol.* **136:**308.

Chandler, J. F., Hogge, G., Endres, M., Jacoby, D. R., Nathanson, N., and Beatty, B., 1991, Reassortment of La Crosse and Tahyna bunyaviruses in *Aedes triseriatus* mosquitoes, *Virus Res.* **20:**181.

Chun, R. W. M., 1983, Clinical aspects of La Crosse encephalitis: Neurological and psychological sequelae, in: *California Serogroup Viruses* (C. H. Calisher and W. H. Thompson, eds.), pp. 193–201, Liss, New York.

Chun, R. W. M., Thompson, W. H., Grabow, J. D., and Matthews, C. G., 1968, California arbovirus encephalitis in children, *Neurology* **18**:369.

Chung, S. I., Livingston, C. W., Jr., Edwards, J. F., Gauer, B. B., and Collisson, E. W., 1990, Congenital malformations in sheep resulting from in utero innoculation of Cache Valley virus, *Am. J. Vet. Res.* **51**:1645.

Clark, G. G., Crabbs, C. L., Bailey, C. L., Calisher, C. H., and Craig, G. B., Jr., 1986, Identification of Aedes campestris from New Mexico: With notes on the isolation of Western equine encephalitis and other arboviruses, *J. Am. Mosq. Control Assoc.* **2**:529.

Crandell, R. A., Livingston, C. W., Jr., and Shelton, M. J., 1989, Laboratory investigation of a naturally occurring outbreak of arthrogryposis–hydranencephaly in Texas sheep, *J. Vet. Diagn. Invest.* **1**:62.

Danielova, V., 1974, The property changes of two Tahyna virus strains by influence of serial passages in various environments, *Z. Bakteriol. Hyg.* **229**:323.

Dargan, D. J., Galt, C. B., and Subak-Sharpe, J. H., 1992, The effect of cicloxolone sodium on the replication in cultured cells of adenovirus type 5, reovirus type 3, poliovirus type 1, two bunyaviruses and Semliki Forest virus, *J. Gen. Virol.* **73**:407.

Davies, F. G., and Jessett, D. M., 1985, A study of the host range and distribution of antibody to Akabane virus (genus Bunyavirus, family Bunyaviridae) in Kenya, *J. Hyg. Cambridge* **95**:191.

DeFoliart, G. R., 1983, Aedes triseriatus: Vector biology in relationship to the persistence of La Crosse virus in endemic foci, in: *California Serogroup Viruses* (C. H. Calisher and W. H. Thompson, eds.), pp. 89–105, Liss, New York.

Deibel, R., Srihongse, S., Grayson, M. A., Grimstad, P. R., Mahdy, M. S., Artsob, H., and Calisher, C. H., 1983, Jamestown Canyon virus: The etiologic agent of an emerging human disease? in: *California Serogroup Viruses* (C. H. Calisher and W. H. Thompson, eds.), pp. 313–328, Liss, New York.

Dixon, K. E., Travassos da Rosa, A. P. A., Travassos da Rosa, J. F., and Llewellyn, C. H., 1981, Oropouche virus. II. Epidemiological observations during an epidemic in Santarem, Para, Brazil in 1975, *Am. J. Trop. Med. Hyg.* **30**:161.

Edwards, J. F., Livingston, C. W., Chung, S.I., and Collisson, E. C., 1989, Ovine arthrogryposis and central nervous system malformations associated with in utero Cache Valley virus infection: Spontaneous disease, *Vet. Pathol.* **26**:33.

Eldridge, B. F., Calisher, C. H., Fryer, J. L., Bright, L., and Hobbs, D. J., 1987, Serological evidence of California serogroup virus activity in Oregon, *J. Wildl. Dis.* **23**:199.

Elliott, R. M., and Wilkie, M. L., 1986, Persistent infection of Aedes albopictus C6/36 cells by Bunyamwera virus, *Virology* **150**:21.

Endres, M. J., Valsamakis, A., Gonzalez-Scarano, F., and Nathanson, N. A., 1990, Neuroattenuated bunyavirus variant: Derivation, Characterization and revertant clones, *J. Virol.* **64**:1927.

Francy, D. B., Karabatsos, N., Wesson, D. M., Moore, C. G., Jr., Lazuick, J. S., Niebylski, M. L., Tsai, T. F., and Craig, G. B., Jr., 1990, A new arbovirus from Aedes albopictus, an Asian mosquito established in the United States, *Science* **250**:1738.

Gauld, L. W., Hanson, R. P., Thompson, W. H., and Sinha, S. K., 1974, Observations on a natural cycle of La Crosse virus (California group) in southwestern Wisconsin, *Am. J. Trop. Med. Hyg.* **23**:983.

Gonzalez-Scarano, F., Shope, R. E., Calisher, C. H., and Nathanson, N., 1982, Characterization of monoclonal antibodies against the G1 and N proteins of La Crosse and Tahyna, two California serogroup bunyaviruses, *Virology* **120**:42.

Gonzalez-Scarano, F., Janssen, R., Najjar, J. A., Pobjecky, N., and Nathanson, N., 1985, An avirulent G1 glycoprotein variant of La Crosse bunyavirus with a defective fusion function, *J. Virol.* **54**:757.

Grady, L. J., Srihongse, S., Grayson, M. A., and Deibel, R., 1983, Monoclonal antibodies against La Crosse virus, *J. Gen. Virol.* **64**:1699.

Grimstad, P. R., Calisher, C. H., Haroff, R. N., Wentworth, B. B., 1984a, Jamestown Canyon virus (California serogroup) is the etiologic agent of widespread infection in Michigan humans, *Am. J. Trop. Med. Hyg.* **35**:376.

Grimstad, P. R., Barrett, R. L., Humphrey, R. L., Sinsko, M. V., 1984b, Serologic evidence for widespread infection with La Crosse and St. Louis encephalitis viruses in the Indiana human population, *Am. J. Epidemiol.* **119**:913.

Griot, C., Gonzalez-Scarano, F., and Nathanson, N., 1993a, Molecular determinants of the virulence and infectivity of California serogroup bunyaviruses, *Annu. Rev. Microbiol.* **47**:117.

Griot, C., Pekosz, A., Lukac, D., Scherer, S. S., Stillmock, K., Schmeidler, D., Endres, M. J., Gonzalez-Scarano, F., and Nathanson, N., 1993b, Polygenic control of neuroinvasiveness in California serogroup bunyaviruses, *J. Virol.* **67**:3861.

Griot, C., Pekosz, A., Davidson, R., Stillmock, K., Hoek, M., Lukac, D., Schmeidler, D., Cobbinah, I., Gonzalez-Scarano, F., and Nathanson, N., 1994, Replication in cultured C2C12 muscle cells correlates with the neuroinvasiveness of California serogroup bunyaviruses, *Virology* **201**:399.

Gundersen, C. B., and Brown, K. L., 1983, Clinical aspects of La Crosse encephalitis: Preliminary report, in: *California Serogroup Viruses* (C. H. Calisher and W. H. Thompson, eds.), pp. 169–177, Liss, New York.

Gundlach, A. L., Grabara, C. S. G., Johnston, G. A. R., and Harper, P. A. W., 1990, Receptor alterations associated with spinal motoneuron degeneration in bovine Akabane disease, *Ann. Neurol.* **27**:513.

Hamblin, C., Anderson, E. C., Jago, M., Mlengeya, T., and Hirji, K., 1990, Antibodies to some pathogenic agents in free-living wild species in Tanzania, *Epidemiol. Infect.* **105**:585.

Hammon, W. McD., and Reeves, W. C., 1952, California encephalitis virus—a newly described agent. I. Evidence of natural infection in man and other animals, *Calif. Med.* **77**:303.

Haziroglu, R., Haritani, M., and Narita, M., 1987, Demonstration of Akabane virus antigen in experimentally infected mice using immunoperoxidase method, *Jpn. J. Vet. Sci.* **49**:133.

Henderson, B. E., and Coleman, P. H., 1971, The growing importance of California arboviruses in the etiology of human disease, *Prog. Med. Virol.* **13**:404.

Hoch, A. L., Pinheiro, F. P., Roberts, D. R., and Gomes, M. D. C., 1987, Laboratory transmission of Oropouche virus by *Culex quinquefasciatus* Say, *PAHO Bull.* **21**:55.

Hoffmann, D., and St. George, T. D., 1985, Growth of epizootic hemorrhagic disease, Akabane, and ephemeral fever viruses in Aedes albopictus cells maintained at various temperatures, *Aust. J. Biol. Sci.* **38**:183.

Iversen, J. O., Wagner, R. J., and Leung, M. K., 1979, Cache Valley virus: Isolations from mosquitoes in Saskatchewan, 1972–1974, *Can. J. Microbiol.* **25**:760.

Iversson, L. B., Travassos da Rosa, A. P. A., Coimbra, T. L. M., Ferreira, I. B., and Nassar, E. D., 1987, Human disease in Ribeira Valley, Brazil caused by Caraparu, a Group C arbovirus—Report of a case, *Rev. Inst. Med. Trop. Sao Paulo* **29**:112.

Jakob, W. L., Davis, T., and Francy, D. B., 1989, Occurrence of Culex erythrothorax in southeastern Colorado and report of virus isolations from this and other mosquito species, *J. Am. Mosq. Control Assoc.* **5**:534.

James, W. S., and Millican, D., 1986, Host-adaptive antigenic variation in Bunyaviruses, *J. Gen. Virol.* **67**:2803.

Janssen, R., Gonzalez-Scarano, F., and Nathanson, N., 1984, Mechanisms of bunyavirus virulence. Comparative pathogenesis of a virulent strain of La Crosse and an avirulent strain of Tahyna virus, *Lab. Invest.* **50**:447.

Janssen, R. S., Nathanson, N., Endres, M. J., and Gonzalez-Scarano, F., 1986, Virulence of La Crosse virus is under polygenic control, *J. Virol.* **59**:1.

Johnson, K. P., and Johnson, R. T., 1968, California encephalitis. II. Studies of experimental infection in the mouse, *J. Neuropathol. Exp. Neurol.* **27**:390.

Jonkers, A. H., Spence, L., Downs, W. G., Aitken, T. H. G., and Tikasingh, E. S., 1968, Arbovirus studies in Bush forest, Trinidad, W. I., September 1959–December 1964, *Am. J. Trop. Med. Hyg.* **17**:276.

Kalfayan, B., 1983, Pathology of La Crosse virus infection in humans, in: *California Serogroup Viruses* (C. H. Calisher and W. H. Thompson, eds.), pp. 179–186, Liss, New York.

Kappus, K. D., Monath, T. P., Kaminski, R. M., and Calisher, C. H., 1983, Reported encephalitis associated with California serogroup virus infections in the United States, 1963–1981, in: *California Serogroup Viruses* (C. H. Calisher and W. H. Thompson, eds.), pp. 31–41, Liss, New York.

Kingsford, L., 1991, Antigenic variance *Curr. Top. Microbiol. Immunol.* **169:**181.

Kingsford, L., Ishizawa, L. D., and Hill, D. W., 1983, Biological activities of monoclonal antibodies reactive with antigenic sites mapped on the G1 glycoprotein of La Crosse virus, *Virology* **129:**443.

Kingsford, L., Boucquey, K. H., and Cardoso, T. P., 1991, Effects of specific monoclonal antibodies on La Crosse virus neutralization: Aggregation, inactivation by Fab fragments, and inhibition of attachment to baby hamster kidney cells, *Virology* **180:**591.

Kirkland, P. D., Barry, R. D., Harper, P. A. W., and Zelski, R. Z., 1988, The development of Akabane virus-induced congenital abnormalities in cattle, *Vet. Rec.* **122:**582.

Kokernot, R. H., Smithburn, K. C., de Meillon, B., and Paterson, H. E., 1958, Isolation of Bunyamwera virus from a naturally infected human being and further isolations from Aedes (Banksinella) Circumluteolus theo, *Am. J. Trop. Med. Hyg.* **7:**579.

Konno, S., Koeda, T., Madarame, H., Ikeda, S., Sasaki, T., Satoh, H., and Nakano, K., 1988, Myopathy and encephalopathy in chick embryos experimentally infected with Akabane virus, *Vet. Pathol.* **25:**1.

Kramer, W. L., Jones, R. H., Holbrook, F. R., Walton, T. E., and Calisher, C. H., 1990, Isolation of arboviruses from Culicoides Midges (Diptera: Ceratopogonidae) in Colorado during an epizootic of vesicular stomatitis New Jersey, *J. Med. Entomol.* **27:**487.

Ksiazek, T. G., and Yuill, T. M., 1977, Viremia and antibody response to La Crosse virus in sentinel gray squirrels (*Sciuris carolinensis*) and chipmunks (*Tamias striatus*), *Am. J. Trop. Med. Hyg.* **26:**815.

Kurogi, H., Inaba, Y., Takahashi, E., Sato, K., Omori, T., Miura, Y., Goto, Y., Fujiwara, Y., Hatano, Y., Kodama, K., Fukuyama, S., Sasaki, N., and Matumoto, M., 1976, Epozootic congenital arthrogryposis–hydranencephaly syndrome in cattle: Isolation of Akabane virus from affected fetuses, *Arch. Virol.* **51:**67.

Kurogi, H., Inaba, Y., Takahashi, E., Sato, K., Satoda, K., Goto, Y., Omori, T., and Matumoto, M., 1977, Congenital abnormalities in newborn calves after inoculation of pregnant cows with Akabane virus, *Infect. Immun.* **17:**338.

Kurogi, H., Inaba, Y., Takahashi, E., Sato, K., Goto, Y., Satoda, K., Omori, T., and Hatakeyama, H., 1978, Development of inactivated vaccine for Akabane disease, *Natl. Inst. Anim. Health Q.* **18:**97.

Kurogi, H., Inaba, Y., Takahashi, E., Sato, K., Akashi, H., Satoda, K., and Omori, T., 1979, An attenuated strain of Akabane virus: A candidate for live virus vaccine, *Natl. Inst. Anim. Health Q.* **19:**12.

LeDuc, J. W., 1987, Epidemiology of Hantaan and related viruses, *Lab. Anim. Sci.* **37:**413.

LeDuc, J. W., Suyemoto, W., Keefe, T. J., Burger, J. F., Eldridge, B. F., and Russell, P. K., 1975a, Ecology of California encephalitis viruses on the Del Mar Va peninsula. I. Virus isolations from mosquitoes, *Am. J. Trop. Med. Hyg.* **24:**118.

LeDuc, J. W., Suyemoto, W., Eldridge, B. F., Russell, P. K., and Barr, A. R., 1975b, Ecology of California encephalitis viruses on the Del Mar Va peninsula. II. Demonstration of transovarial transmission, *Am. J. Trop. Med. Hyg.* **24:**124.

Ludwig, G. V., Israel, B. A., Christensen, B. M., Yuill, T. M., and Schultz, K. T., 1991, Monoclonal antibodies directed against the envelope glycoproteins of La Crosse virus, *Microb. Pathogen.* **11:**411.

Lyons, M. J., and Heyduk, J., 1973, Aspects of the developmental morphology of California encephalitis virus in cultured vertebrate and arthropod cells and in mouse brain, *Virology* **54:**37.

McClure, S., McCullagh, P., Parsonson, I. M., McPhee, D. A., Della-Porta, A. J., and Orsini, A., 1988, Maturation of immunological reactivity in the fetal lamb infected with Akabane virus, *J. Comp. Pathol.* **99:**133.

McLean, D. M., Gubash, S. M., Grass, P. N., Miller, M. A., Petric, M., and Walters, T. E., 1975,

California encephalitis virus development in mosquitoes as revealed by transmission studies, immunoperoxidase staining, and electron microscopy, *Can. J. Microbiol.* **21**:453.

McLean, D. M., Grass, P. N., Judd, B. D., and Stolz, K. J., 1979, Bunyavirus development in arctic and *Aedes aegypti* mosquitoes as revealed by glucose oxidase staining and immunofluorescence, *Arch. Virol.* **62**:313.

McLean, R. G., Calisher, C. H., and Parham, G. L., 1987, Isolation of Cache Valley virus and detection of antibody for selected arboviruses in Michigan horses in 1980, *Am. J. Vet. Res.* **48**:1039.

Main, A. J., and Crans, W. J., 1986, Cache Valley virus from Aedes sollicitans in New Jersey, *J. Am. Mosq. Control Assoc.* **2**:95.

Malkova, D., 1971, Virulence of Tahyna virus in mice and its relation to thermosensitivity and character of plaque population markers, *Acta Virol.* **15**:473.

Malkova, D., and Marhoul, Z., 1971, Plaquing of different strains of Tahyna virus in GMK cells, *Acta Virol.* **15**:227.

Mangiafico, J. A., Sanchez, J. L., Figueiredo, L. T., LeDuc, W., and Peters, C. J., 1988, Isolation of a newly recognized Bunyamwera serogroup virus from a febrile human in Panama, *Am. J. Trop. Med. Hyg.* **39**:593.

Miller, B. R., 1983, A variant of La Crosse virus attenuated for Aedes triseriatus mosquitoes, *Am. J. Trop. Med. Hyg.* **32**:1422.

Mitchell, C. J., Lvov, S. D., Savage, H. M., Calisher, C. H., Smith, G. C., Lvov, D. K., and Gubler, D. J.;, 1993, Vector and host relationships of California serogroup viruses in western Siberia, *Am. J. Trop. Med. Hyg.* **49**:53.

Monath, T. P. C., Nuckolls, J. G., Berall, J., Bauer, H., Chappell, W. A., and Coleman, P. H., 1970, Studies of California encephalitis in Minnesota, *Am. J. Epidemiol.* **92**:40.

Murphy, F. A., Whitfield, S. G., Coleman, P. H., and Calisher, C. H., 1968, California group arboviruses: Electron microscopic studies, *Exp. Mol. Pathol.* **9**:44.

Narita, M., and Kawashima, K., 1993, Detection of Akabane viral antigen and immunoglobulin-containing cells in ovine fetuses by use of immunoperoxidase staining, *Am. J. Vet. Res.* **54**:420.

Neitzel, D. F., and Grimstad, P. R., 1991, Serological evidence of California group and Cache Valley virus infection in Minnesota white-tailed deer, *J. Wildl. Dis.* **27**:230.

Newton, S. E., Short, N. J., and Dalgarno, L., 1981, Bunyamwera virus replication in cultured Aedes albopictus (mosquito) cells: Establishment of a persistent viral infection, *J. Virol.* **38**:1015.

Oya, A., Okuno, T., Ogata, T., Kobayashi, I., and Matsuyama, T., 1961, Akabane, a new arbovirus isolated in Japan, *Jpn. J. Med. Sci. Biol.* **14**:101.

Pantuwatana, S., Thompson, W. H., Watts, D. M., and Hanson, R. P., 1972, Experimental infection of chipmunks and squirrels with La Crosse and Trivittatus viruses and biological transmission of La Crosse virus by *Aedes triseriatus*, *Am. J. Trop. Med. Hyg.*, **21**:476.

Parsonson, I. M., and McPhee, D. A., 1985, Bunyavirus pathogenesis, *Adv. Virus Res.* **30**:279.

Parsonson, I. M., Della-Porta, A. J., and Snowdon, W. A., 1977, Congenital abnormalities in newborn lambs after infection of pregnant sheep with Akabane virus, *Infect. Immun.* **15**:254.

Parsonson, I. M., McPhee, D. A., Della-Porta, A. J., McClure, S., and McCullagh, P., 1988, Transmission of Akabane virus from the ewe to the early fetus (32 to 53 days), *J. Comp. Pathol.* **98**:215.

Patrican, L. A., DeFoliart, G. R., and Yuill, T. M., 1985, La Crosse viremias in juvenile, subadult and adult chipmunks (Tamias striatus) following feeding by transovarially-infected Aedes triseriatus, *Am. J. Trop. Med. Hyg.* **34**:596.

Pekosz, A., Griot, C., Nathanson, N., and González-Scarano, F., 1996, Tropism of Bunyaviruses: Evidence for a G1 glycoprotein mediated entry pathway common to the California serogroup, *Virology*, in press.

Pekosz, A., Griot, C., Stillmock, K., Nathanson, N., and González-Scarano, F., 1995, Protection from La Crosse virus encephalitis with recombinant glycoproteins: Role of neutralizing anti-G1 antibodies, *J. Virol.* **69**:3475–3481.

Pinheiro, F. P., Hoch, A. L., Gomes, M. D. C., and Roberts, D. R., 1981a, Oropouche virus. IV. Laboratory transmission by *Culicoides paraensis, Am. J. Trop. Med. Hyg.* **30:**172.

Pinheiro, F. P., Travassos da Rosa, A. P. A., Travassos da Rosa, J. F. S., Ishak, R., Freitas, R. B., Gomes, M. L. C., LeDuc, J. W., and Oliva, O. F. P., 1981b, Oropouche virus: A review of clinical, epidemiological, and ecological findings, *Am. J. Trop. Med. Hyg.* **30:**149.

Pinheiro, F. P., Travassos da Rosa, A. P. A., Gomes, M. L. C., LeDuc, J. W., and Hoch, A. L., 1982, Transmission of Oropouche virus from man to hamster by the midge *Culicoides paraensis, Science* **215:**1251.

Reeves, W. C., and Hammon, W. McD., 1962, Epidemiology of the arthropod-borne viral encephalitides in Kern County, California, 1943–1952, *Univ. Calif. Publ. Public Health* **4:**1.

Reeves, W. C., Emmons, R. W., and Hardy, J. L., 1983, Historical perspective on California encephalitis virus in California, in: *California Serogroup Viruses* (C. H. Calisher and W. H. Thompson, eds.), pp. 19–30, Liss, New York.

Roberts, A., Rossier, Kolakofsky, D., Nathanson, N., and Gonzáles-Scarano, F., 1995, Completion of the La Crosse virus genome sequence and genetic comparisons of the L proteins of the Bunyaviridae, *Virology* **206:**742–745.

Roberts, D. R., Hoch, A. L., Dixon, K. E., and Llewellyn, C. H., 1981, Oropouche virus. III. Entomological observations from three epidemics in Para, Brazil, 1975, *Am. J. Trop. Med. Hyg.* **30:**165.

Rodhain, F., Gonzalez, J. P., Mercier, E., Helynck, B., Larouze, B., and Hannoun, C., 1989, Arbovirus infections and viral haemorrhagic fevers in Uganda: A serological survey in Karamoja district, 1984, *Trans. R. Soc. Trop. Med. Hyg.* **83:**851.

Rowley, W. A., Wong, Y. W., Dorsey, D. C., Hausler, R. V., and Currier, R. W., 1983, California serogroup viruses in Iowa, in: *California Serogroup Viruses* (C. H. Calisher and W. H. Thompson, eds.), pp. 237–246, Liss, New York.

Schopen, S., Labuda, M., and Beaty, B., 1991, Vertical and veneral transmission of California group viruses by *Aedes triseriatus* and *Culiseta inornata* mosquitoes, *Acta Virol.,* **35:**373.

Sellers, R. F., and Pedgley, D. E., 1985, Possible windborne spread to western Turkey of bluetongue virus in 1977 and of Akabane virus in 1979, *J. Hyg. Cambridge* **95:**149.

Seymour, C., Amundson, T. E., Yuill, T. M., and Bishop, D. H. L., 1983, Experimental infection of chipmunks and snowshoe hares with La Crosse and snowshoe hare viruses and four of their reassortants, *Am. J. Trop. Med. Hyg.* **32:**1147.

Shope, R. E., 1985, Bunyaviruses, in: *Virology* (B. N. Fields, ed.), pp. 1055–1082, Raven Press, New York.

Smith, G. C., and Francy, D. B., 1991, Laboratory studies of a Brazilian strain of *Aedes albopictus* as a potential vector of Mayaro and Oropouche viruses, *J. Am. Mosq. Control Assoc.* **7:**89.

Smithburn, K. C., Haddow, A. J., and Mahaffy, A. F., 1946, A neurotropic virus isolated from Aedes mosquitos caught in the Semliki Forest, *Am. J. Trop. Med.* **26:**189.

Srihongse, S., Grayson, M. A., and Diebel, R., 1984, California serogroup viruses in New York State: The role of subtypes in human infections, *Am. J. Trop. Med. Hyg.* **33:**1218.

Sundin, D. R., Beaty, B. J., Nathanson, N., and Gonzalez-Scarano, F., 1987, A G1 glycoprotein epitope of La Crosse virus: A determinant of infection of Aedes triseriatus, *Science* **235:**591.

Tesh, R. B., and Beaty, B. J., 1983, Localization of California serogroup viruses in mosquitoes, in: *California Serogroup Viruses* (C. H. Calisher and W. H. Thompson, eds.), pp. 67–75, Liss, New York.

Theiler, M., and Downs, W. G., 1973, Bunyamwera supergroup, in: *The Arthropod-borne Viruses of Vertebrates,* pp. 209–262, Yale University Press, New Haven, CT.

Thompson, W. H., 1983, Vector–virus relationships, in: *California Serogroup Viruses* (C. H. Calisher and W. H. Thompson, eds.), pp. 57–66, Liss, New York.

Thompson, W. H., and Beaty, B. J., 1978, Venereal transmission of La Crosse virus from male to female Aedes triseriatus, *Am. J. Trop. Med. Hyg.* **27:**187.

Thompson, W. H., Kalfayan, B., and Anslow, R. O., 1965, Isolation of California encephalitis group virus from a fatal human illness, *Am. J. Epidemiol.* **81:**245.

Tignor, G. H., Burrage, T. G., Smith, A. L., and Shope, R. E., 1983, California serogroup gene structure–function relationships; virulence and tissue tropisms, in: *California Serogroup Viruses* (C. H. Calisher and W. H. Thompson, eds.), pp. 129–138, Liss, New York.

Traavik, T., Mehl, R., and Wiger, R., 1978, California encephalitis group viruses isolated from mosquitoes collected in southern and arctic Norway, *Acta Pathol. Microbiol. Scand.* **86:**335.

Traavik, T., Mehl, R., and Wiger, R., 1985, Mosquito-borne arboviruses in Norway: Further isolations and detection of antibodies to California encephalitis viruses in human, sheep and wildlife sera, *J. Hyg. Cambridge* **94:**111.

Turell, M. J., and Hardy, J. L., 1980, Carbon dioxide sensitivity of mosquitoes infected with California encephalitis virus, *Science* **209:**1029.

Turell, M. J., and LeDuc, J. W., 1983, The role of mosquitoes in the natural history of California serogroup viruses, in: *California Serogroup Viruses* (C. H. Calisher and W. H. Thompson, eds.), pp. 43–56, Liss, New York.

Turell, M. J., Gargan, T. P., and Bailey, C. L., 1984, Replication and dissemination of Rift Valley fever virus in Culex pipiens, *Am. J. Trop. Med. Hyg.* **33:**176.

Tyler, K. L., McPhee, D. A., and Fields, B. N., 1986, Distinct pathways of viral spread in the host determined by reovirus S1 gene segment, *Science* **233:**770.

Wallnerova, A., 1969, Fluorescent antibody technique in study of experimental Tahyna and La Crosse virus infections in suckling mice, in: *Arboviruses of the California Complex and the Bunyamwera Group* (V. Bardos, ed.), pp. 237–246, Slovak Academy of Sciences, Bratislava.

Watts, D. M., Pantuwatana, S., DeFoliart, G. R., Yuill, T. M., and Thompson, W. H., 1973, Transovarial transmission of La Crosse virus (California encephalitis group) in the mosquito, Aedes triseriatus, *Science* **182:**1140.

Way, H. J., Bowen, E. T. W., and Platt, G. S., 1976, Comparative studies of some African arboviruses in cell culture in mice, *J. Gen. Virol.* **30:**123.

Whitman, L., and Shope, R. E., 1962, The California complex of arthropod-borne viruses and its relationship to the Bunyamwera group through Guaroa virus, *Am. J. Trop. Med. Hyg.* **11:**691.

Woodruff, B. A., Baron, R. C., and Tsai, T. F., 1992, Symptomatic La Crosse virus infections of the central nervous system: A study of risk factors in an endemic area, *Am. J. Epidemiol.* **136:**320.

Woodruff, P. W. R., Morrill, J. C., Burans, J. P., Hyams, K. C., and Woody, J. N., 1988, A study of viral and rickettsial exposure and causes of fever in Juba, southern Sudan, *Trans. R. Soc. Trop. Med. Hyg.* **82:**761.

Worth, C. B., Paterson, H. E., and de Meillon, B., 1961, The incidence of arthropod-borne viruses in a population of culicine mosquitoes in Tongaland, Union of South Africa (January, 1956, through April, 1960), *Am. J. Trop. Med. Hyg.* **10:**583.

Yuill, T. M., 1983, The role of mammals in the maintenance and dissemination of La Crosse virus, in: *California Serogroup Viruses* (C. H. Calisher and W. H. Thompson, eds.), pp. 77–88, Liss, New York.

CHAPTER 10

Epidemiology and Pathogenesis of Hemorrhagic Fever with Renal Syndrome

Ho Wang Lee

I. EPIDEMIOLOGY

The outbreak of deadly hantavirus pulmonary syndrome (HPS) in the United States in 1993 (CDC, 1993) changed the concept of hemorrhagic fever with renal syndrome (HFRS) (WHO, 1983) because the clinical manifestations of this newly described hantavirus infection (Duchin *et al.*, 1994) are quite different from the known typical forms of HFRS caused by Hantaan, Seoul, and Puumala viruses and because it occurred in the United States, where hantavirus disease had not been known to exist previously. Recent data on hantavirus disease show that HFRS occurs in the Old World and HPS, caused by novel hantaviruses, occurs in the New World. Epidemiological studies of HFRS progressed rapidly after the discovery of Hantaan virus, the etiologic agent of Korean hemorrhagic fever (Lee and Lee, 1976; Lee *et al.*, 1978) through seroepidemiologic surveys of hantavirus infections in humans and in animals. It was long believed that HFRS only occurs in certain limited rural areas of Eurasia, such as Korea, China, Far East of Russia, and Northern Europe (Lee, 1982). Recent findings (Lee and van der Groen, 1989; Lee *et al.*, 1990b), however, have shown that HFRS patients can be found not only in rural areas but also in urban cities in many parts of the world.

HO WANG LEE • Asan Institute for Life Sciences, Asan Medical Center, Seoul, Korea.

The Bunyaviridae, edited by Richard M. Elliott, Plenum Press, New York, 1996.

A. Geographical Distribution of HFRS and Hantaviruses

HFRS, especially the mild form of HFRS, called nephropathia epidemica (Myhrman, 1951), has been recorded for more than 80 years in Europe and large outbreaks of HFRS occurred among soldiers stationed in the field during past wars as shown in Table I. The history of HFRS is useful and interesting in understanding the epidemiology of the disease. HFRS was a major military problem during the American Civil War (Jellison, 1971) and the First World War (Abercrombie, 1916; Bradford, 1916), and also for Russian, Japanese, and German troops during the Second World War (Ishii *et al.*, 1942; Jellison, 1971; Smorodintsev *et al.*, 1944). However, it was the incidence of Korean hemorrhagic fever among UN troops during the Korean War (Smadel, 1953) that attracted world attention.

There are at least four clinical forms of hantavirus infection in humans: fatal, severe, moderate, and mild. The fatal form, HPS (Duchin *et al.*, 1994; Hjelle *et al.*, 1994) with over 50% mortality, is caused by the newly identified Four Corners virus (also called Muerto Canyon or Sin Nombre virus; C. J. Peters, personal communication), has so far been found in the Americas: the United States, Canada, and Brazil (Chapter 11). The severe form, HFRS, has 4–15% fatality, is caused by Hantaan and Belgrade viruses (Gligic *et al.*, 1992), and is common in Asia and the Balkan countries: Korea, China, Mon-

TABLE I. War and Hemorrhagic Fever with Renal Syndrome

Date	Name of war	Name of disease	Number of patients (fatality)	Local hantavirus
1861–1866	American Civil War	Epidemic nephritis	14,187	Sin Nombre Seoul Prospect Hill
1914–1918	First World War (British troops)	War nephritis, trench nephritis, epidemic nephritis	12,000	Puumala Seoul
1939–1945	Second World War			
	1. Japanese troops in China	Epidemic hemorrhagic fever	12,500 (15–30%)	Hantaan Seoul
	2. Soviet troops in the Far East	Hemorrhagic nephrosonephritis	8,000 (10–20%)	Hantaan Puumala
	3. German and Finnish troops in Lapland	Nephropathia epidemica, field fever	10,000	Puumala
	4. German war prisoners in Yugoslavia	Epidemic nephritis, virus glomerulonephritis	6,000	Puumala Hantaan Belgrade
1951–1954	Korean War (UN troops)	Korean hemorrhagic fever	3,256 (5–15%)	Hantaan Seoul

golia, Russia, Hong Kong, Myanmar, Greece, and the former Yugoslavia. The moderate form caused by Seoul virus (Lee *et al.*, 1990b) occurs in Japan, Korea, China, and Southeast Asia. The majority of cases reported in 17 European countries are of the mild form, caused by Puumala virus in Russia, Finland, Sweden, Norway, Denmark, Bulgaria, Hungary, Albania, Germany, Belgium, the Netherlands, Switzerland, England, Portugal, the former Yugoslavia, and Greece; and also in one African country, the Central African Republic.

Worldwide, the total number of hospitalized HFRS patients is about 60,000–150,000 annually, as shown in Table II. Most of these cases occur in Asia and Europe: China, Russia, Korea, Japan, Finland, Sweden, Norway,

TABLE II. Hemorrhagic Fever with Renal Syndrome and
Hantavirus Pulmonary Syndrome Patients by Country

Continent	Country	No. of HFRS patients[a]
Asia	China	60,000–150,000/year
	Russia	hundreds to thousands/year
	S. Korea	500–2000/year
	N. Korea	hundreds/year
	Japan	316 (1961–1987)
	Hong Kong	7 (1985–1987)
	Malaysia	6 (1985)
	Sri Lanka	4 (1987)
	Myanmar	3 (1994)
	Singapore	1 (1991)
Europe	Finland	hundreds since 1940
	Sweden	hundreds since 1954
	Yugoslavia	hundreds since 1957
	Norway	hundreds/year since 1952
	Bulgaria	hundreds
	France	hundreds
	Greece	hundreds
	Hungary	136 (1952–1980)
	Germany	46 (1984–1991)
	Netherlands	25 since 1981
	Italy	14 (1984–1987)
	Great Britain	7 (1984–1987)
	Belgium	4 (1978)
	Austria	1 (1990)
	Switzerland	1 (1990)
	Portugal	1 (1994)
Americas	USA	95 (1993–1994)
	Canada	3 (1994)
	Brazil	1 (1994)
Africa	Central African Republic	1 (1986–1987)

[a]There are about 150,000 hospitalized HFRS patients with 3–10% fatality each year worldwide and about 100 hantavirus pulmonary syndrome patients with 70% fatality in the Americas.

Belgium, and France (Lee *et al.*, 1990b). Seroepidemiological surveys show that hantavirus infections are distributed throughout much of the world (Lee *et al.*, 1990b; Nuti and Lee, 1991; Puthavathana *et al.*, 1992), and humans seropositive for hantaviruses have been confirmed worldwide (Fig. 1). Sixteen countries in Asia and Oceania are focally enzootic for hantaviruses: Japan, South Korea, North Korea, China, Mongolia, Russia, Taiwan, Fiji, Hong Kong, Malaysia, India, Myanmar, Indonesia, Singapore, Australia, and Sri Lanka; 20 countries in Europe: Russia, Finland, Sweden, Norway, Denmark, Bulgaria, Hungary, Czechoslovakia, Poland, Albania, Germany, France, Portugal, Belgium, Netherlands, Switzerland, Italy, England, the former Yugoslavia, and Greece; 1 country in Africa, the Central African Republic (Couland *et al.*, 1987); and 6 countries in the Americas: Canada, the United States, Mexico, Panama, Brazil, and Argentina.

Rodents sampled from the field infected with hantaviruses were demonstrated in Asia, Europe, Africa and the Americas, specifically Korea, Japan, China, Hong Kong, Thailand, Sri Lanka, Russia, Sweden, Finland, Norway, Germany, Belgium, Netherlands, France, Italy, the former Yugoslavia, Greece, Egypt, the United States, Brazil, and Argentina. Urban rats infected with hantaviruses have been trapped in 12 Asian countries: Japan, Korea, China, Hong Kong, Malaysia, Thailand, Sri Lanka, India, Singapore, Fiji, Indonesia, Philippines; in 3 European countries: Belgium, Germany, Italy; and in 3 countries in the Americas: Canada, the United States, and Brazil.

Laboratory rats infected with hantaviruses have been found in 11 countries around the world: Japan, Korea, China, Russia, Belgium, England, Malaysia, Hong Kong, Singapore, the United States, and Argentina. Infections of laboratory workers with Hantaan, Puumala, and Seoul viruses have been reported from Russia, Korea, Japan, China, Belgium, England, and France (Desmyter *et al.*, 1983; Lee and Johnson, 1982; Lloyd *et al.*, 1984; Umenai *et al.*, 1979). In Asia, 1 out of 149 laboratory-acquired cases of HFRS resulted in death, as shown in Table III. Laboratory infection with hantaviruses is very dangerous and it is strongly recommended that laboratory animals be screened against Hantaan, Seoul, and Puumala viruses before use in experiments. The importation and exportation of hantavirus-free small laboratory animals from one country to another should be controlled under internationally adopted animal quarantine rules.

B. Epidemiologic Type

Originally it was thought that HFRS only occurred in endemic rural areas of Asia, with farmers and soldiers being the most likely victims. Areas infected with Hantaan and Puumala viruses were also thought to be limited to certain geographical regions of Eurasia, infecting only their inhabitants and travelers. Recent findings (Lee *et al.*, 1990b), however, have shown that HFRS patients can be found not only in rural areas but also in urban areas (Tamura, 1964). Three epidemiologic patterns of HFRS are now recognized

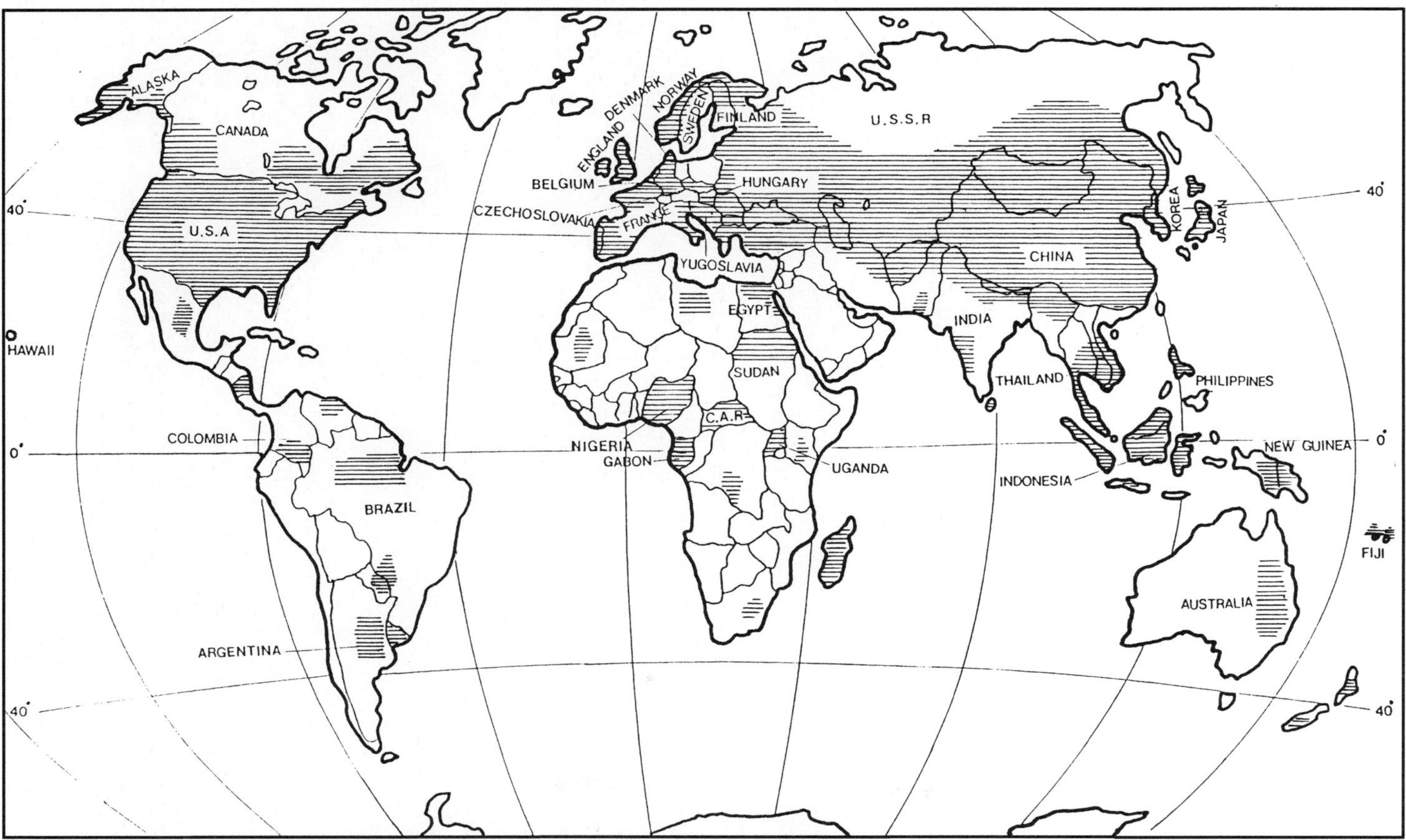

FIGURE 1. Global distribution of hantaviruses based on demonstration of immunofluorescent antibodies against the viruses among humans and rodents.

TABLE III. Laboratory-Acquired
Infections with Hantaviruses

Country	No. of patients
Russia	113 (1962)
Japan	149 (1 death) since 1964 at 24 institutes
Korea	9 since 1980 at 8 institutes
China	Many
Belgium	4
France	2
Netherlands	2
England	1

according to the location of outbreak and the reservoir hosts of the disease: rural, urban, and animal room.

1. Rural Type

The reservoirs for viruses causing HFRS in the rural endemic areas of the world are field mice (Lee and van der Groen, 1989). These rodents live mainly in fields but will invade houses during the winter season. Rural-type cases occur annually in the endemic areas and there are two seasonal peaks, in late spring and in the fall when the incidence of the infected *Apodemus* mice is high in Asia and in the Balkans, as shown in Fig. 2. Outbreaks of HFRS occur in the winter season in Russia and in Scandinavian countries. Victims are primarily farmers, soldiers, and construction workers, ranging in age from 20 to 50, who are working or stationed in the field. There is increasing evidence that many species of wild rodents and insectivores are also reservoirs of hantaviruses in Russia (Tkachenko *et al.*, 1983, 1987), China, and the former Yugoslavia (Diglisic *et al.*, 1994).

2. Urban Type

The reservoir for the hantavirus responsible for the urban type of HFRS is the house rat. This type of HFRS poses a major threat worldwide. One hundred thirty cases of HFRS were reported among residents of urban areas of Osaka, Japan, during the 1960s (Tamura, 1964), and recently over 100 cases annually of HFRS have been reported in metropolitan areas of Seoul and other large cities in Korea where the patients had never been outside the city limits, but who had histories of contact with house rats (Lee, 1989). Cases of the urban type occur throughout the year, but tend to be more frequent in the fall and winter seasons.

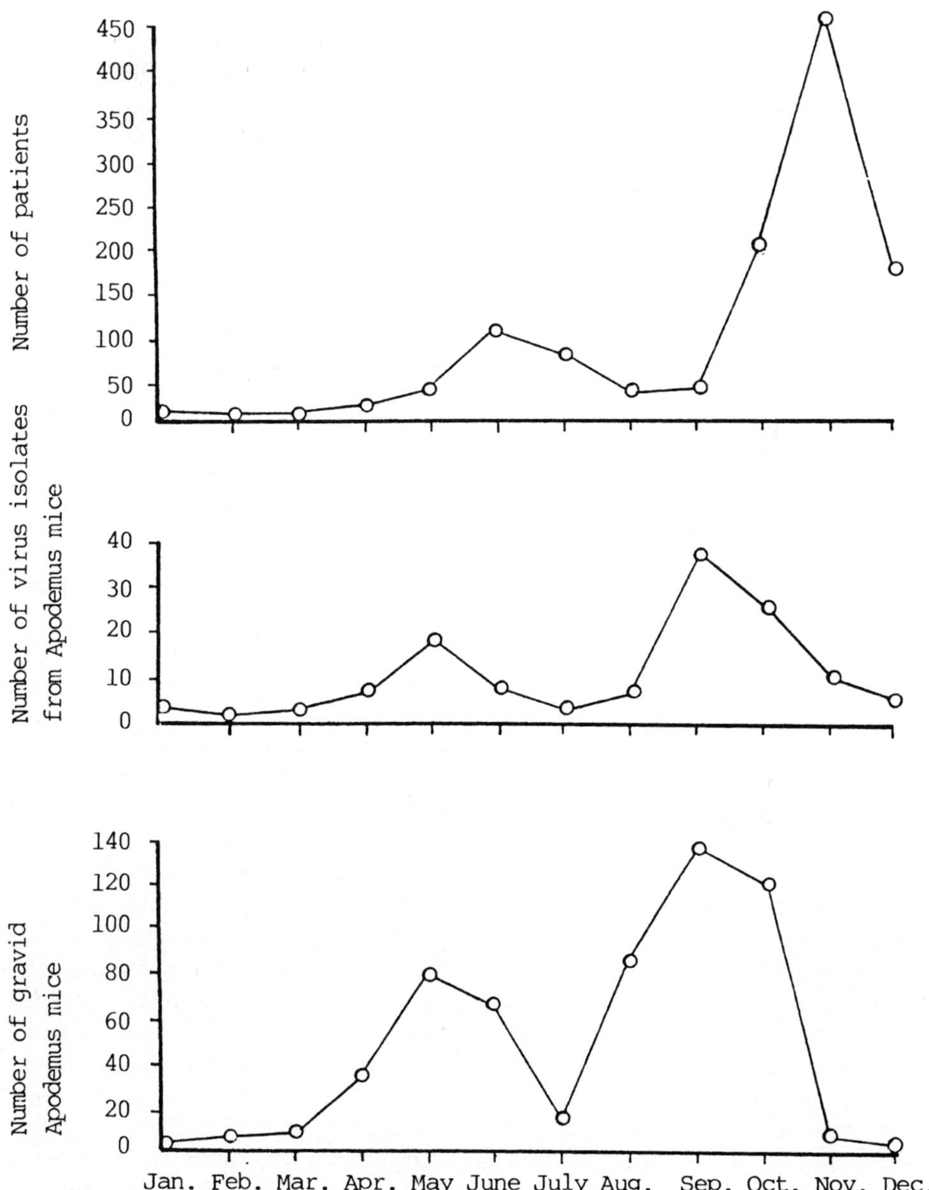

FIGURE 2. Cumulative seasonal prevalence of HFRS, and number of gravid and of infected *Apodemus agrarius* with Hantaan virus in the endemic area of Korea.

3. Animal Room Type

Animal room HFRS can be acquired from colonized experimental rats and hamsters (Lee and van der Groen, 1989), as proven by the demonstration of antibody and subsequent isolation of the virus from these animals. There were 33 outbreaks of HFRS from 1976 to 1985 among personnel in animal rooms of medical centers in Korea and Japan where Hantaan or Seoul virus experiments had *not* been conducted (Kawamata *et al.*, 1987; Lee *et al.*, 1981a, 1986). The victims numbered over 150, one of whom died. Animal room HFRS may occur at any time of the year, but a series of outbreaks in Korea and Japan occurred during the winter season in nonventilated animal rooms when the air in the rooms was dry. One hundred sixteen cases of HFRS occurred in an institute in Moscow after the introduction of wild mice, caught in areas endemic for HFRS, into the animal rooms in 1962 and a few cases have occurred in animal rooms of research institutes in France, Belgium, the Netherlands, and the United Kingdom (Desmyter *et al.*, 1983; Dournon *et al.*, 1984; Lloyd *et al.*, 1984; Osterhaus *et al.*, 1984). These incidents demonstrate that exports, imports, and exchanges of animals between research institutes can be potentially dangerous.

C. Host–Hantavirus Relationship

1. Reservoirs

The main reservoir of Hantaan virus in the rural endemic areas of Korea and China is the field mouse *Apodemus agrarius* (Lee *et al.*, 1981b; Song *et al.*, 1983); in Finland and west of the Ural mountains, the reservoir of Puumala virus is the bank vole *Clethrionomys glareolus* and field mice (Tkachenko *et al.*, 1983, 1987; Niklasson and LeDuc, 1987); and in the urban areas of Korea, Japan, and China, *Rattus* and *R. norvegicus* are reservoirs of Seoul virus (Lee *et al.*, 1990b). The reservoirs of new hantaviruses in the United States are the deer mouse *Peromyscus maniculatus* and the white-footed mouse *Peromyscus leucopus* (Nernkar *et al.*, 1993; Song *et al.*, 1994). Hantavirus antigens have been detected in 16 different rodent species and 4 different insectivorous species in the USSR and China (Gavrilovskaya *et al.*, 1983; Song *et al.*, 1983; Tkachenko *et al.*, 1983, 1987). Recently, a Seoul virus was isolated from a Syrian hamster purchased at a local animal farm in Korea (Lee *et al.*, 1990b). Hantaan virus was isolated from a resident bird in Siberia (Slonova *et al.*, 1992) and from native bats in Korea (Kim *et al.*, 1994), though the role of birds and bats as reservoirs of hantaviruses has to be confirmed and requires further investigation. The reservoirs of hantaviruses do not show clinical signs of illness but excrete the virus in the urine and saliva for a long period of time (Lee *et al.*, 1981a,b; Yanagihara *et al.*, 1985). The lungs of reservoir hosts contain larger amounts of virus than other tissues. Hantaviruses are species specific and there is no animal model to mimic the disease in humans.

2. Vectors

Although the identification of Hantaan group viruses as members of the family *Bunyaviridae* (Schmaljohn *et al.*, 1985) suggested that they may be arthropod-transmitted, there is no experimental evidence to confirm this. It has been hypothesized that Hantaan virus can be transmitted by ecto-parasites harbored by various field mice (Kasahara *et al.*, 1944), but Hantaan virus has not been isolated from arthropods nor has it been successfully cultivated in arthropod-derived cell lines.

3. Seasonal Incidence

HFRS occurs throughout the year, but the incidence is greatest during dry seasons. There are two seasonal peaks in Korea and China, a small one in late spring and a larger peak in late fall, as shown in Fig. 2. Cases usually appear singly, but small outbreaks do develop particularly when groups of susceptible persons are exposed to a contaminated focus such as digging burrows of infected rodents (Lee and van der Groen, 1989).

4. Age and Sex

The disease appears to affect most frequently those 20–50 years old; cases in children under age 10 are rare. Although HFRS occurs in both sexes, the accumulated incidence figures thus far show that twice as many males as females are infected. The victims are primarily farmers and soldiers stationed in the field and there is evidence of an increase in the number of patients among hunters, campers, and golfers in Asia.

D. Transmission

Large quantities of virus are excreted in the saliva, urine, and feces of infected *Apodemus* mice (Lee *et al.*, 1981b) and *Clethrionomys* voles (Yanagihara *et al.*, 1985). Excretion of Hantaan virus in *Apodemus* mice persists in the saliva and feces for at least 1 month and in the urine for several months, and horizontal transmission of the virus among *Apodemus* mice has been demonstrated. Noninfected *Apodemus* mice caged with infected *Apodemus* mice for several days acquired infection beginning 10 days after initial exposure. The results were the same whether the mice carried ectoparasites or had been cleaned of parasites. The main route of infection between *Apodemus* mice is via the respiratory tract, although it can be transmitted by the saliva, urine, or feces. Transmission of the virus among house rats and experimental rats was similar to that among *Apodemus* mice but the virus was detected in the excreta for a shorter time, about 1–2 weeks (Lee *et al.*, 1986). In contrast to transmission between rodents, there is no evidence of direct human-to-human transmission of the virus among hospitalized patients.

E. Groups at Risk

The following groups are at increased risk of infection by hantaviruses.

1. Personnel working in laboratories where research into the disease is being conducted
2. Animal room workers
3. Animal breeders
4. Rodent control officers
5. Mammalogists
6. Soldiers
7. Farmers
8. Construction workers
9. Hunters, hikers, and campers

Table IV shows the infection rate of people with hantaviruses in South Korea: telephone line construction employees and golf course employees are high-risk groups.

F. Risk Factors

The number of hantavirus-infected rodents and the coincidence of breeding seasons of field rodents with the dry windy periods in the endemic areas (April–May and September–October) are important risk factors. The activity of *Apodemus* mice increases during breeding seasons and epidemics of HFRS occur right after the breeding seasons in China, Korea, and Far East of Russia, as indicated in Fig. 2. Soil, foods, and grass near the burrows and

TABLE IV. Infection Rate of People with Hantaviruses in Korea

Group by profession or locality	IF antibody-positive rate to Hantaan virus ($n = 150–950$)
Residents of cities (Seoul, Pusan, Incheon)	1.0%
Blood donors (Seoul)	1.2%
Soldiers (ROK and U.S. Army)	1.1%
Residents of the rural endemic areas of HFRS (Tongducheon)	3.8%
Golf course employees (caddies, field workers)	4.2%
Telephone line construction employees	9.0%

paths of rodents become contaminated with rodent excreta containing virus. Planting and harvesting crops by farmers and military exercises by soldiers during dry windy epidemic seasons in the endemic areas increase the risk of infection by the respiratory route (Nuzum *et al.*, 1988). Poor ventilation of the animal rooms containing infected animals in the winter season is a decisive factor in laboratory infection.

G. Prevention and Control

Control of laboratory animal infection by hantaviruses is only possible by eradication of all infected animals from an animal facility and introduction of uninfected clean laboratory animals. Successful eradication of hantavirus from a contaminated animal facility by application of cesarean section and foster mother techniques has been demonstrated. Consequently, continuous monitoring of laboratory animals by the immunofluorescence technique and/or ELISA test for hantavirus antibodies is required both in animal facilities and in animal breeding farms. If an animal laboratory worker demonstrates anti-hantavirus antibodies, one should suspect that hantavirus-infected animals may be present in the animal laboratory.

Rodent control measures in rural areas are expensive and difficult to maintain over long periods since it is impossible to eradicate the reservoir of the virus from nature. In urban areas, however, rodent control is feasible and should be encouraged. Basically, control depends on reducing the contact between humans and rodent excreta; soldiers, farmers, and workers in endemic areas should not dig mouse burrows and should keep away from the rodent burrows and paths.

Recently, Lee *et al.* (1990a) developed a formalin-inactivated Hantaan virus vaccine which is commercially available in Korea. Some results on the persistence of antibodies after immunization with this vaccine suggest that a booster vaccination at 1 year is necessary for the maintenance of protective antibodies against Hantaan virus in humans (Lee *et al.*, 1992; Kim *et al.*, 1994).

II. PATHOGENESIS

Little is known about the pathogenesis of HFRS in humans, and so far a suitable animal model for the disease in humans has not been reported. Human specimens from autopsied patients with Korean hemorrhagic fever (KHF) and biopsied specimens from nephropathia epidemica (NE) patients, and samples from naturally infected wild rodents and experimentally infected mice and rats have been examined by immunohistological and histopathological methods. Hantavirus antigens were detected in the pituitary, brain, spleen, kidneys, lungs, and liver in human tissues (Kurata *et al.*, 1987).

In naturally infected animals, viral antigens were demonstrated in the lungs, liver, kidneys, and salivary glands (Lee *et al.*, 1981a). Histopathological changes in the kidney vary according to the stage of KHF. As the disease progresses, vascular congestion occurs in the intertubular spaces and is followed by progressive tubular damage, the kidneys become swollen and on section reveal a pale cortex overlying an extremely congested, often hemorrhagic, medulla, which occasionally exhibits necrosis of the pyramids (Yanagihara and Gajdusek, 1987).

Renal biopsy material from NE cases in Western Europe showed conspicuous cellular infiltrates predominantly in the medulla. The infiltrates were associated with medullary interstitial hemorrhages together with hyperemic peritubular capillaries. Tubular lesions, present in all cases, included patterning and differentiation of tubular epithelium with disappearance of the brush border and vacuolization of epithelial cells (Laehdevirta, 1971).

Several groups have attempted to infect various rodents with hantaviruses. Infection of newborn rodents is fatal, but infection of adult animals with these viruses results in prolonged infection without any clinical symptoms. In experimentally infected rodents, viral antigens were demonstrated in almost all organs, especially in nerve tissues and in endothelial cells (Lee *et al.*, 1981a; Yanagihara and Gajdusek, 1987). Recently, it was shown that both humoral and cellular immunity play a role in protection against Hantaan virus infection in mice and that T cells possessing L3 TQ4 Lyt 2^+ markers on the cell surface are especially important in elimination of infectious virus *in vivo* (Asasa *et al.*, 1987). However, considerably more research is needed into all biological and clinical aspects of infection with this group of viruses.

III. REFERENCES

Abercrombie, R. G., 1916, Observations on the acute phase of five hundred cases of war nephritis, *J. R. Army Med. Corps* **27**:10.

Asasa, J. H., Tamura, M., and Kondo, K., 1987, Role of T-lymphocyte subset in protection and recovery from Hantaan virus infection in mice, *J. Gen. Virol.* **68**:1961.

Bradford, J. R., 1916, Nephritis in the British troops in Flanders, *Q. J. Med.* **9**:445.

Centers for Disease Control, 1993, Update: Outbreak of hantavirus infection—Southwestern United States, *MMWR* **42**:421.

Couland, X., Chonaib, E., and Georges, A. J., 1987, First human case of hemorrhagic fever with renal syndrome in the Central African Republic, *Trans. R. Soc. Trop. Med. Hyg.* **81**:686.

Desmyter, J., LeDuc, J. W., Johnson, K. M., Bresseur, F., Deckers, C., and van Ypersele de Strihou, C., 1983, Laboratory rat associated outbreak of hemorrhagic fever with renal syndrome due to Hantaan-like virus in Belgium, *Lancet* **2**:1445.

Diglisic, G., Xiao, S., Gligic, A., Otradovic, M., Stojanovic, R., Velimirovic, D., Lukac, V., Rossi, C. A., and LeDuc, J. W., 1994, Isolation of a Puumala-like virus from *Mus musculus* captured in Yugoslavia and its association with severe hemorrhagic fever with renal syndrome, *J. Infect. Dis.* **169**:204.

Dournon, E., Moriniere, B., Mattheson, S., Girard, P. M., Gonzales, J. P., Hirsh, F., and McCormick, J. B., 1984, HFRS after a wild rodent bite in the Haute Savoie and risk of exposure to Hantaan-like virus in a Paris laboratory, *Lancet* **1:**676.

Duchin, J. S., Koster, F. T., Peters, C. J., Simpson, G. L., Tempest, B., Zaki, S. R., Ksiazek, T. G., Rollin, P. E., Nichol, S., Umland, E. T., Moolenaar, R. L., Reef, S. E., Nolte, K. B., Gallaher, M. M., Butler, J. C., Rreiman, R. F., and the Hantavirus Study Group, 1994, Hantavirus pulmonary syndrome: A clinical description of 17 patients with a newly recognized disease, *N. Engl. J. M.* **330:**949.

Gavrilovskaya, I. N., Apekina, N. S., Miasnikov, Y. A., Yu, A., Bernshtein, A. D., Ryltseva, E. V., Gorbachkova, E. A., and Chumakov, M. P., 1983, Features of circulation of hemorrhagic fever with renal syndrome (HFRS) virus among small mammals in the European USSR, *Arch. Virol.* **75:**313.

Gligic, A., Dimcovic, N., Xiao, S. Y., Yanagihara, R., and Gajdusek, D. C., 1992, Belgrade virus, a new hantavirus causing severe hemorrhagic fever with renal syndrome in Yugoslavia, *J. Infect. Dis.* **166:**113.

Hjelle, B., Jenison, S., Torrez-Martinez, N., Yamada, T., Notte, K., Zumwalt, R., Macinnes, K., and Meyers, G., 1994, A novel hantavirus associated with an outbreak of fatal respiratory disease in the southwestern United States: Evolutionary relationships to known hantaviruses, *J. Virol.* **68:**592.

Ishii, S., Ando, K., Watanabe, N., Murakami, R., Nagayama, T., and Ishikawa, T., 1942, Studies on Songo fever, *Jpn. Army Med. J.* **355:**1755.

Jellison, W. H., 1971, Korean hemorrhagic fever and related diseases; a critical review and hypothesis, Mountain Press Publ. Co., p. 12.

Kasahara, S., Kitano, M., Kikuchi, H., Sakuyama, M., Kanazawa, K., Nezu, N., Yoshimara, M., and Kudo, T., 1944, Studies on pathogen of epidemic hemorrhagic fever, *J. Jpn. Pathol.* **34:**1.

Kawamata, J., Yamanouchi, T., Miyamoto, H., Tanakashi, M., Yamanishi, K., Kurata, T., Dohmae, K., and Lee, H. W., 1987, Control of laboratory acquired hemorrhagic fever with renal syndrome (HFRS) in Japan, *Lab. Anim. Sci.* **37:**431.

Kim, G. R., Lee, Y. T., and Park, C. H., 1994, A new natural reservoir of hantavirus; isolation of hantaviruses from lung tissues of bats, *Arch. Virol.* **134:**85.

Kim, Y. S., and Lee, H. W., 1994, Persistence of antibodies after immunization with inactivated vaccine against HFRS among children and teenagers, *J. Korean Clin. Pharmacol.* **2:**48.

Kurata, T., Sata, T., and Yamanishi, T., 1987, Viral pathology of hemorrhagic fever with renal syndrome virus infections in human and rodents, (abstract) 29th Int. Coll. Hantaviruses, Antwerp, p. 8.

Laehdevirta, J., 1971, Nephropathia epidemica in Finland. A clinical histological and epidemiological study, *Ann. Clin. Res.* **3(Suppl. 8):**1.

Lee, C. H., Byun, K. S., Kim, W. J., Woo, Y. D., and Lee, H. W., 1992, Persistence of antibodies after immunization with an inactivated vaccine against HFRS in humans, *J. Korean Soc. Microbiol.* **22:**239.

Lee, H. W., 1982, Korean hemorrhagic fever, *Prog. Med. Virol.* **28:**96.

Lee, H. W., 1989, Hemorrhagic fever with renal syndrome in Korea, *Rev. Infect. Dis.* **2(Suppl. 4):** S864.

Lee, H. W., and Johnson, K. M., 1982, Laboratory-acquired infection with Hantaan virus, *J. Infect. Dis.* **146:**638.

Lee, H. W., and Lee, P. W., 1976, Korean hemorrhagic fever. I. Demonstration of causative antigen and antibodies, *Korean J. Intern. Med.* **19:**371.

Lee, H. W., and van der Groen, G., 1989, Hemorrhagic fever with renal syndrome, *Prog. Med. Virol.* **36:**62.

Lee, H. W., Lee, P. W., and Johnson, K. M., 1978, Isolation of the etiologic agent of Korean hemorrhagic fever, *J. Infect. Dis.* **137:**298.

Lee, H. W., French, G. R., Lee, P. W., Baek, L. J., Tsuchiya, K., and Foulke, R. S., 1981a, Observations on natural and laboratory infection of rodents with the etiologic agent of Korean hemorrhagic fever, *Am. J. Trop. Med. Hyg.* **30:**477.

Lee, H. W., Lee, P. W., Baek, L. J., Song, C. K., Seong, I. W., and Kim, J. H., 1981b, Intraspecific transmission of Hantaan virus, the etiologic agent of Korean haemorrhagic fever in the rodent *Apodemus agrarius*, *Am. J. Trop. Med. Hyg.* **30**:1106.

Lee, H. W., Baek, L. J., and Kim, H. D., 1986, Studies on laboratory rat infection with Hantavirus in animal rooms of institutes in Seoul, *J. Korean Soc. Virol.* **16**:113.

Lee, H. W., An, C. N., Song, J. W., Baek, L. J., Seo, T. J., and Park, S. C., 1990a, Field trial of an inactivated vaccine against hemorrhagic fever with renal syndrome in humans, *Arch. Virol.* **(Suppl. 1)**:35.

Lee, H. W., Lee, P. W., Baek, L. J., and Chu, Y. K., 1990b, Geographical distribution of hemorrhagic fever with renal syndrome and hantaviruses, *Arch. Virol.* **(Suppl. 1)**:5.

Lloyd, G., Bowen, E. T. W., Jones, N., and Pendry, A., 1984, HFRS outbreak associated with laboratory rats in UK, *Lancet* **1**:1175.

Myhrman, G., 1951, Nephropathia epidemica, a new infectious disease in northern Scandinavia, *Acta Med. Scand.* **140**:52.

Nernkar, V. R., Song, K., Gajdusek, D. C., and Yanagihara, R., 1993, Genetically distinct hantavirus in deer mice, *Lancet* **342**:1058.

Niklasson, B., and LeDuc, J. W., 1987, Epidemiology of nephropathia epidemic in Sweden, *J. Infect. Dis.* **155**:269.

Nuti, M., and Lee, H. W., 1991, Serological evidence of hantavirus infection in some tropical populations, *Trans. R. Soc. Trop. Med. Hyg.* **85**:297.

Nuzum, E. O., Rossi, C. A., Stephenson, E. H., and LeDuc, J. W., 1988, Aerosol transmission of Hantaan and related viruses to laboratory rats, *Am. Soc. Trop. Med. Hyg.* **38**:636.

Osterhaus, A., Spijkers, Van Steenis, B., and van der Groen, G., 1984, Hantavirus infection in Nederland, *Ned. Tijdschr. Geneeskd.* **128**:2461.

Puthavathana, P., Lee, H. W., and Kang, C. Y., 1992, Typing of hantaviruses from five continents by polymerase chain reaction, *Virus Res.* **26**:1.

Schmaljohn, C. S., Hasty, S. E., Dalrymple, J. M., LeDuc, J. W., Lee, H. W., von Bonsdorff, C. H., Brummer-Korvenkontio, M., Vaheri, A., Tsai, T. F., Regnery, H. L., Goldgaber, D., and Lee, P. W., 1985 Antigenic and genetic properties of viruses linked to hemorrhagic fever with renal syndrome, *Science* **227**:1041.

Slonova, R. A., Tkachenko, E. A., Knshnarev, E. L., Dzagurova, T. K., and Astakova, T. I., 1992, Hantavirus isolation from birds, *Acta Virol.* **36**:493.

Smadel, J. E., 1953, Epidemic hemorrhagic fever, *Am. J. Public Health* **43**:1327.

Smorodintsev, A. A., Altchuler, I. S., Dunaevskii, M. I., Kiselev, M. V., Churilov, A. V., and Darshkevich, V., 1944, Etiology and clinical observation of hemorrhagic nephrosonephritis, Medgiz, Moscow, p. 38.

Song, G., Liao, H. X., Fu, J. L., Hang, C. S., Ni, D. S., Xia, Z. G., Lee, Z. Z., Su, T., Qiu, H. L., Zhang, Q. F., Gao, G. Z., and Zhao, Y. S., 1983, Etiologic studies of epidemic hemorrhagic fever (hemorrhagic fever with renal syndrome), *J. Infect. Dis.* **147**:654.

Song, J., Baek, L. J., Gajdusek, D. C., Yanagihara, R., Gavrilovskaya, I., Luft, R. J., Machow, E. R., and Hjelle, B., 1994, Isolation of pathogenic hantavirus from white-footed mouse, *Lancet* **344**:1637.

Tamura, M., 1964, Occurrence of epidemic hemorrhagic fever in Osaka city: First cases found in Japan with characteristic features of marked proteinuria, *Biken J.* **7**:79.

Tkachenko, E. A., Ivanov, A. P., Donets, M. A., Miasnikov, Y. A., Ryltseva, E. V., Gaponova, L. K., Bashkirtsev, V. N., Okulova, N. M., and Drozdov, G. S., 1983, Potential reservoir and vectors of hemorrhagic fever with renal syndrome (HFRS) in the USSR, *Ann. Soc. Belg. Med. Trop.* **63**:267.

Tkachenko, E. A., Ryltseva, E. V., and Miasnikov, Y. A., 1987, Study of circulation of hemorrhagic fever with renal syndrome virus among small mammals in the USSR, *Probl. Virol.* **6**:709.

Tsai, T. F., 1987, Hemorrhagic fever with renal syndrome mode of transmission to humans, *Lab. Anim. Sci.* **37**:428.

Umenai, T., Lee, H. W., Lee, P. W., Saito, T., Toyoda, T., Hongo, M., Yoshinage, K., Nobunaga, T.,

Horiuchi, T., and Ishida, N., 1979, Korean hemorrhagic fever in staff in an animal laboratory, *Lancet* **1:**1314.
WHO, 1983, Hemorrhagic fever with renal syndrome, memorandum from a WHO meeting, *Bull. WHO* **61:**269.
Yanagihara, R., and Gajdusek, D. C., 1987, Hemorrhagic fever with renal syndrome: Global epidemiology and ecology of hantavirus infection, *Med. Virol.* **6:**171.
Yanagihara, R., Amyx, H. L., and Gajdusek, D. C., 1985, Experimental infection with Puumala virus, the etiologic agent of nephropathia epidemica, in bank voles (*Clethrionomys glareolus*), *J. Virol.* **55:**34.

CHAPTER 11

Hantavirus Pulmonary Syndrome and Newly Described Hantaviruses in the United States

STUART T. NICHOL, THOMAS G. KSIAZEK, PIERRE E. ROLLIN, AND C. J. PETERS

I. INTRODUCTION

In May, 1993, an outbreak of an acute undifferentiated febrile illness, followed by rapid development of respiratory failure, shock, and, in many cases, death, was identified in the Four Corners region of the southwestern United States (Centers for Disease Control and Prevention, 1993). Specimens from the patients were tested for a wide variety of agents at the University of New Mexico (UNM), the state health departments of Arizona and New Mexico, and the Centers for Disease Control and Prevention (CDC). In early June, the first evidence of the cause of illness was found. Detection of IgG and IgM antibodies to several hantaviruses by enzyme-linked immunosorbent assay (ELISA) and indirect immunofluorescence (IFA) suggested that the illness was caused by a serologically cross-reactive hantavirus. Within a few days,

STUART T. NICHOL, THOMAS G. KSIAZEK, PIERRE E. ROLLIN, AND C. J. PETERS • Special Pathogens Branch, Division of Viral and Rickettsial Diseases, National Center for Infectious Diseases, Centers for Disease Control and Prevention, Atlanta, Georgia 30333.

The Bunyaviridae, edited by Richard M. Elliott, Plenum Press, New York, 1996.

other laboratory and epidemiologic information confirmed the preliminary results. Antigen was detected in endothelial cells of lungs and other organs by immunohistochemistry, using a broadly reactive hantavirus nucleoprotein monoclonal antibody (Zaki *et al.*, 1995). Nichol and colleagues (Nichol *et al.*, 1993), using nested reverse transcriptase–polymerase chain reaction (RT-PCR), independently confirmed the serologic results by detecting hantavirus genetic sequences in patient tissues. Furthermore, the sequence was unique and represented a new hantavirus most closely related to Prospect Hill virus, a North American hantavirus not known to cause human disease. Like previously discovered hantaviruses, this newly recognized virus was found to be carried by rodents, and the deer mouse (*Peromyscus maniculatus*) was incriminated (Childs *et al.*, 1994). This rodent species had the highest antibody prevalence among animals tested, and RT-PCR products from rodents captured at case households were identical or closely related to those from the human patients. The concurrence of antibody and virus RNA detection in nearly all (96%) of the deer mice tested by both techniques suggested a chronic infection of this host, leading to the conclusion that *P. maniculatus* was the primary reservoir of this newly recognized hantavirus.

In common with previously characterized hantaviruses, successful cell culture isolation was difficult and required several months. The causative virus was isolated from *P. maniculatus* tissue in Vero cells only after blind passage in laboratory-raised *P. maniculatus* (Elliott *et al.*, 1994). The agent of this newly described disease, hantavirus pulmonary syndrome (HPS), was registered as Sin Nombre (SN) virus. Investigation of HPS cases occurring outside of the geographical range of *P. maniculatus* has led to the description of several other distinct hantaviruses (see below) also associated with human disease.

II. CLINICAL MANIFESTATIONS

HPS begins with a nonspecific prodromal phase, lasting for 3 to 6 days, always including fever and myalgia, and frequently including headache, and gastrointestinal symptoms (e.g., nausea, vomiting, abdominal pain); dizziness and cough are also seen in some patients. This prodrome is followed by a cardiopulmonary phase with progressive cough, dyspnea, and shortness of breath, usually resulting in the patient requiring emergency medical attention. Tachypnea, tachycardia, fever, and hypotension are the most common findings at admission (Duchin *et al.*, 1994). No hemorrhagic signs have been noted. Pulmonary radiographic findings have shown evidence of interstitial edema, and extensive bibasilar or perihilar airspace disease, normal heart size, and pleural effusions are frequently present during the course of the disease (Duchin *et al.*, 1994; Ketai *et al.*, 1994). Initial clinical laboratory findings included hemoconcentration, thrombocytopenia, increased proportion of immature granulocytes and atypical lymphocytes on peripheral blood smears, sometimes marked leukocytosis (>25,000 cells/mm^3), and elevated levels of serum lactate dehydrogenase and aspartate aminotransferase. Renal failure is not a feature of HPS, but mild to moderate proteinuria is frequently observed, and elevated creatininemia has been occasionally noticed.

Progressive pulmonary edema and hypoxemia almost always require intubation and mechanical ventilation. The hemodynamic profile of patients hospitalized at the UNM medical center showed a drop in the cardiac index, with raised systemic vascular resistance; the pulmonary edema has been related to increased permeability of the capillary bed and not to increased hydrostatic forces (Levy and Simpson, 1994). Severe cardiopulmonary dysfunctions correlate with a poor prognosis. Severe hypotension followed by cardiac arrhythmias have been seen in the patients who died. Recovery in survivors was as rapid as the initial decline with markedly improved oxygenation and hemodynamic functions (Levy and Simpson, 1994; Duchin *et al.*, 1994).

Treatment of the patients consists of maintaining adequate oxygenation and careful monitoring and support of hemodynamic function. Intravenous fluids should be used with extreme caution to prevent exacerbation of the pulmonary edema. An open-label clinical trial using an antiviral drug (ribavirin) was carried out on patients with suspected HPS, but no conclusion about the benefit of ribavirin could be determined (Chapman *et al.*, 1994). This trial was discontinued and will be replaced by a placebo-controlled trial to determine any beneficial effect of ribavirin therapy (Butler and Peters, 1994; Chapman *et al.*, unpublished data).

III. DIAGNOSTIC TESTS

Rapid serologic diagnostic methods include ELISA, using homologous recombinant nucleoprotein antigen for IgG (Feldmann, *et al.*, 1993) and native inactivated Sin Nombre virus–cell suspension antigen employing an IgM capture technique (Ksiazek *et al.*, 1995). Researchers at UNM developed a Western blot assay in which expressed nucleoprotein and glycoprotein are used (Jenison *et al.*, 1994). RT-PCR can be done on patient blood clot or serum (Hjelle *et al.*, 1994b) or necropsy tissues (Nichol *et al.*, 1993). When only fixed tissues are available, immunohistochemistry is the simplest means to confirm the diagnosis (Zaki *et al.*, 1994) and remains an invaluable technique to study the pathogenesis of the disease (Zaki *et al.*, 1995). RT-PCR detection of virus RNA in fixed tissues has also been successful; while this procedure is more difficult and time-consuming than other methods of detection, it has the advantage of genetic typing of the virus by sequence analysis of amplified PCR products (Schwartz *et al.*, 1995).

IV. EPIDEMIOLOGY IN RODENT RESERVOIRS AND HUMANS

During the initial outbreak investigation, deer mice were found to be the most abundant species in the region, and also had the highest prevalence of antibody to Sin Nombre virus. Since then, a large national rodent survey has been conducted, with more than 25,000 small mammals tested. Antibody to Sin Nombre virus among *P. maniculatus* has been found in most states within the geographic distribution of this rodent species in the United States. Antibody to Sin Nombre virus has also been found in other species, genera, or families of rodents (Childs *et al.*, 1994; Ksiazek *et al.*, unpublished data).

Most human HPS cases identified in Canada (Canada Health and Welfare, 1994) and in the United States have been within the range of *P. maniculatus*. From 1959, the year for which the first case was retrospectively confirmed, through October 18, 1995, 122 persons with illness meeting the surveillance case definition have been reported to and confirmed by CDC. Sixty-one (50%) cases were fatal. Patients have ranged in age from 11 to 69 years (median age, 35 years), and 57 (55.3%) have been male. HPS patients have been reported from 23 states (Fig. 1). Most have lived in rural areas. A case–control study during the initial outbreak (Zeitz *et al.*, 1995) confirmed that, as expected, contact with rodents (e.g., trapping, handling) in households was an important risk factor. A study of rodent workers (Armstrong *et al.*, 1994) showed that past infection with Sin Nombre virus was correlated with the number of rodents handled by mammalogists and field biologists, but the overall seroprevalence was less than 1%.

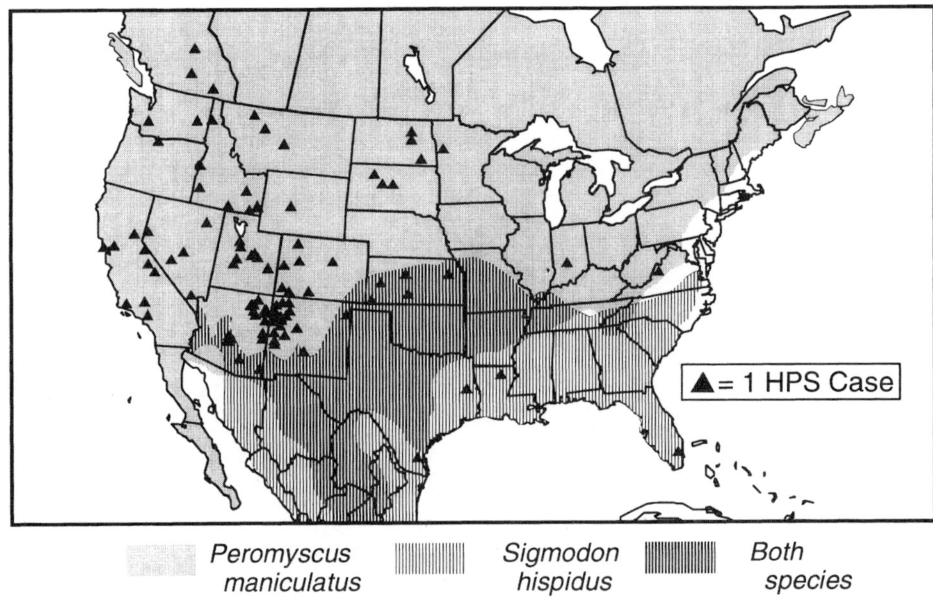

FIGURE 1. Distribution of known rodent hosts for hantaviruses and location of HPS cases as of March 22, 1995, United States and Canada. Rodent distributions from Burt and Grossenheider, 1980.

V. ETIOLOGIC AGENT

Genetic diversity among the four previously defined classic hantavirus serotypes (see Chapter 3)—Hantaan (HTN), Seoul (SEO), Puumala (PUU), and Prospect Hill (PH) viruses—is quite high, with approximately 50% identity at the nucleotide sequence level. However, HTN and SEO viruses are more closely related to one another (approximately 70% nucleotide identity), as are PUU and PT viruses. Two sets of deoxyoligonucleotide primers (targeting the G2 coding region of the virus M segment) were initially designed for sensitive and specific detection of low levels of hantavirus of either HTN–SEO serotypes or PUU–PH serotypes by an RT-PCR assay (Nichol *et al.*, 1993). Nested RT-PCR of total RNA extracted from HPS patient tissues generated specific DNA products only with the PUU/PH primer set. Use of previously described primer sets for hantaviruses did not yield products when applied directly to clinical material. Analysis of the nucleotide sequences indicated that the DNA bands gave hantavirus-like sequences corresponding to the predicted target region of virus genome M segment, but that this virus represented a new and unique member of the hantavirus genus. An improved PCR method (involving the use of specific primers deduced from the originally obtained sequence) facilitated a comparison of virus RNA present in human autopsy and infected rodent tissues from the Four Corners region. A direct genetic link between the virus in infected *Peromyscus* rodents and the virus in the human HPS cases was demonstrated.

Heterologous reassortment has never been reported among hantaviruses although such reassortment has been seen among other members of the family *Bunyaviridae* (Murphy and Pringle, 1987). Complete sequence characterization of the three RNA segments of Sin Nombre virus (Spiropoulou *et al.*, 1994, Chizhikov *et al.* 1995) has shown no evidence of genetic reassortment with previously recognized hantaviruses; each RNA segment is unique and approximately 30% different at the nucleotide level from the segments of PH virus, the most closely related hantavirus. These results argue against RNA segment reassortment between different previously known hantaviruses playing a role in the apparent emergence of the newly recognized SN virus. However, evidence of reassortment in nature among different lineages of SN virus has been found (Li *et al.*, 1995; Henderson *et al.*, unpublished data).

Despite the fact that the G2 coding region of the genome targeted for amplification is generally highly conserved among hantaviruses, a surprisingly high level of genetic diversity was observed among specimens obtained from human HPS patients and rodents collected within the broad geographic range. Up to 10% sequence variation was seen among hantaviruses from New Mexico, Arizona, and Colorado, where most HPS activity was first observed. Even greater variation was noted when the analysis included viruses from other regions. This finding is also consistent with the hypothesized lengthy association of SN viruses with rodent populations

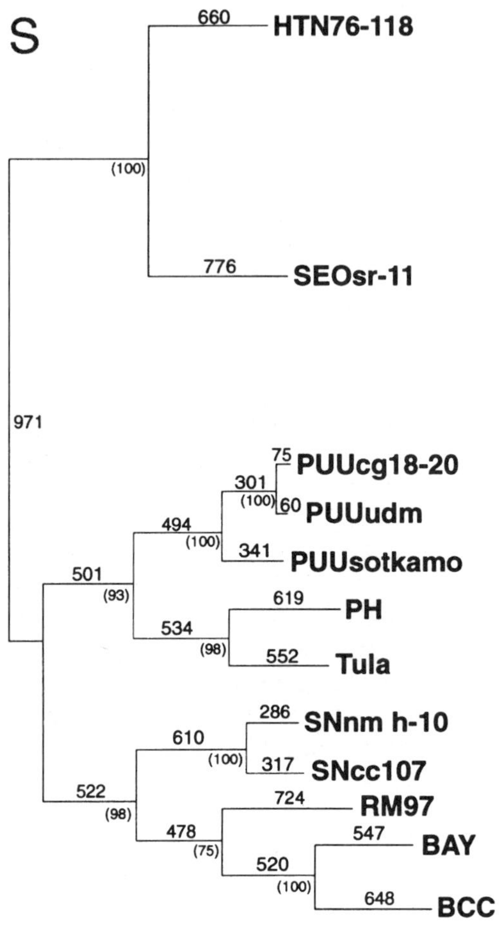

FIGURE 2.

across the country. It appears highly unlikely that SN virus variants independently maintained in rodent populations across the United States would simultaneously and independently acquire the ability to cause the severe pulmonary disease of humans. It seems much more likely that this disease and associated hantavirus have existed unnoticed for many years, and the retrospective identification of cases by immunohistochemistry or serology have confirmed this suspicion. These data further support suggestions of a long-term association of SN virus with *P. maniculatus,* and coevolution of hantaviruses and their respective primary rodent hosts (Childs *et al.,* 1994; Xiao *et al.,* 1994).

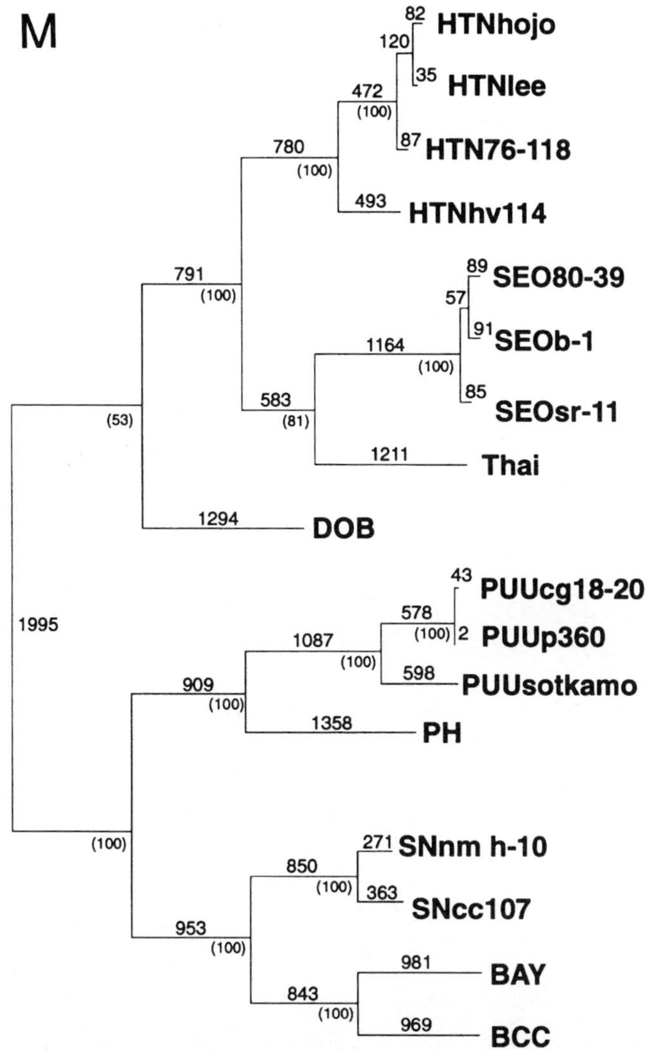

FIGURE 2. Phylogenetic relationship of hantavirus entire S, M, and partial L genome segment nucleotide sequences. (Reproduced from Ravkov *et al.*, 1995.) Phylogenetic analysis of nucleotide sequence differences was carried out by the weighted maximum parsimony method, using the PAUP 3.1.1 software program (Swofford, 1991). A transversion:transition weighting of 4:1 was used based on earlier analyses (Spiropoulou *et al.*, 1994; Morzunov *et al.*, 1995). See original legend for all virus sequence references. Viruses include: Thailand (Thai) from *Bandicotta indica* (Xiao *et al.*, 1994), Dobrava (DOB) from *Apodemus flavicollis* (Avsic-Zupanc and Schmaljohn, unpublished), SN virus (Four Corners) strain CC107 (Li *et al.*, 1995); Tula from *Microtis arvalis* (26); El Moro Canyon (RM-97) from *Reithrodontomys megalotis* (Hjelle *et al.*, 1994a). A single most parsimonious tree was obtained for each segment analyzed. Horizontal lengths are proportional to nucleotide step differences (indicated above lines). Vertical distances are for graphic representation only. Bootstrap confidence limits were calculated from 500 heuristic search replicates and limits in excess of 50% are indicated in parentheses.

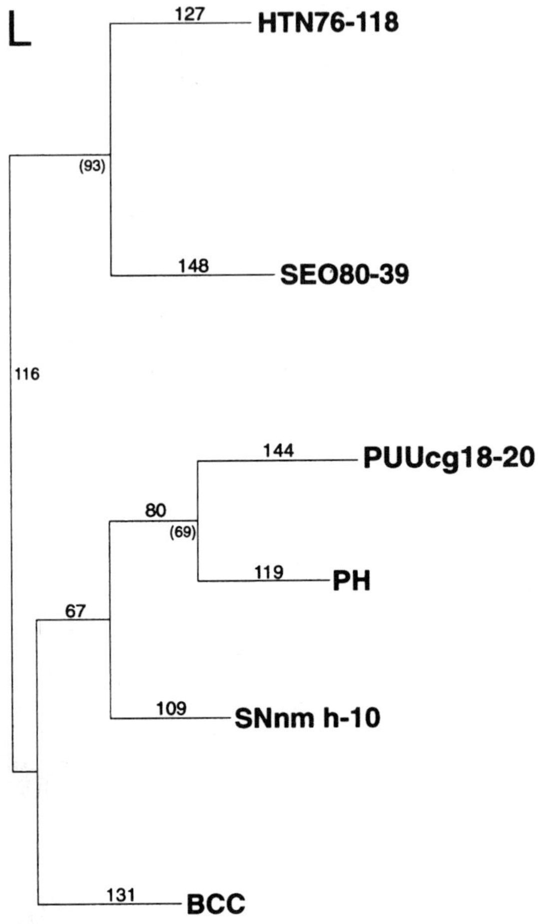

FIGURE 2. (*Continued*)

VI. OTHER NEWLY RECOGNIZED HANTAVIRUSES

During HPS case investigations or related rodent studies, other hantaviruses have been isolated or detected by RT-PCR. Bayou (BAY) virus was detected in human autopsy tissues from a patient who died of HPS in Louisiana (Khan *et al.*, 1995). Nucleotide sequence analysis of the entire S and M segments of the virus genome indicated that this virus was related to Sin Nombre virus but was clearly distinct, differing by approximately 30% at the nucleotide level (Morzunov *et al.*, 1995). This infection occurred outside of the range of *P. maniculatus* and the BAY virus reservoir has now been identified as the rice rat (Oryzomys palustris). Black Creek Canal (BCC) virus was isolated from a cotton rat (*Sigmodon hispidus*) captured in Florida and was also shown to be associated with a severe HPS case (Rollin *et al.*, 1995).

The analysis of the complete S, M, and partial L segments of the virus genome confirmed that this virus was also related to SN and BAY viruses but, again, was quite distinct, differing by approximately 25% at the nucleotide identity level (Ravkov *et al.*, 1995) BAY and BCC viruses form two new lineages among the New World hantaviruses (Fig. 2). Interestingly, both patients infected by these viruses in Louisiana and Florida had renal involvement, which is not prominent in HPS cases caused by SN virus (Khan *et al.*, 1995). Renal involvement is, however, a prominent feature of hemorrhagic fever with renal syndrome (HFRS) caused by HTN, SEO, and PUU viruses. More recently, an SN-like virus has been associated with a fatal HPS case in the northeastern United States that was linked to the white-footed mouse (*P. leucopus*) (Fig. 3). This virus has been successfully passed in laboratory rodents (Song *et al.*, 1994), and may represent another newly recognized hantavirus (NY 1) or an eastern variant of SN virus found predominantly in *P. leucopus*, but also in *P. maniculatus* (Mather *et al.*, unpublished).

Several additional unique hantavirus sequences have been detected in tissues from the western harvest mouse (*Reithrodontomys megalotis*) (Hjelle *et al.*, 1994a) (Fig. 2), mountain vole (*Microtus montanus*), meadow vole (*M. pennsylvanicus*), and prairie vole (*M. ochrogaster*) (Rowe *et al.*, unpublished) in various areas of the western United States (Fig. 3). It appears that there are still additional hantaviruses present in the region with unknown potential to cause human illness. In addition, discovery of a unique hantavirus associated with an HPS case in Brazil, detection of a BCC-like virus in *S. alstoni* in Venezuela (Fulhorst, unpublished data) (Fig. 3), and serological evidence for hantavirus infection in typical HPS patients in Argentina indicate that HPS-related hantaviruses are not restricted to North America.

VII. REMAINING QUESTIONS

Long-term ecological studies of reservoir rodents have been initiated to explore the effect of rodent population fluctuations on virus maintenance and transmission in the field. Educational tools were developed to reduce rodent–human contact. Other viruses in additional rodent reservoirs are actively being sought. Models of complex virus–rodent host interaction in the laboratory are also being developed. The broad correspondence between hantavirus phylogeny and rodent host phylogeny strongly supports the concept of their coevolution and ancient association (Nichol *et al.*, 1993; Childs *et al.*, 1994; Spiropoulou *et al.*, 1994; Xiao *et al.*, 1994).

Analysis of the entire S, M, and partial L segment nucleotide sequences and the deduced nucleocapsid and glycoprotein amino acid sequences reveals no simple explanation for the unusually high mortality (greater than 50%) associated with HPS or for the novel pulmonary disease features of HPS virus infection. Obvious differences in virus characteristics, such as genetic constellation via segment reassortment, G1/G2 cleavage sites, potential protein N-glycosylation sites, that could correlate with altered virus pathogenicity

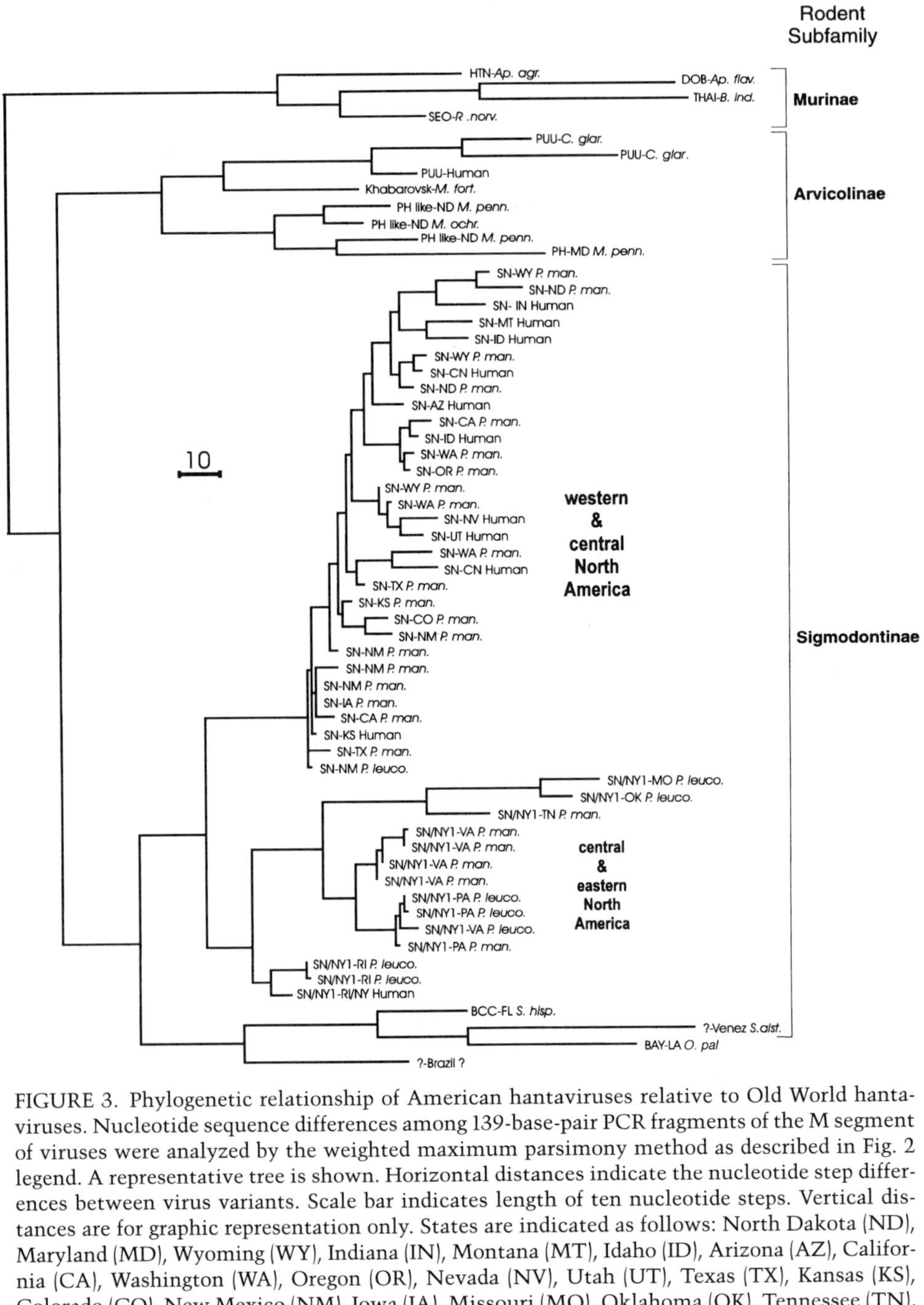

FIGURE 3. Phylogenetic relationship of American hantaviruses relative to Old World hantaviruses. Nucleotide sequence differences among 139-base-pair PCR fragments of the M segment of viruses were analyzed by the weighted maximum parsimony method as described in Fig. 2 legend. A representative tree is shown. Horizontal distances indicate the nucleotide step differences between virus variants. Scale bar indicates length of ten nucleotide steps. Vertical distances are for graphic representation only. States are indicated as follows: North Dakota (ND), Maryland (MD), Wyoming (WY), Indiana (IN), Montana (MT), Idaho (ID), Arizona (AZ), California (CA), Washington (WA), Oregon (OR), Nevada (NV), Utah (UT), Texas (TX), Kansas (KS), Colorado (CO), New Mexico (NM), Iowa (IA), Missouri (MO), Oklahoma (OK), Tennessee (TN), Virginia (VA), Pennsylvania (PA), Rhode Island (RI), New York (NY), Florida (FL), and Louisiana (LA). Canada and Venezuela are abbreviated CN and Venez, respectively. Rodent host species are indicated as follows: *Apodemus agrarius* (*Ap. agr.*) and *flavicollis* (*Ap. flav.*); *Bandicotta indica* (*B. ind.*); *Rattus norvegicus* (*R. norv.*); *Clethrionomys glareolus* (*C. glar.*); *Microtus fortis* (*M. fort.*), *pennsylvanicus* (*M. penn.*) and *ochrogaster* (*M. ochr.*); *Peromyscus maniculatus* (*P. man.*) and *leucopus* (*P. leuco.*); *Sigmodon hispidus* (*S. hisp.*) and *alstoni* (*S. alst.*).

are not readily apparent. Investigation of the virulence of the virus would be possible with the development of suitable animal model systems, together with the genetic manipulation of virus RNA and protein features through molecular cloning and expression systems. A nonhuman primate model for HPS would ideally permit pathogenic and therapeutic studies, and effects of modulation of the immune response on the early events of viral replication.

VIII. REFERENCES

Armstrong, L. R., Khabbaz, R. F., Childs, J. E., Rollin, P. E., Martin, M. L., Clarke, M., Holman, R. C., Peters, C. J., and Ksiazek, T. G., 1994, In: *Abstracts of the 43rd Annual Meeting of the American Society of Tropical Medicine and Hygiene*, abstract #4, Cincinnati, November 13–17.

Burt, W. H., and Grossenheider, R. P., 1980, *A Field Guide to the Mammals*, 3rd ed., Houghton Mifflin, Boston.

Butler, J. C., and Peters, C. J., 1994, Hantaviruses and hantavirus pulmonary syndrome, *Clin. Infect. Dis.* **19:**387.

Canada Health and Welfare, 1994, First reported cases of hantavirus pulmonary syndrome in Canada, *C.C.D.R.* **20:**121.

Centers for Disease Control and Prevention, 1993, Outbreak of acute illness—Southwestern United States, 1993, *MMWR* **42:**421.

Chapman, L. E., Mertz, G., Khan, A. S., Hart, D. C., Peters, C. J., Koster, F., Ksiazek, T. G., Rollin, P. E., Wilson, L. J., Baum, K. F., Pavia, A. T., Christenson, J. C., Allen, S., Rubin, P. J., Goad, D., and the Ribavirin Study Group, 1994, In: *Abstracts of the 34th Annual Interscience Conference on Antimicrobial Agents and Chemotherapy*, abstract H-111, Orlando, FL, October 4–7.

Childs, J. E., Ksiazek, T. G., Spiropoulou, C. F., Krebs, J. W., Morzunov, S., Maupin, G. O., Gage, K. L., Rollin, P. E., Sarisky, J., Enscore, R. E., Frey, J. K., Peters, C. J., and Nichol, S. T., 1994, Serologic and genetic identification of *Peromyscus maniculatus* as the primary rodent reservoir for a new hantavirus in the southwestern United States, *J. Infect. Dis.* **169:**1271.

Chizhikov, V. E., Spiropoulou, C. F., Morzunov, S. P., Monroe, M. C., Peters, C. J., and Nichol, S. T., 1995, Complete genetic characterization and analysis of isolation of Sin Nombre virus, *J. Virol.* **69:**8132.

Duchin, J. S., Koster, F., Peters, C. J., Simpson, G., Tempest, B., Zaki, S., Ksiazek, T. G., Rollin, P. E., Nichol, S., Umland, E., Moolenaar, R. L., Reef, S. E., Nolte, K. B., Gallaher, M. M., Butler, J. C., Breiman, R. F., and the Hantavirus Study Group, 1994, Hantavirus pulmonary syndrome: A clinical description of 17 patients with a newly recognized disease, *N. Engl. J. Med.* **330:**949.

Elliott, L. H., Ksiazek, T. G., Rollin, P. E. Spiropoulou, C. F., Morzunov, S., Monroe, M., Goldsmith, C. S., Humphrey, C. D., Zaki, S. R., Krebs, J. W., Maupin, G., Gage, K., Childs, J. E., Nichol, S. T., and Peters, C. J., 1994, Isolation of the causative agent of hantavirus pulmonary syndrome, *Am. J. Trop. Med. Hyg.* **51:**102.

Feldmann, H., Sanchez, A., Morzunov, S., Piropoulou, C. F., Rollin, P. E., Ksiazek, T. G., Peters, C. J., and Nichol, S. T., 1993, Utilization of autopsy RNA for the synthesis of the nucleocapsid antigen of a newly recognized virus associated with hantavirus pulmonary syndrome, *Virus Res.* **30:**351.

Hjelle, B., Chavez-Giles, F., Torrez-Martinez, N., Yates, T., Sarisky, J., Webb, J., and Ascher, M., 1994a, Genetic identification of a novel hantavirus of the harvest mouse *Reithrodontomys megalotis*, *J. Virol.* **68:**6751.

Hjelle, B., Spiropoulou, C. F., Torrez-Martinez, N., Morzunov, S., Peters, C. J., and Nichol, S. T., 1994b, Detection of Muerto Canyon virus RNA in peripheral blood mononuclear cells from patients with hantavirus pulmonary syndrome, *J. Infect. Dis.* **170:**1013.

Jenison, S., Yamada, T., Morris, C., Anderson, B., Torrez-Martinez, N., Keller, N., and Hjelle, B., 1994, Characterization of human antibody responses to Four Corners hantavirus infections among patients with hantavirus pulmonary syndrome, *J. Virol.* **68:**3000.

Ketai, L. H., Williamson, M. R., Telepak, R. J., Levy, H., Koster, F. T., Nolte, K. B., and Allen, S. E., 1994, Hantavirus pulmonary syndrome—Radiographic findings in 16 patients, *Radiology* **191**:665.

Khan, A. S., Spiropoulou, C. F., Morzunov, S., Zaki, S., Kohn, M., Nawas, S., McFarland, L., and Nichol, S. T., 1995, A fatal illness associated with a new hantavirus in Louisiana, *J. Med. Virol.* **46**:281.

Ksiazek, T. G., Peters, C. J., Rollin, P. E., Zaki, S., Nichol, S. T., Spiropoulou, C., Morzunov, S., Feldmann, H., Sanchez, A., Khan, A. S., Mahy, B. W. J., Wachsmuth, K., and Butler, J. C., 1995, Identification of a new North American hantavirus that causes acute pulmonary insufficiency, *Am. J. Trop. Med. Hyg.* **52**:1017.

Levy, H., and Simpson, S. Q., 1994, Hantavirus pulmonary syndrome, *Am. J. Resp. Crit. Care Med.* **149**:1710.

Li, D., Schmaljohn, A. L., Anderson, K., and Schmaljohn, C. S., 1995, Complete nucleotide sequences of the M and S segments of two hantavirus isolates from California: Evidence for reassortment in nature among viruses related to hantavirus pulmonary syndrome, *Virology* **206**:973.

Morzunov, S., Feldmann, H., Spiropoulou, C. F., Semenova, V. A., Rollin, P. E., Ksiazek, T. G., Peters, C. J., and Nichol, S. T., 1995, A newly recognized virus associated with a fatal case of hantavirus pulmonary syndrome in Louisiana, *J. Virol.* **69**:1980.

Murphy, J., and Pringle, C. R., 1987, Bunyavirus mutants: Reassortment group assignment and G1 protein variants, in: *The Biology of Negative Strand Viruses* (B. W. J. Mahy and D. Kolakofsky, eds.), pp. 357–362, Elsevier, Amsterdam.

Nichol, S. T., Spiropoulou, C. F., Morzunov, S., Rollin, P. E., Ksiazek, T. G., Feldmann, H., Sanchez, A., Childs, J., Zaki, S., and Peters, C. J., 1993, Genetic identification of a hantavirus associated with an outbreak of acute respiratory illness, *Science* **262**:914.

Ravkov, E. V., Rollin, P. E., Ksiazek, T. G., Peters, C. J., and Nichol, S. T., 1995, Genetic and serologic analysis of Black Creek Canal virus and its association with human disease and *Sigmodon hispidus* infection, *Virology* **210**:482.

Rollin, P. E., Ksiazek, T. G., Elliott, L. H., Ravkov, E. V., Martin, M. L., Morzunov, S., Livingstone, W., Monroe, M., Glass, G., Ruo, S., Khan, A. S., Childs, J. E., Nichol, S., and Peters, C. J., 1995, Isolation of Black Creek Canal virus, a new hantavirus from *Sigmodon hispidus* in Florida, *J. Med. Virol.* **46**:35.

Schwarz, T. F., Zaki, S. R., Morzunov, S., Peters, C. J., and Nichol, S. T., 1995, Detection and sequence confirmation of Sin Nombre virus RNA in paraffin-embedded human tissues using one-step RT-PCR, *J. Virol. Methods* **51**:349.

Song, J., Baek, L., Gajdusek, D. C., Yanagihara, R., Gavrilovskaya, I., Luft, B. J., Mackow, E. R., and Hjelle, B., 1994, Isolation of pathogenic hantavirus from white-footed mouse (*Peromyscus leucopus*), *Lancet* **344**:1637.

Spiropoulou, C. F., Morzunov, S., Feldmann, H., Sanchez, A., Peters, C. J., and Nichol, S. T., 1994, Genome structure and variability of a virus causing hantavirus pulmonary syndrome, *Virology* **200**:715.

Swofford, D. L., 1991, PAUP: Phylogenetic Analysis Using Parsimony, Version 3.1.1. Computer program. Illinois Natural History Survey, Champaign.

Xiao, S. Y., Leduc, J. W., Chu, Y. K., and Schmaljohn, C. S., 1994, Phylogenetic analyses of virus isolates in the genus *Hantavirus* family *Bunyaviridae*, *Virology* **198**:205.

Zaki, S. R., Albers, R. C., Greer, P. W., Coffield, L. M., Armstrong, L. R., Khan, A. S., Khabbaz, R., and Peters, C. J., 1994, Retrospective diagnosis of a 1983 case of fatal hantavirus pulmonary syndrome, *Lancet* **343**:1037.

Zaki, S., Greer, P. W., Coffield, L. M., Goldsmith, C. S., Nolte, K. B., Foucar, K., Feddersen, R. M., Zumwalt, R., Miller, G. L., Rollin, P. E., Ksiazek, T. G., Nichol, S. T., Mahy, B. W. J., and Peters, C. J., 1995, Hantavirus pulmonary syndrome: Pathogenesis of an emerging infectious disease, *Am. J. Pathol.* **146**:552.

Zeitz, P. S., Butler, J. C., Cheek, J. E., Samuel, M. C., Childs, J. E., Shands, L. A., Turner, R. E., Vorhees, R. E., Sarisky, J., Rollin, P. E., Ksiazek, T. G., Chapman, L., Reef, S. E., Komatsu, K. K., Krebs, J. W., Gage, K., Maupin, G. O., Sewell, C. M., Breiman, R. F., and Peters, C. J., 1995, A case–control study of hantavirus pulmonary syndrome during an outbreak in the southwestern United States, *J. Infect. Dis.* **171**:864.

CHAPTER 12

Epidemiology and Pathogenesis of Rift Valley Fever and Other Phleboviruses

J. C. MORRILL AND D. J. McCLAIN

I. INTRODUCTION

Rift Valley fever (RVF) virus is unique among the phleboviruses in its pathogenicity for humans and domestic animals, its various routes of infection, and its wide host range. Since the first isolation and detailed description of the disease following an epizootic in sheep in the Rift Valley of Kenya in 1930 (Daubney *et al.*, 1931), there have been significant epizootics in South Africa (Swanepoel and Coetzer, 1995), Egypt in 1977–1979 (Meegan and Hoogstraal, 1979; Meegan, 1979), West Africa in 1987 (Digoutte and Peters, 1989), Madagascar in 1990 (Morvan *et al.*, 1992), and most recently a reintroduction into Egypt in 1993 (Arthur *et al.*, 1993). Presently, virologic and serologic evidence suggests that the virus exists throughout sub-Saharan Africa and Madagascar and, in light of its recurrence in Egypt, may be extending its range even farther though, to date, no outbreaks have been reported outside Africa.

Modern transportation provides a potential means for global transmission via infected animals and humans incubating the virus as well as infected

The views of the authors do not purport to reflect the position of the Department of the Army or the Department of Defense.

J. C. MORRILL AND D. J. McCLAIN • U.S. Army Medical Research Institute of Infectious Diseases, Fort Detrick, Frederick, Maryland 21702-5011.

The Bunyaviridae, edited by Richard M. Elliott, Plenum Press, New York, 1996.

vectors carried as stowaways. Hence, international travelers to endemic areas should be aware of the disease and its public health and agricultural importance.

Typically the virus is associated with pastoral regions where habitat conducive to the maintenance of arthropod vectors is present. Natural hosts for RVF virus include mosquitoes, sheep, cattle, buffalo, camels, goats, other ruminants, and humans. Newborn lambs, calves, and puppies are highly susceptible. Mice, lambs less than 7 days old, and baby hamsters are the most susceptible experimental animals. Several species of wild rodents are susceptible to RVF but the epidemiologic significance of their role in virus maintenance and transmission is not known. Various avian and reptilian species have been tested for susceptibility and are refractory to RVF virus.

Epidemics of RVF typically center around regions where there are large concentrations of sheep and cattle. Explosive epidemics occur periodically and are associated with periods of heavy rainfall producing localized flooding and dense or expanding vector populations (Davies *et al.*, 1985). Transovarially infected floodwater *Aedes* eggs hatch producing infected adults which feed extensively on cattle (Linthicum *et al.*, 1985). Other mosquito species feeding on the infected livestock ingest viremic blood meals and, if those mosquitoes are efficient vectors, develop disseminated infections and become competent secondary vectors. Secondary vectors include mosquitoes of many species of the genera *Aedes*, *Anopheles*, *Culex*, *Eretmapodites*, and *Mansonia*. *Culicoides* spp. and sand flies may play limited roles in biological transmission and, along with other arthropods, mechanical transmission. *Culex pipiens* was an important mosquito vector in the Egyptian epizootic in 1977 (Meegan *et al.*, 1980).

In the absence of epidemics, a cycle of enzootic circulation exists in many regions of Africa. Livestock infections, probably acquired by the bite of infected mosquitoes, result in low rates of disease and abortion that are undiagnosed because of confusion with other livestock diseases as well as a lack of diagnostic capabilities. Reservoirs for RVF virus are unidentified, though there is strong evidence of interepidemic maintenance via transovarial transmission in certain *Aedes* mosquitoes (Logan *et al.*, 1991). The infected eggs are deposited and may remain dormant in depressions, called "dambos" in East Africa or "pans" in South Africa, that are subject to inundation. When flooding occurs, the eggs hatch and infected larvae emerge and develop into infected adults. Through monitoring of changes in vegetation, satellite remote sensing is being used to identify areas of flooding which may trigger hatching of floodwater *Aedes* and provide breeding sites for secondary vectors (Linthicum *et al.*, 1987).

Though aerosol transmission between infected and susceptible livestock appears less important than mosquito transmission, humans may be infected by aerosol in the laboratory and during slaughter of viremic animals (Hoogstraal *et al.*, 1979). Blood, serum, and the products of abortion from RVF virus-infected animals are sources for infection of humans in at-risk occupations such as abattoir workers, farmers, veterinarians, and laboratory technicians (Peters and Meegan, 1981). The major natural means of RVF transmission is by bite of infected mosquitoes, though mechanical transmission by arthropods is possible. Consumption of milk or meat from infected animals does not appear to be a common means of transmission.

High concentrations of virus may be found in amniotic fluid and the serosanguinous fluid in the thorax of aborted lamb carcasses and may provide a source of environmental contamination as well as diagnostic material.

II. CLINICAL SIGNS

In Africa, the disease in animals seems to be limited to domestic ruminants, with imported European animals being more severely affected than native African breeds. Sheep and cattle are the primary domestic ruminant species affected by RVF virus, with goats being involved to a lesser extent.

Clinical signs vary considerably and are related to the species and age of the animal involved. Disease progression and severity of disease are generally inversely proportional to age (Easterday *et al.*, 1962a,b; Coackley *et al.*, 1967; Kaschula, 1957; McIntosh *et al.*, 1973). Adult cattle and sheep may suffer mortality rates of 10–30% or higher, depending on the nutritional state of the animal; but in animals less than 7 days old, fatality rates may approach 100%. The disease is characterized by a short incubation period, fever, hepatitis, abortion, and death. Widespread abortion, infertility, and rapidly fatal neonatal disease are typical of outbreaks among cattle and sheep. Fulminant neonatal disease may be the first indication of RVF in areas where abortion rates are high as a result of other abortogenic agents. Newborn lambs and kids are highly susceptible to infection with RVF virus and may suffer 90–100% mortality rates. Experimentally infected calves and lambs experience an acute febrile response, often exceeding 42°C, accompanied by viremia, followed by collapse and death within 24 to 48 hr. Experimentally infected pregnant ewes experience a fever of up to 42°C for 1 to 4 days, followed by recovery or prostration and death; abortion occurs 5 to 20 days later in survivors. Other overt signs are inconsistent, but include congestion of mucous membranes, injected conjunctiva, hyperemia of the oral mucosa, mucopurulent nasal discharge, salivation, vomiting, anorexia, general weakness, an unsteady gait, fetid diarrhea, and a rapid decrease in milk production. A definite leukopenia, most severe in younger animals, which corresponds to maximal viremia and temperature response, is seen, often followed by leukocytosis in later stages of the disease. Elevated serum AST, GGT, and LDH values are common. Experimentally infected animals are viremic for 2 to 5 days with titers often in excess of 10^8 PFU/ml. No long-term carrier state in animals has been identified. Central nervous system involvement, evidenced by encephalitis, occurs periodically in experimentally infected rodents surviving a week or more after experiencing a brief episode of low viremia or in animals with high viremias that have been treated with antiviral drugs or passive antibodies (Peters *et al.*, 1986). Weanling gerbils, *Meriones unguiculatus*, appear to be refractory to liver disease and uniformly develop fatal encephalitis, providing a unique model to study RVFV-induced encephalitis (Anderson *et al.*, 1988). While the incidence of encephalitis in cattle naturally infected with RVF virus is not known, there is a single report of RVF viral encephalomyelitis in an experimentally infected calf (Rippy *et al.*, 1992). The animal appeared to recover after viremia and pyrexia but became moribund and was euthanized 9 days after virus inoculation. The calf was no longer viremic but RVF virus was isolated from the brain and multifocal necrotizing encephalitic lesions were observed on pathologic examination.

The disease in humans is usually a temporarily incapacitating illness. Infection results in fever, malaise, headache, and myalgia, often with other constitutional symptoms developing, followed by complete recovery. Probably 1% or less of human infections progress to the more severe and often fatal complications of hemorrhagic disease, encephalitis, or retinal disease (Jouan *et al.*, 1989; McIntosh *et al.*, 1980; Peters and LeDuc, 1991; Van Velden *et al.*, 1977). The determinants of these different syndromes are unknown. However, during the RVF virus outbreak in Egypt in 1993, a presumptive case definition of ocular disease characterized by macular and paramacular retinal lesions, frequently with hemorrhage and edema, following a febrile episode was established (Arthur *et al.*, 1993). This clinical presentation was quite different from the previous outbreak in 1977–1978 in which the hemorrhagic form was frequently seen and accounted for nearly 600 human deaths (Laughlin *et al.*, 1979; Meegan, 1979).

The introduction of RVF virus into Egypt in 1977 produced the largest recorded RVF epidemic. Prior to this epidemic, only 4 human deaths attributable to RVF had been reported. The sudden and unexpected appearance of this previously geographically limited sub-Saharan virus and the unprecedented numbers of encephalitic, ocular, and fatal hemorrhagic disease remain an enigma. Introduction via importation of diseased animals from the south or wind-borne arthropods are unproven possibilities. The epidemic centered in the fertile Nile Delta region harboring an essentially naive population of human, livestock, and arthropod hosts and vectors. The demography of the region coupled with an alteration in virulence of the virus, perhaps through reassortment, and the presence of endemic hepatotropic organisms, like *Schistosoma mansoni*, may have contributed to this devastating epizootic. Although extensive epidemiologic data could not be collected, an estimated 18,000 to > 200,000 clinical cases in humans occurred with 598 fatalities and about 800 cases of ocular disease associated with RVF virus infection (Meegan, 1979). Animal losses resulting from abortion and mortality were high and impacted significantly on the availability and cost of animal protein in Egypt.

The hemorrhagic fever syndrome seen in an estimated 1% of human cases has a case fatality of approximately 50% and is manifested during the course of acute illness. Complications of encephalitis and retinal disease usually develop as the acute illness fades or during the recovery period.

The presence of serum antibody to RVF virus seems to be the major immunologic defense mechanism in recovery. In rodents and monkeys the outcome of RVF virus infection appears to be regulated by serum antibody and interferon and the early appearance of serum interferon may be a contributory factor in limiting viremia and preventing clinical disease (Peters *et al.*, 1986; Morrill *et al.*, 1989, 1990).

III. PATHOLOGY

The most consistent pathologic changes in all species affected involve the liver. The liver appears to be the primary site of virus replication and initial mild hepatocellular changes rapidly progress to final massive necrosis. As the disease progresses in neonates, the necrotic foci may enlarge to 2 mm in diameter and the liver becomes friable, irregularly congested, and may become mottled brown or yellow in color. As these necrotic areas enlarge, extensive destruction of normal hepatic architecture occurs. Hepatic lesions in adult ruminants are not as severe as those found in neonates, but multiple necrotic areas may be present. In some animals, only small, microscopic necrotic areas with varying degrees of visceral and serosal hemorrhages are seen. Coagulated blood may be found in the lumen of the gallbladder in those cases with marked hemorrhage in the liver. Hemorrhages are seen infrequently in the abomasum and intestinal tract.

The rhesus monkey provides a realistic model for human infection with hepatic lesions occurring in the characteristic midzonal pattern seen in humans and other animals. Experimentally infected rhesus monkeys experience a transient viremia, often exceeding 10^7 PFU/ml serum. Usually the viremic phase is followed by an uncomplicated recovery though a variety of clinical symptoms including diminished food intake, lethargy, cutaneous petechiae, and occasional vomiting may be observed. The disease, in approximately 20% of infected monkeys, progresses to a fatal hemorrhagic form which is thought to be mediated by disseminated intravascular coagulation (Cosgriff *et al.*, 1989; Peters *et al.*, 1989). Death is preceded by epistaxis, petechial to purpuric cutaneous lesions, anorexia, and vomiting. Microscopically, extensive moderate to severe centrolobular and midzonal coagulative necrosis occurs in all lobes of the liver.

IV. DIAGNOSIS

An epidemiologic pattern suggestive of RVF includes: short incubation period; high mortality in lambs, calves, and kids that are less than 1 week of age; illness in adult sheep and cattle; high abortion rate among cows and ewes; liver lesions at necropsy; an acute febrile disease in humans; and the presence of dense populations of arthropod vectors.

In the laboratory, a characteristic histopathologic finding of liver necrosis in all susceptible animals often provides the first clue that the disease is RVF. A definitive diagnosis of RVF is accomplished by isolating and identifying the virus or by observing a fourfold rise in specific, neutralizing antibody titer between acute and convalescent sera (Peters and Meegan, 1981). During past epizootics, the most common material used for virus isolation included whole blood or serum collected from animals at the peak of pyrexia. Fresh specimens of liver from animals dying of the illness and the products of abortion are also excellent diagnostic materials. Infected humans are also a source of diagnostic material; and, if possible, suspected mosquito vectors should be collected for virus-isolation studies.

RVF virus may be isolated in laboratory rodents as well as in a number of common cell culture systems; however, virus isolation should not be attempted unless adequate personal protection, such as vaccination, can be assured or Biosafety Level 3 (BSL-3) containment facilities are available (Peters and Meegan, 1981; Eddy *et al.*, 1981). Laboratory animals of choice for isolation are suckling mice, adult mice, and hamsters. RVF virus is one of the few viruses that will kill adult mice and hamsters within 1 to 4 days after intraperitoneal inoculation (Wood *et al.*, 1990).

Serologic techniques used to demonstrate RVF virus antibody in domestic animals and humans include HI, CF, IFA, agar gel diffusion, plaque-reduction neutralization, and ELISA tests. A quick and practical means of diagnosis is the simultaneous application of the IgM and IgG antibody ELISA to human and animal sera (Ksiazek *et al.*, 1989). The IgM response provides a measure of recent infection since IgM antibodies decline within several months after infection and the IgG response is a measure of lifetime exposure. The ELISA is also used to demonstrate viral antigen in suspect tissue and serum. Nucleic acid hybridization and enzyme immunochemistry techniques for detection of viral antigen have been useful but are less sensitive than virus isolation. Polymerase chain reaction (PCR) methodology is exquisitely sensitive and specific and its utility as a diagnostic tool for RVF virus is being evaluated.

V. TREATMENT AND CONTROL

No specific treatments are currently available. RVF virus is sensitive to several antiviral agents and interferon *in vitro*. Experimental studies in RVF virus-infected rhesus macaques show that ribavirin and recombinant interferon alpha are effective prophylactic drugs; however, chemotherapeutic efficacy for the disease has not been demonstrated (Peters *et al.*, 1986; Morrill *et al.*, 1989). Passive antibody therapy, by administration of immune plasma or serum, may be effective but impractical in an epizootic. Neonatal calves have been shown to be completely protected against experimental challenge with virulent virus through ingestion of colostrum from immune dams (Mebus, 1992).

Relocation of animals to an altitude where mosquitoes are absent or application of residual insecticides to animals and their pens and barns has been suggested, though movement of animals during an epizootic is undesirable and rarely practical, and effectiveness of residual insecticides in animal holding areas is dependent on vector habits. Limiting amplification of virus in domestic animals will probably block extensive human disease and mass vaccination is the method of choice in controlling RVF during an epizootic.

Effective live attenuated and killed veterinary vaccines for RVF are in use in many African countries (Assad *et al.*, 1983). The live attenuated Smithburn strain provides long-lasting immunity but is abortogenic in pregnant ewes. The live-virus vaccines should be used only in enzootic areas of Africa or to control an epizootic.

Killed vaccines are recommended for use outside enzootic areas of Africa. A formalin-inactivated vaccine is safe for pregnant ewes but provides only short-term immunity and requires booster inoculations to maintain a durable immunity. Stringent production controls are necessary to ensure the absence of residual live virus.

The only vaccine cleared for human use is a killed product available only from the United States Army Medical Research and Materiel Command (USAMRMC). This vaccine is in limited supply and requires an initial three-dose series for protective immunity with annual booster inoculations required to maintain that immunity.

A live attenuated vaccine (MP-12) developed for use in livestock and humans is being tested (Caplen *et al.*, 1985). Extensive laboratory studies have shown this vaccine to be safe and efficacious against virulent virus challenge in pregnant cows and ewes as well as neonatal calves and lambs (Morrill *et al.*, 1987, 1991; Hubbard *et al.*, 1991; Mebus *et al.*, 1989). Under experimental conditions, the vaccine does not induce fetal damage in sheep or cattle. Limited field studies in Senegal have shown the MP-12 vaccine to be safe, immunogenic, and nonabortogenic (R. Lancelot, 1993, personal communication). More extensive field studies are anticipated in the future. The MP-12 vaccine is less neurovirulent than Smithburn strain in rhesus monkeys and, since the genome of this virus has at least one attenuating lesion on each of the three segments, reversion to virulence is unlikely and reassortment with wild-type virus would produce attenuated progeny. The American Committee on Arthropod-Borne Viruses' (ACAV) Subcommittee on Arbovirus Laboratory Safety (SALS) has determined that the MP-12 vaccine strain may be handled at BSL-2, providing additional safety to humans involved in vaccine production and vaccination procedures (*Biosafety*, 1993). Presently the current lot of MP-12 vaccine is undergoing human testing at the United States Army Medical Research Institute of Infectious Diseases (USAMRIID).

If MP-12 proves to be a safe and effective immunogen, it would provide a measure of protection to those in high-risk professions with the advantages of single-dose immunization providing a rapid and durable immunity and a low cost per dose.

Suggested specific measures to control an RVF epidemic include:

1. Implement an animal vaccination program using a live attenuated RVF vaccine (inactivated vaccine for pregnant animals and neonatal lambs).
   a. Establish a vaccine barrier between known affected areas and unaffected areas.
   b. Positively identify vaccinated animals with an ear tag, tattoo, or other means of identification not easily counterfeited or duplicated.
   c. Prohibit the use of common needles to immunize herds or flocks.
   d. Prohibit the movement of nonvaccinated animals from affected areas.
   e. Employ integrated vector control measures in the areas of active virus transmission and elsewhere as practical and appropriate. Use caution when using insecticides to prevent destruction of mosquito predator species as well as contamination of water and food supplies.
   f. Use personal protective measures such as insect sprays, repellents, and bednets.
2. Implement active surveillance for human and animal disease as well as seroprevalence outside the area of active virus transmission.
   a. Inform human health care providers and veterinarians of the present epizootic/epidemic and be alert for cases exhibiting common signs and sequelae of the disease.
   b. Alert those in high-risk occupations (farmers, herdsmen, and abattoir workers) to the potential hazard of aerosol and parenteral infection through the slaughtering of sick animals and assisting with abortions or handling of products of abortion from ruminants.
   c. Public awareness of the threat through radio, newspaper, and television broadcasts and instructions in personal protective measures. This information should be truthful, accurate, and informative and care must be taken to instruct and not induce panic or over reaction.
3. Vaccination should be sought for certain professionals such as veterinarians, physicians, health care providers, biomedical researchers, and laboratory technicians who are at greatest risk of infection through their attendance to patients or processing of laboratory specimens.

VI. OTHER PHLEBOVIRUSES

Sandfly fever (SF) is a self-limited febrile viral illness transmitted by biting insects of the genus *Phlebotomus* (Sabin, 1952). It occurs in Africa,

Europe, and Asia with seasonal incidence peaking between April and October (Tesh *et al.*, 1976). The virus is transovarially transmitted in sand flies (*Phlebotomus papatasi*), a phenomenon that is important to the maintenance of endemic disease. Humans may also serve as viremic vertebrate hosts in a human–*Phlebotomus*–human cycle. Although serosurveys have suggested that small mammals may have antibody to certain viral strains, the significance of this finding to the maintenance of the disease is uncertain (Le-Lay-Rogues *et al.*, 1983).

SF has a wide geographic distribution in those parts of Europe, Africa, and Asia between 20 and 45° north latitude, reflecting the range of *P. papatasi* (Tesh *et al.*, 1976; Saidi *et al.*, 1977). The disease persists mainly in the lower altitudes of these subtropical and tropical countries in which there are long periods of hot, dry weather. Its occurrence is distinctly seasonal, with the highest incidence occurring during the late spring and summer months, depending on the prevailing temperatures and timing of the rainy season (Sabin, 1944). There are more than 20 viral isolates from phlebotomine flies in both hemispheres that are antigenically related to SF viruses, and which cause rare cases of human disease (Tesh *et al.*, 1974a,b, 1982). However, sandfly fever Sicilian (SFS) and sandfly fever Naples (SFN) are the most important epidemiologically.

Field isolates of SF virus have demonstrated poor infectivity and lack of pathogenicity for various laboratory animals, hence undermining attempts to understand the pathogenic mechanisms of disease (Sabin, 1952). These studies have consisted of the inoculation of hamsters, mice, rats, rabbits, guinea pigs, and monkeys. Pathogenicity for suckling mice has been demonstrated only after serial blind passage and adaptation to mouse brain.

The clinical illness is well-defined by human volunteer studies as self-limited (hence the eponym three-day fever), with no mortality or sequelae and a very predictable clinical course (Hertig and Sabin, 1964). After intravenous inoculation of human volunteers with the virus, the incubation period has been shown to be $1\frac{1}{2}$ to 2 days, followed by a temperature of $> 102°F$ in two-thirds of subjects. The duration of fever is from 1 to 4 days and is accompanied by a frontal or retroorbital headache, malaise, myalgias, anorexia, and lymphopenia. In addition, many patients will also have low back pain, photophobia, and nausea. A small percentage may suffer from arthralgias, odynophagia, and/or vomiting. Infrequently, a patient may experience abdominal pain lasting 1–2 days. On physical examination, persons with SF appear flushed, and often have conjunctival injection. The heart rate is usually elevated early in the course of the disease in association with the fever.

The most distinctive laboratory feature of SF is leukopenia, which occurs in approximately 90% of infected subjects within 2–3 days of resolution of fever (Hertig and Sabin, 1964). Slight decrements in volunteers' platelet counts are occasionally noted, but never below normal limits and always with spontaneous recovery. Mild elevations of the liver transaminases and alkaline phosphatase (2–3× normal values) may also occur during the febrile period, and routinely return to normal within the following week.

Diagnostic techniques for SF are similar to those employed for RVF and include the isolation of the virus from febrile patients or the demonstration of a fourfold increase in neutralizing antibody. Enzyme immunoassays have also been used to detect IgM and IgG antibodies in serum.

Other phleboviruses have been discovered which are serologically related to Naples and Sicilian types, and which are broadly distributed in Eurasia and the Americas (Tesh *et al.*, 1974b; Travassos da Rosa *et al.*, 1983; Shope *et al.*, 1980). These viruses have been recovered from phlebotomine flies and mosquitoes. Several of these phleboviruses appear to cause a disease in humans similar to SF: Chagres, Alenquer, Candiru, and Punta Toro viruses. Toscana virus represents a strain that is distinct albeit related to SF Naples, and which has been reported as a cause of aseptic meningitis in the Tuscany region of Italy and in Portugal (Verani *et al.*, 1980, 1984; Ehrnst *et al.*, 1985). Although transovarial transmission of viruses in *Phlebotomus* spp. has been demonstrated, decline of virus infection rates in successive generations suggests that these agents may not be maintained indefinitely by such a mechanism (Ciufolini *et al.*, 1985, 1989; Endris *et al.*, 1983).

VII. REFERENCES

Anderson, G. W., Slone, T. W., Jr., and Peters, C. J., 1988, The gerbil, *Meriones unguiculatus*, a model for Rift Valley fever viral encephalitis, *Arch. Virol.* **102:**187.

Arthur, R. R., Said El-Sharkawy, M., Cope, S. E., Botros, B. A., Oun, S., Morrill, J. C., Shope, R. E., Hibbs, R. G., Darwish, M. A., and Imam, I. Z. E., 1993, Recurrence of Rift Valley fever in Egypt, *Lancet* **342:**1149.

Assad, F., Davies, F. G., Eddy, G. A., El Karamany, R., Meegan, J. M., Ozawa, Y., Shimshony, A., Shope, R. E., Walker, J., and Yedloutschnig, R. J., 1983, The use of veterinary vaccines for prevention and control of Rift Valley fever: Memorandum from a WHO/FAO meeting, *Bull. WHO* **61:**261.

Biosafety in Microbiological and Biomedical Laboratories, 1993, HHS Publication No. (CDC) 93-8395, U.S. Department of Health and Human Services, 3rd ed., May, p. 130.

Caplen, H., Peters, C. J., and Bishop, D. H. L., 1985, Mutagen-directed attenuation of Rift Valley fever virus as a method for vaccine development, *J. Gen. Virol.* **66:**2271.

Ciufolini, M. G., Maroli, M., and Verani, P., 1985, Growth of two phleboviruses after experimental infection of their suspected sand fly vector, *Phlebotomus perniciosus* (Diptera: Psychodidae), *Am. J. Trop. Med. Hyg.* **34:**174.

Ciufolini, M. G., Maroli, M., Guandalini, E., Marchi, A., and Verani, P., 1989, Experimental studies on the maintenance of Toscana and Arbia viruses (Bunyaviridae: *Phlebovirus*), *Am. J. Trop. Med. Hyg.* **40(6):**669.

Coackley, W., Pini, A., and Gosden, D., 1967, Experimental infection of cattle with pantropic Rift Valley fever virus, *Res. Vet. Sci.* **8:**399.

Cosgriff, T. M., Morrill, J. C., Jennings, G. B., Hodgson, L. A., Slayter, M. V., Gibbs, P. H., and Peters, C. J., 1989, The hemostatic derangement produced by Rift Valley fever virus in rhesus monkeys, *Rev. Infect. Dis.* **11(Suppl. 4):**S807.

Daubney, R., Hudson, J. R., and Garnham, P. C., 1931, Enzootic hepatitis or Rift Valley fever: An undescribed virus disease of sheep, cattle, and man from East Africa, *J. Pathol. Bacteriol.* **34:**545.

Davies, F. G., Linthicum, K. J., and James, A. D., 1985, Rainfall and epizootic Rift Valley fever, *Bull. WHO* **63:**941.

Digoutte, J. P., and Peters, C. J., 1989, General aspects of the 1987 Rift Valley fever epidemic in Mauritania, *Res. Virol.* **140**:27.

Easterday, B. C., Murphy, L. C., and Bennett, D. G., 1962a, Experimental Rift Valley fever in lambs and sheep, *Am. J. Vet. Res.* **23**:1231.

Easterday, B. C., Murphy, L. C., and Bennett, D. G., 1962b, Experimental Rift Valley fever in calves, goats, and pigs, *Am. J. Vet. Res.* **23**:1224.

Eddy, G. A., Peters, C. J., Meadors, G., and Cole, F. E., Jr., 1981, Rift Valley fever vaccine for humans, in: *Contributions to Epidemiology and Biostatistics* (N. Goldblum, T. A. Swartz, and M. A. Klingberg, eds.), Vol. 3, pp. 124–141, Karger, Basel.

Ehrnst, A., Peters, C. J., Niklasson, B., Svedmyr, A., and Holmgren, B., 1985, Neurovirulent Toscana virus (a sandfly fever virus) in Swedish man after visit to Portugal [letter], *Lancet* **2**:1212.

Endris, R. G., Tesh, R. B., and Young, D. G., 1983, Transovarial transmission of Rio Grande virus (Bunyaviridae: *Phlebovirus*) by the sandfly, *Lutzomyia anthrophora*, *Am. J. Trop. Med. Hyg.* **32**:862.

Hertig, M., and Sabin, A. B., 1964, Sandfly fever (papataci, phlebotomus, three-day fever), in: *Preventive Medicine in World War II*, Vol. 7 (E. C. Hoff, ed.), 6:109–174, Office of the Surgeon General, U.S. Department of the Army, Washington, D.C.

Hoogstraal, H., Meegan, J. M., Khalil, G. M., and Adham, F. K., 1979, The Rift Valley fever epizootic in Egypt 1977–1978. 2. Ecological and entomological studies, *Trans. R. Soc. Trop. Med. Hyg.* **73**:624.

Hubbard, K. A., Baskerville, A., and Stephenson, J. R., 1991, Ability of a mutagenized virus variant to protect young lambs from Rift Valley fever, *Am. J. Vet. Res.* **52**:302.

Jouan, A., Coulibaly, I., Adam, F., Philippe, B., Riou, O., Leguenno, B., Christie, R., Ould Merzoug, N., Ksiazek, T., and Digoutte, J. P., 1989, Analytical study of a Rift Valley fever epidemic, *Res. Virol.* **140**:175.

Kaschula, V. R., 1957, Rift Valley fever as a veterinary and medical problem, *J. Am. Vet. Med. Assoc.* **131**:219.

Ksiazek, T. G., Jouan, A., Meegan, J. M., LeGuenno, B., Wilson, M. L., Peters, C. J., Digoutte, J. P., Guillaud, M., Merzoug, N. O., and Touray, E. M., 1989, Rift Valley fever among domestic animals in the recent West African outbreak, *Res. Virol.* **140**:67.

Laughlin, L. W., Meegan, J. M., Strausbaugh, L. J., Morens, D. M., and Watten, R. H., 1979, Epidemic Rift Valley fever in Egypt: Observations of the spectrum of human illness, *Trans. R. Soc. Trop. Med. Hyg.* **73**:630.

Le-Lay-Rogues, G., Valle, M., Chastel, C., and Beaucournu, J. C., 1983, Small wild mammals and arboviruses in Italy, *Bull. Soc. Pathol. Exot. Filiales* **76**:333.

Linthicum, K. J., Kaburia, H. F. A., Davies, F. G., and Lindquist, K. J., 1985, A blood meal analysis of engorged mosquitoes found in Rift Valley fever epizootic areas in Kenya, *J. Am. Mosq. Control Assoc.* **1**:93.

Linthicum, K. J., Bailey, C. L., Davies, F. G., and Tucker, C. J., 1987, Detection of Rift Valley fever viral activity in Kenya by satellite remote sensing imagery, *Science* **235**:1656.

Logan, T. M., Linthicum, K. J., Davies, F. G., Binepal, Y. S., and Roberts, C. R., 1991, Isolation of Rift Valley fever virus from mosquitoes collected during an outbreak in domestic animals in Kenya, *J. Med. Entomol.* **28**:293.

McIntosh, B. M., Dickinson, D. B., and Dos Santos, I., 1973, Rift Valley fever. 3. Viremia in cattle and sheep, *J. S. Afr. Vet. Assoc.* **44**:167.

McIntosh, B. M., Russel, D., Dos Santos, I., and Gear, J. H. S., 1980, Rift Valley fever in humans in South Africa, *S. Afr. Med. J.* **58**:803.

Mebus, C. A., 1992, *Rift Valley Fever. Foreign Animal Diseases.* U.S. Animal Health Assoc., Richmond, VA, pp. 311–317.

Mebus, C. A., Morrill, J. C., Breeze, R. G., and Peters, C. J., 1989, Safety and immunity of a mutagen attenuated Rift Valley fever virus in pregnant cattle, *Proc. 93rd Annul. Meet. U.S. Anim. Health Assoc.* pp. 132–136.

Meegan, J. M., 1979, The Rift Valley fever epizootic in Egypt 1977–78, 1. Description of the epizootic and virological studies, *Trans. R. Soc. Trop. Med. Hyg.* **73**:618.

Meegan, J. M., and Hoogstraal, H., 1979, An epizootic of Rift Valley fever in Egypt in 1977, *Vet. Rec.* **105**:124.

Meegan, J. M., Khalil, G. M., Hoogstraal, H., and Adham, F. K., 1980, Experimental transmission and field isolation studies implicating *Culex pipiens* as a vector of Rift Valley fever virus in Egypt, *Am. J. Trop. Med. Hyg.* **29**:1405.

Morrill, J. C., Jennings, G. B., Caplen, H., Turell, M. J., Johnson, A. J., and Peters, C. J., 1987, Pathogenicity and immunogenicity of a mutagen-attenuated Rift Valley fever virus immunogen in pregnant ewes, *Am. J. Vet. Res.* **48**:1042.

Morrill, J. C., Jennings, G. B., Cosgriff, T. M., Gibbs, P. H., and Peters, C. J., 1989, Prevention of Rift Valley fever in rhesus monkeys with interferon-α, *Rev. Infect. Dis.* **11(Suppl. 4)**:S815.

Morrill, J. C., Jennings, G. B., Johnson, A. J., Cosgriff, T. M., Gibbs, P. H., and Peters, C. J., 1990, Pathogenesis of Rift Valley fever in rhesus monkeys: Role of interferon response, *Arch. Virol.* **110**:195.

Morrill, J. C., Carpenter, L., Dalrymple, J., and Peters, C. J., 1991, Further evaluation of ZH-548 strain MP12 as a live attenuated Rift Valley fever vaccine for sheep, *Vaccine* **9**:35.

Morvan, J., Rollin, P. E., Laventure, S., Rakotoarivony, I., and Roux, J., 1992, Rift Valley fever epizootic in the central highlands of Madagascar, *Res. Virol.* **143**:407.

Peters, C. J., and LeDuc, J. W., 1991, Bunyaviridae: Bunyaviruses, phleboviruses, and related viruses, in: *Textbook of Human Virology*, 2nd ed. (B. Belshe, ed.), pp. 571–614, Mosby Year Book, St. Louis.

Peters, C. J., and Meegan, J. M., 1981, Rift Valley fever, in: *Handbook Series in Zoonoses, Section B, Viral Zoonoses* (J. N. Steele, ed.), Vol. 1, pp. 403–420, CRC Press, Boca Raton, FL.

Peters, C. J., Reynolds, J. A., Slone, T. W., Jones, D. E., and Stephen, E. L., 1986, Prophylaxis of Rift Valley fever virus with antiviral drugs, immune serum, and interferon inducer, and a macrophage activator, *Antiviral Res.* **6**:285.

Peters, C. J., Liu, C. T., Anderson, G. W., Jr., Morrill, J. C., and Jahrling, P. B., 1989, Pathogenesis of viral hemorrhagic fever: Rift Valley fever and Lassa fever contrasted, *Rev. Infect. Dis.* **11(Suppl. 4)**:S743.

Rippy, M. K., Topper, M. J., Mebus, C. A., and Morrill, J. C., 1992, Rift Valley fever virus-induced encephalomyelitis and hepatitis in calves, *Vet. Pathol.* **29**:495.

Sabin, A. B., 1944, Experimental studies on phlebotomus (papataci, sandfly) fever during World War II, *Arch. Gesamte Virusforsch.* **4**:367.

Sabin, A. B., 1952, Phlebotomus fever, in: *Viral and Rickettsial Infections of Man*, 2nd ed. (T. M. Rivers, ed.), pp. 569–577, Lippincott, Philadelphia.

Saidi, S., Tesh, R., Javadian, E., Sahabi, Z., and Nadim, A., 1977, Studies on the epidemiology of sandfly fever in Iran. II. The prevalence of human and animal infection with five phlebotomus fever virus serotypes in Isfahan province, *Am. J. Trop. Med. Hyg.* **26**:288.

Shope, R. E., Peters, C. J., and Walker, J. S., 1980, Serological relation between Rift Valley fever virus and viruses of phlebotomus fever serogroup, *Lancet* **1**:886.

Swanepoel, R., and Coetzer, J. A. W., 1995, Rift Valley fever, in: *Infectious Diseases of Livestock in Southern Africa* (J. A. W. Coetzer, G. R. Thomson, R. C. Tushn, and N. P. J. Kriek, eds.), Oxford University Press Southern Africa, Cape Town.

Tesh, R. B., Chaniotis, B. N., Peralta, P. H., and Johnson, K. M., 1974a, Ecology of viruses isolated from Panamanian phlebotomine sandflies, *Am. J. Trop. Med. Hyg.* **23**:258.

Tesh, R. B., Peralta, P. H., Shope, R. E., Chaniotis, B. N., and Johnson, K. M., 1974b, Antigenic relationships among phlebotomus fever group arboviruses and their implications for the epidemiology of sandfly fever, *Am. J. Trop. Med. Hyg.* **24**:135.

Tesh, R. B., Saidi, S., Gajdamovic, S. J., Rodhain, F., and Vesenjak-Hirjan, J., 1976, Serological studies on the epidemiology of sandfly fever in the Old World, *Bull. WHO* **54**:663.

Tesh, R. B., Peters, C. J., and Meegan, J. M., 1982, Studies on the antigenic relationship among Phleboviruses, *Am. J. Trop. Med. Hyg.* **31**:149.

Travassos da Rosa, A. P. A., Tesh, R. B., Pinheiro, F. P., Travassos da Rosa, J. F. S., and Peterson, N. E., 1983, Characterization of eight new phlebotomus fever serogroup arboviruses (Bunyaviridae: *Phlebovirus*) from the Amazon region of Brazil, *Am. J. Trop. Med. Hyg.* **32**:1164.

Van Velden, D. J. J., Meyer, J. D., Olivier, J., Gear, J. H. S., and McIntosh, B., 1977, Rift Valley fever affecting humans in South Africa: A clinicopathological study, *S. Afr. Med. J.* **51:**867.

Verani, P., Lopes, M. C., Nicoletti, L., and Balducci, M., 1980, Studies on *Phlebotomus*-transmitted viruses in Italy I: Isolation and characterization of a sandfly fever Naples-like virus, in: *Arboviruses in the Mediterranean Countries* (J. Vesenjak-Hirjan *et al.*, eds.), pp. 195–201, Gustav Fischer Verlag, Stuttgart.

Verani, P., Nicoletti, L., and Ciufolini, M. G., 1984, Antigenic and biological characterization of Toscana virus, new phlebotomus fever group virus isolated in Italy, *Acta Virol.* **28:**39.

Wood, O. L., Meegan, J. M., Morrill, J. C., and Stephenson, E. H., 1990, Rift Valley fever virus, in: *Virus Infections of Ruminants* (Z. Dinter and B. Morein, eds.), pp. 481–494, Elsevier, Amsterdam.

The *Bunyaviridae*

Concluding Remarks and Future Prospects

RICHARD M. ELLIOTT

The preceding chapters of this volume have given detailed information on the biology and molecular characteristics of viruses representing the five genera in the family *Bunyaviridae*. In this final chapter I will briefly review comparative data on the family, discuss some aspects of the evolution of the *Bunyaviridae*, and suggest new approaches for future studies on these viruses.

I. COMPARISON OF GENOME CODING STRATEGIES

The *Bunyaviridae* are united by a few familial characteristics such as a tripartite negative or ambisense single-stranded RNA genome, cytoplasmic site of replication with maturation at the Golgi, and viral mRNA transcription primed by capped oligonucleotides cleaved from the 5' ends of host cell mRNAs. In view of the numbers of viruses that meet these less than stringent criteria, it is not surprising that considerable diversity exists at the biological level in terms of hosts and vectors infected, and at the molecular level in terms of genome coding and replication strategies. Most viruses in the family are transmitted by arthropods and one can generalize as to which

RICHARD M. ELLIOTT • Institute of Virology, University of Glasgow, Glasgow G11 5JR, Scotland.

The Bunyaviridae, edited by Richard M. Elliott, Plenum Press, New York, 1996.

arthropods are used by viruses in a particular genus: bunyaviruses are usually transmitted by mosquitoes or biting midges, nairoviruses by ticks, and phleboviruses by sand flies or, the Uukuniemi group, by ticks. Table I in Chapter 1 lists the exceptions to these. Tospoviruses are transmitted to plants by thrips. No arthropod vector has been identified for hantaviruses, which are maintained in nature as persistent infections of rodents and are transmitted to humans via infected rodent secretions; the term "robovirus" (*rodent-borne* virus) has been coined by analogy with the term "arbovirus."

Complete genome sequences are now available for at last one representative of each of the five genera (Table I). Across the family a common coding pattern for the four structural proteins emerges: the L RNA segment encodes the L protein, the M RNA segment encodes the virion glycoproteins, and the S RNA segment encodes the nucleocapsid protein. However, much diversity exists in the coding strategies that have evolved for different viruses, and in addition some viruses also encode nonstructural proteins (i.e., proteins not incorporated into the mature virus particle). The coding strategies are compared schematically in Fig. 1. All L proteins are encoded in a negative-sense manner by the L RNA. Bunya-, hanta-, and phleboviruses have similarly sized L proteins, about 250 kDa, whereas the L proteins of the plant-infecting tospoviruses (332 kDa) and of the nairoviruses (459 kDa) are significantly larger. No data are available that might suggest why the latter L proteins are bigger or what additional domains they might contain.

In contrast to the uniformity in L segment coding strategy, a plethora of strategies are diplayed in the M and S segments. The bunya-, hanta-, and phlebovirus M segments encode, using a negative-sense strategy, a precursor polyprotein of G1 and G2 (and perhaps a nonstructural protein, NSm) which is probably cotranslationally cleaved. The position of NSm in the precursor varies between the different genera: sandwiched between G2 and G1 in bunyaviruses, at the N-terminus in some phleboviruses (but not in the uukuviruses), and is absent in hantaviruses. The nairovirus M segment is also negative-sense and encodes a glycoprotein precursor, but processing occurs over a period of several hours, i.e., it is posttranslational, and has yet to be fully resolved (Marriott *et al.*, 1992; Chapter 4). Tospoviruses display further diversity in using an ambisense coding strategy for the M segment (Kormelink *et al.*, 1992; Law *et al.*, 1992; Chapter 6). The structural glycoproteins are encoded as a precursor in the viral complementary-sense RNA (i.e., in a negative-sense manner) but an equivalent of NSm is encoded in the same sense as the 5' end of the genome RNA. These proteins are translated from two subgenomic mRNAs, the message for NSm being transcribed from the complement of the genome RNA.

Nairo- and hantavirus S segments encode a single product, the N protein, in the complementary-sense RNA. Bunya-, phlebo-, and tospoviruses encode two proteins, N (25–30 kDa) and a nonstructural protein, NSs. Both bunyavirus N and NSs proteins are translated from the same mRNA, the result of alternative initiation of translation at different AUG codons. For phlebo- and tospoviruses N is translated from a subgenomic complementary-sense mRNA, whereas NSs is translated from a subgenomic genome sense RNA, i.e., in an ambisense arrangement.

Here a word should be said about the nomenclature of *Bunyaviridae* proteins, which at best provides facile working designations and at worst is potentially misleading. The naming of the glycoproteins and nonstructural proteins is of particular concern. The designations G1 and G2 were based on the relative electrophoretic mobility of the glycoproteins in SDS–polyacryl-amide gels, with the slower-migrating glycoprotein being designated G1 and the faster G2. Now that molecular cloning and sequencing studies have elucidated the coding strategies of the M segment in different genera, confu-sion can arise, even within the same genus: the G1 of one virus being equiva-lent to G2 of another, and vica versa (e.g., Punta Toro and Rift Valley fever virus glycoproteins; Chapter 5). Also, evidence for the isofunctionality of the first or N-terminally encoded glycoprotein has now been obtained for three genera, where it has been shown that this protein contains the Golgi target-ting and retention signal (Chapter 7). Hence, it seems more appropriate to rename the glycoproteins according to their position within the polyprotein, i.e., GN for the N-terminal glycoprotein and GC for the C-terminal glycopro-tein (Lappin *et al.*, 1994). Turning to the nonstructural proteins, the present nomenclature merely reflects the segment that encodes the protein and the fact that it is not apparently contained in mature virions. In view of the considerable variation in size of these similarly named proteins, and that some are indeed encoded in an ambisense fashion and hence subject to different temporal control in the virus life cycle, renaming of these proteins should be undertaken when some idea of their function is forthcoming.

II. EVOLUTIONARY RELATIONSHIPS OF *BUNYAVIRIDAE*

A. Mechanisms of Evolution

The large number of individual viruses classified into the family *Bunya-viridae* (>300; Chapter 1) is taken as evidence of their evolutionary success and potential. The two major paths of evolution for RNA viruses with segmented genomes are genetic drift—the acquisition of genome changes through point mutations, insertions, deletions, inversions, etc.,—and ge-netic shift—the generation of reassortant viruses following genome segment exchange during the course of a mixed infection. These have been discussed in detail in Chapter 8. Viral RNA-dependent RNA polymerases lack proof-reading functions and have a high error rate (10^{-2} to 10^{-4}), which gives rise to permanent sequence heterogeneity in the population: the quasispecies con-cept (reviewed by Domingo and Holland, 1988). The mutation rate *in vivo* is influenced by many factors such as immune pressure, but an important consideration for the evolutionary potential of arboviruses is the ability to

TABLE I. Compilation of Complete Genome Segment Sequences of *Bunyaviridae*

Genus (serogroup)	Virus (strain)	Seg-ment	Accession number	Reference
Bunyavirus				
(Simbu)	Aino	S	M22011	Akashi *et al.* (1984)
(Bunyamwera)	Batai, BAT	S	X73464	Dunn *et al.* (1994)
	Bunyamwera, BUN	S	D00379	Elliott (1989a)
	Bunyamwera	M	M11852	Lees *et al.* (1986)
	Bunyamwera	L	X14383	Elliott (1989b)
	Cache Valley, CV	S	X73465	Dunn *et al.* (1994)
	Germiston, GER	S	M19420	Gerbaud *et al.* (1987a)
	Germiston	M	M21951	Pardigon *et al.* (1988)
	Guaroa, GRO	S	X73466	Dunn *et al.* (1994)
	Kairi, KRI	S	X73467	Dunn *et al.* (1994)
	Maguari, MAG	S	D00380	Elliott and McGregor (1989)
	Main Drain, MD	S	X73496	Dunn *et al.* (1994)
	Northway, NOR	S	X73470	Dunn *et al.* (1994)
(California)	California encephalitis, CE (BFS283)	S	U12797	Bowen *et al.* (1995)
	California encephalitis (E6071)	S	U12800	Bowen *et al.* (1995)
	Jamestown Canyon, JC (61V2235)	S	U12796	Bowen *et al.* (1995)
	Jamestown Canyon (DAV28)	S	U12799	Bowen *et al.* (1995)
	Jerry Slough, JS	S	U12798	Bowen *et al.* (1995)
	Keystone, KEY	S	U12801	Bowen *et al.* (1995)
	La Crosse, LAC	S	K00610	Akashi and Bishop (1983)
			K00108	Cabridilla *et al.* (1983)
	La Crosse	M	M87664	Jacoby *et al.* (unpublished)
	La Crosse (74-328213)	M	D10370	Grady *et al.* (1987)
	La Crosse (30928-31)	M	U18979	Huang *et al.* (1995)
	La Crosse (22988-89)	M	U18980	Huang *et al.* (1995)
	La Crosse	L	U12396	Roberts *et al.* (1995)
	Lumbo, LUM	S	X73468	Dunn *et al.* (1994)
	Melao, MEL	S	U12802	Bowen *et al.* (1995)
	snowshoe hare, SSH	S	J02390	Bishop *et al.* (1982)
	snowshoe hare	M	K02539	Eshita and Bishop (1984)
	Trivittatus, TVT	S	U12803	Bowen *et al.* (1995)
Hantavirus				
(Hantaan)	Hantaan, HTN (76-118)	S	M14626	Schmaljohn *et al.* (1986)
	Hantaan (76-118)	M	M14627	Schmaljohn *et al.* (1987)
			Y00386	Yoo and Kang (1987)
	Hantaan (76-118)	L	X55901	Schmaljohn (1990)
	Hantaan (HV114)	M	L08753	Xiao *et al.* (1993a)
	Hantaan (Hojo)	M	D00376	Schmaljohn *et al.* (1988)
	Hantaan (Lee)	M	D00377	Schmaljohn *et al.* (1988)
	Hantaan (cl-1)	S	D25530	Isegawa *et al.* (1994)
	Hantaan (cl-1)	M	D25529	Isegawa *et al.* (1994)
	Hantaan (cl-1)	L	D25528	Isegawa *et al.* (1994)
	Hantaan (cl-2)	S	D25533	Isegawa *et al.* (1994)
	Hantaan (cl-2)	M	D25532	Isegawa *et al.* (1994)

(*continued*)

TABLE I. (Continued)

Genus (serogroup)	Virus (strain)	Segment	Accession number	Reference
(Hantaan) (*cont.*)	Hantaan (c1-2)	L	D25531	Isegawa *et al.* (1994)
	Dobrova, DOB (3970/87)	M	L33685	Avsic-Zupanc and Schmaljohn (unpublished)
(Seoul)	Seoul, SEO (Biken-1)	M	X53861	Isegawa *et al.* (1990)
	Seoul (KI-83-262)	M	D17592	Kariwa *et al.* (1994)
	Seoul (KI-85-1)	M	D17593	Kariwa *et al.* (1994)
	Seoul (KI-88-5)	M	D17594	Kariwa *et al.* (1994)
	Seoul (80-39)	M	S47716	Antic *et al.* (1991a)
	Seoul (80-39)	L	X56492	Antic *et al.* (1991b)
	Seoul (R22)	M	S68035	Xu *et al.* (1991)
	Seoul (SR-11)	S	M34881	Arikawa *et al.* (1990)
	Seoul (SR-11)	M	M34882	Arikawa *et al.* (1990)
	Thailand, THAI (749)	M	L08756	Xiao *et al.* (1994)
(Puumala)	Puumala, PUU (Bashkiria/Cg18-20/84)	S	M32750	Stohwasser *et al.* (1990)
	Puumala (Bashkiria/Cg18-20/84)	M	M29979	Giebel *et al.* (1989)
	Puumala (Bashkiria/P360/84)	S	L11347	Xiao *et al.* (1993b)
	Puumala (Bashkiria/P360/84	M	L08755	Xiao *et al.* (1993b)
	Puumala (Bashkiria/K27/84)	S	L08804	Xiao *et al.* (1993b)
	Puumala (Bashkiria/K27/84)	M	L08754	Xiao *et al.* (1993b)
	Puumala (Evo/13Cg/93)	S	Z30703	Plyusnin *et al.* (unpublished)
	Puumala (Evo/14Cg/93)	S	Z30704	Plyusnin *et al.* (unpublished)
	Puumala (Evo/15Cg/93)	S	Z30705	Plyusnin *et al.* (unpublished)
	Puumala (Sotkamo/V-2969/81)	S	X61035	Vapalahti *et al.* (1992)
	Puumala (Sotkamo/V-2969/81)	M	X61034	Vapalahti *et al.* (1992)
	Puumala (Bashkiria/Cg18-20/84)	L	M63194	Stohwasser *et al.* (1991)
	Puumala (Udmurtia/894Cg/91)	S	Z21497	Plyusnin *et al.* (1994b)
	Puumala (Udmurtia/444Cg/88)	S	Z30706	Plyusnin *et al.* (unpublished)
	Puumala (Udmurtia/458Cg/88)	S	Z30707	Plyusnin *et al.* (unpublished)
	Puumala (Udmurtia/338Cg/92)	S	Z30708	Plyusnin *et al.* (unpublished)
	Puumala (Vranica)	S	U14137	Reip *et al.* (unpublished)
	Puumala (Vranica)	M	U14136	Reip *et al.* (unpublished)
	Tula (23Ma/87)	S	Z30941	Plyusnin *et al.* (1994a)
	Tula (53Ma/87)	S	Z30942	Plyusnin *et al.* (1994a)

(continued)

TABLE I. (*Continued*)

Genus (serogroup)	Virus (strain)	Seg-ment	Accession number	Reference
Puumala (*cont.*)	Tula (76Ma/87)	S	Z30943	Plyusnin *et al.* (1994a)
	Tula (175Ma/87)	S	Z30944	Plyusnin *et al.* (1994a)
	Tula (249Mr/87)	S	Z30945	Plyusnin *et al.* (1994a)
(Prospect Hill)	Prospect Hill (PH-1)	S	Z49098	Plyusnin *et al.* (unpublished)
	Prospect Hill (MP-20)	S	X55128	Parrington and Kang (1990)
	Prospect Hill (PHV-1)	M	X55128	Parrington *et al.* (1991)
(Hantavirus pulmonary syndrome)	Bayou	S	L36929	Morzunov *et al.* (1995)
	Bayou	M	L36930	Morzunov *et al.* (1995)
	Black Creek Canal	S	L39949	Ravkov *et al.* (1995)
	Black Creek Canal	M	L39950	Ravkov *et al.* (1995)
	Convict Creek 107	S	L33683	Li *et al.* (1995)
	Convict Creek 107	M	L33474	Li *et al.* (1995)
	Convict Creek 74	S	L33816	Li *et al.* (1995)
	Convict Creek 74	M	L33684	Li *et al.* (1995)
	Sin Nombre (Muerto Canyon)	S	L25784	Spiropoulou *et al.* (1994)
	Sin Nombre (Muerto Canyon)	M	L25783	Spiropoulou *et al.* (1994)
	harvest mouse hantavirus-1 (El Moro Canyon; RM-97)	S	U11427	Hjelle *et al.* (1994)
	harvest mouse hantavirus-2 (Rio Segundo; RMx-Costa-1)	S	U18100	Hjelle *et al.* (1995)
Nairovirus	Crimean-Congo hemorrhagic fever, CCHF (C68031)	S	M86625	Marriott and Nuttall (1992)
	Crimean-Congo hemorrhagic fever (AP92)	S	U04958	Marriott *et al.* (1994)
	Dugbe, DUG	S	M25150	Ward *et al.* (1990)
	Dugbe	M	M94133	Marriott *et al.* (1992)
	Dugbe	L	U15018	Marriott and Nuttall (unpublished)
	Hazara, HAZ (JC280)	S	M86624	Marriott and Nuttall (1992)
Phlebovirus	Rift Valley fever, RVF (ZH-548M12)	S	X53771	Giorgi *et al.* (1991)
	Rift Valley fever (ZH-501)	M	M11157	Collett *et al.* (1985)
	Rift Valley fever (ZH-548M12)	M	M25276	Takehara *et al.* (1989)
	Rift Valley fever (ZH-548M12)	L	X56464	Müller *et al.* (1994)
	Punta Toro, PT	S	K02736	Ihara *et al.* (1984)
	Punta Toro	M	M11156	Ihara *et al.* (1985)
	Sandfly fever, Sicilian, SFS	S	J04418	Marriott *et al.* (1989)

(continued)

TABLE I. (*Continued*)

Genus (serogroup)	Virus (strain)	Seg-ment	Accession number	Reference
Phlebovirus (*cont.*)	Toscana, TOS	S	X53794	Giorgi *et al.* (1991)
	Toscana	L	X68414	Accardi *et al.* (1993)
	Uukuniemi, UUK	S	33551	Simons *et al.* (1990)
	Uukuniemi	M	M17417	Rönnholm and Pettersson (1987)
	Uukuniemi	L	D10759	Elliott *et al.* (1992)
Tospovirus	tomato spotted wilt, TSW (BR-01)	S	D00645	de Haan *et al.* (1990)
	tomato spotted wilt (B)	S	L12048	Pang *et al.* (1993)
	tomato spotted wilt (BL)	S	L20953	Pang *et al.* (1992)
	tomato spotted wilt (L3)	S	D13926	Maiss *et al.* (1991)
	tomato spotted wilt (serogroup IV)	S	Z46419	Heinze *et al.* (1994)
	tomato spotted wilt (BR-01)	M	S48091	Kormelink *et al.* (1992)
	tomato spotted wilt (BR-01)	L	D10066	de Haan *et al.* (1991)
	impatiens necrotic spot, INS	M	M74904	Law *et al.* (1992)
	impatiens necrotic spot	S	X66972	de Haan *et al.* (1992)

replicate in cells of disparate phylogeny. Usually the viruses cause inapparent lifelong persistent infections of the vector in contrast to the often lytic effect on vertebrate cells, an effect that can be reproduced in tissue culture cells (Scallan and Elliott, 1992). The persistent infection of the vector host is an excellent opportunity to promote genomic changes, as exemplified by the appearance of temperature-sensitive and plaque morphology mutants in the population of virus shed from mosquito cells persistently infected with Bunyamwera virus (Newton *et al.*, 1981; Elliott and Wilkie, 1986). Persistent infection of the vector *in vivo* allows the accumulation of point mutations in the viral genome, and subsequent transmission of the virus could lead to the emergence of a variant with altered pathogenicity or tropism. If the vector fed on separate vertebrates viremic with different viruses, the possibility of genetic interactions in the vector could be enhanced. Clearly the vector can exert an extreme influence on evolution of the virus, and of course ongoing evolution of the vector itself could also have an effect: a change in the behavior of the vector could lead to new vertebrate species being infected and the subsequent adaptation of the virus to its novel environment.

The capability of the *Bunyaviridae* to exchange genome segments greatly enhances their evolutionary potential. Most data on reassortment have come from laboratory experiments, which are fully discussed in Chapter 8, but there is some evidence for reassortment in nature. A classic example is provided by analysis of six group C bunyaviruses isolated in the Utinga Forest in Brazil (Casals and Whitman, 1961; Shope and Causey, 1962). By

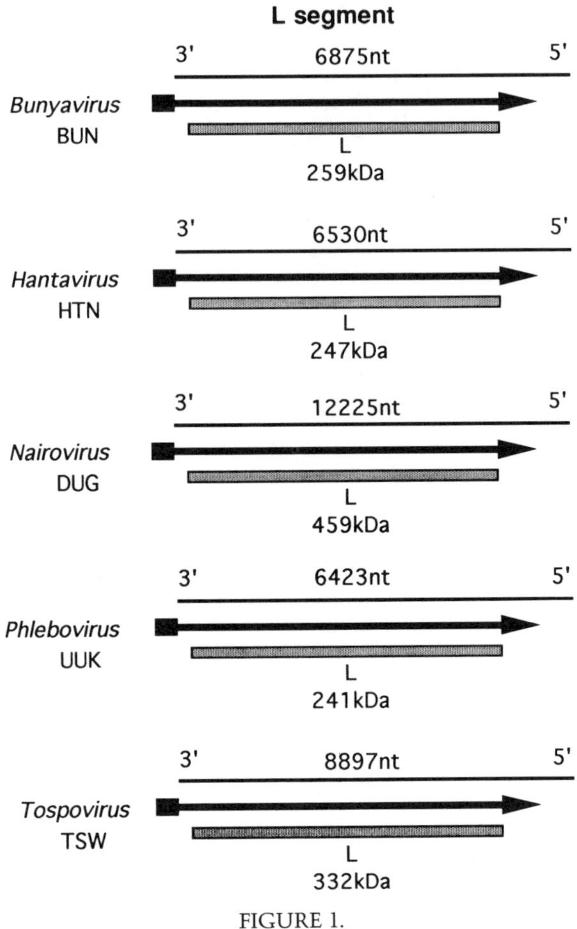

FIGURE 1.

hemagglutination-inhibition and neutralization tests (which assay M segment gene products) the viruses segregate into three groups. However, by complement fixation (which assays the S segment gene products) the viruses fall into three alternate groups. These results are indicative of reassortment occurring between these viruses and a detailed discussion of the natural history of these viruses is given by Shope (1985). High-frequency reassortment of bunyaviruses occurs in mosquitoes following oral or intrathoracic infection and reassortant viruses have been found in the ovaries of dually infected mosquitoes; the newly generated reassortants could be transmitted transovarially to mosquito progeny and orally to mice (Beaty *et al.*, 1981, 1985; Beaty and Bishop, 1988; Chandler *et al.*, 1990). Obviously genome segment reassortment is a major means by which viruses can evolve having

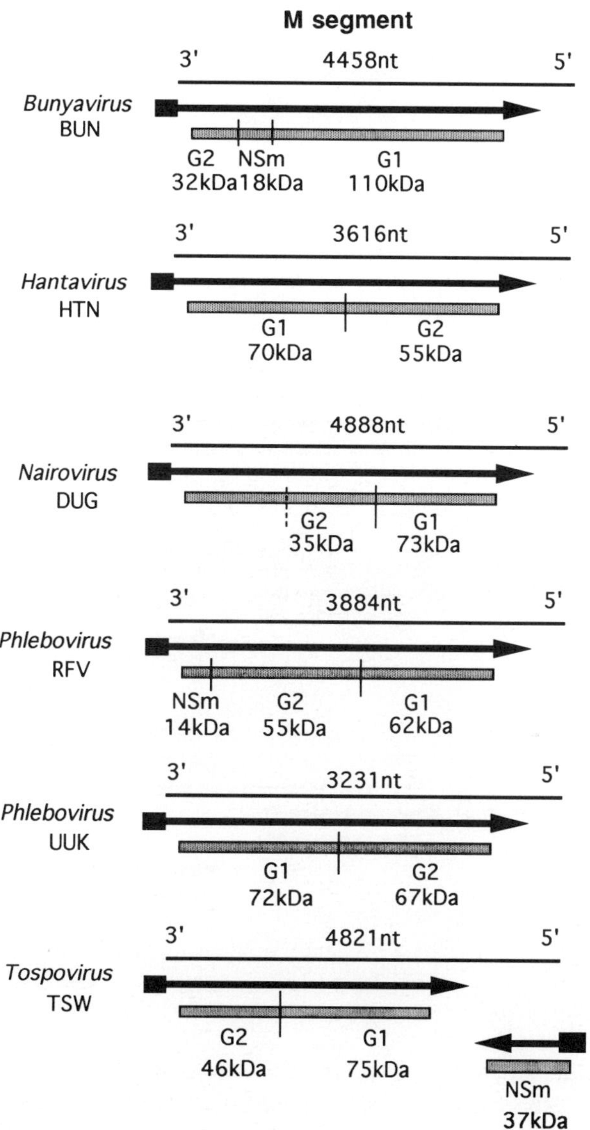

FIGURE 1. Coding strategies of *Bunyaviridae* genome segments. Genomic RNAs are represented by thin lines (the number of nucleotides is given above) and mRNAs are shown as arrows (■ indicates host-derived primer sequence at 5' end, ▶ indicates 3' end). Gene products, with their size in kilodaltons, are represented by solid rectangles. Two examples of phlebovirus M segments are given which differ with respect to the presence or absence of NSm. Virus abbreviations are given in Table I.

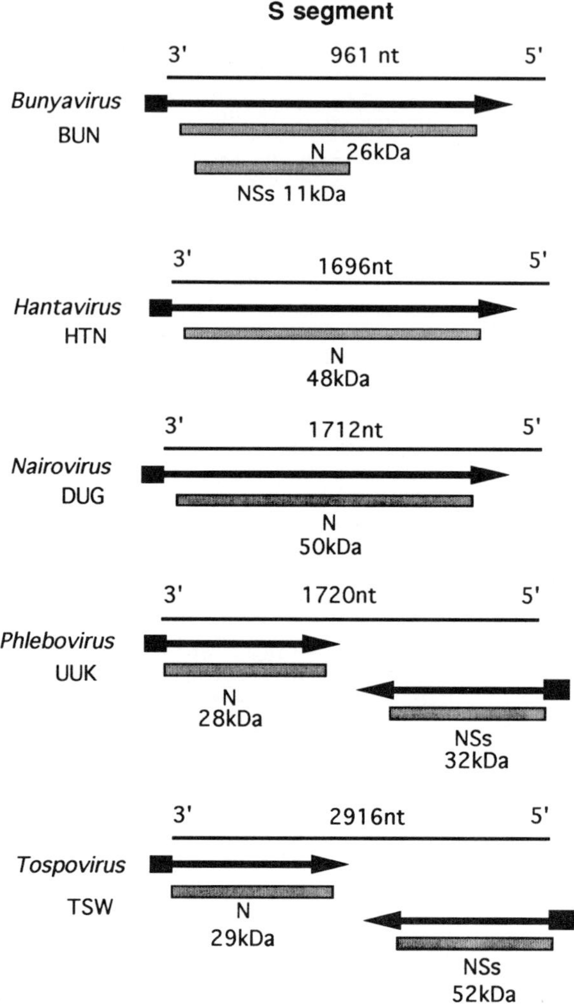

FIGURE 1. (*Continued*)

altered host range, pathogenicity, virulence, tropisms, etc., though the extent
to which this occurs for the *Bunyaviridae* in nature remains to be assessed.

B. Tenuiviruses

Before moving on to discuss some observations derived from sequence
comparisons of *Bunyaviridae* RNAs and proteins, it is relevant to introduce
the tenuiviruses. This is a small group of plant-infecting viruses which, from
recent biochemical evidence, shows evolutionary relatedness to the *Bunya-*

viridae. Five tenuiviruses are currently recognized: rice stripe virus (RSV, the type member), maize stripe virus (MStV), rice hojo blanca virus (RHBV), European wheat striate mosaic virus (EWSMV), and rice grassy stunt virus (RGSV). Tenuiviruses have an ssRNA genome comprising four (RSV, RHBV, and RGSV) or five (MStV) segments, and the RNAs are encapsidated in a nucleocapsid (NC) protein to form a ribonucleoprotein (RNP) complex; an RNA-dependent RNA polymerase activity is associated with the RNP. No spherical enveloped particles have been detected in infected plant tissue but infectivity is associated with filamentous RNPs which bear some resemblance to *Bunyaviridae* nucleocapsids. The tenuiviruses infect plants of the *Graminae* family, and RSV and RHBV can cause economically significant losses of rice yields in Asia and tropical America. The viruses are transmitted in a persistent manner by delphacid planthoppers; the viruses can replicate in the vector and can be passed transovarially to successive generations. A recent review of tenuiviruses is provided by Ramirez and Haenni (1994).

The complete genome of RSV has been cloned and sequenced, and the sequences of RNA segments 3 and 4 of RHBV and segments 3, 4, and 5 of MStV have also been determined; these data and properties of the predicted protein species they encode are summarized in Table II. The coding strategies of the genome segments are schematized in Fig. 2. Segment 1 and segment 5 encode single gene products in a negative-sense manner, whereas segments 2, 3, and 4 each encode two proteins in an ambisense arrangement. Details of protein homologies with *Bunyaviridae* proteins are given in the following section, but two other similarities between tenuiviruses and *Bunyaviridae* are mentioned here. First, the terminal sequences of the tenuivirus RNA segments are conserved, the 3' and 5' ends show inverted complementarity (Takahashi *et al.*, 1990), and circular forms of the RNAs (presumably panhandle forms) have been observed by electron microscopy. Second, the mode of transcription of tenuiviruses is by cap-snatching as the 5' ends of viral mRNAs contain 10–15 additional, capped, nontemplated heterogeneous nucleotides (Huiet *et al.*, 1993; Ramirez *et al.*, 1995).

C. Terminal Nucleotide Sequences

A feature of segmented genome negative- or ambisense RNA viruses is that the terminal sequences of the viral genome segments are conserved and the 3' and 5' ends are complementary; consensus sequences specific for each genus or group of viruses can be derived (Fig. 3). It is assumed that the complementarity accounts for the circular or panhandle forms of RNA and RNPs observed by electron microscopy (Bouloy *et al.*, 1973/74; Obijeski *et al.*, 1976; Samso *et al.*, 1975, 1976; Hewlett *et al.*, 1977; Pardigon *et al.*, 1982). Further, the complementary sequences may be involved in packaging of the genome by the N protein (Raju and Kolakofsky, 1987, 1989) and may provide signals for recognition by the virus-coded polymerase. Some sim-

TABLE II. Tenuivirus Genomes: Compilation of Available Complete Segment Sequences and Their Encoded Proteins

Virus	RNA segment	Length (nt)	Accession number	Encoded protein[a]	Designation	Function	Reference
Rice stripe virus (RSV), isolate T	1	8970	D31879	vc:336.8k	Pol	RNA polymerase	Toriyama *et al.* (1994)
	2[b]	3514	D13176	vc:94k	NSvc2	?	Takahashi *et al.* (1993)
				v:22.8k	NS2	?	
	3	2504	X53563	vc:35.1k	coat	Nucleocapsid protein	Zhu *et al.* (1991)
				v:23.9k	NS3	Disease-specific?	
	4	2157	D01164	vc:32.5k	NS4vc	?	Zhu *et al.* (1992)
				v:20.5k	S	Stripe disease-specific	
Rice stripe virus (RSV), isolate M	3	2475	D01094	vc:35.1k	coat	Nucleocapsid protein	Kakutani *et al.* (1991)
				v:23.8k	NS3	?	
	4	2137	D01039	vc:32.4k	NSvc4	?	Kakutani *et al.* (1990)
				v:20.5k	S	Stripe disease-specific	
Maize stripe virus (MStV)	3	2357	M57426	vc:34.5k	N	Nucleocapsid protein	Huiet *et al.* (1991)
				v:22.7k	NS3	?	
	4	2227	L13438	vc:31.9k	NS4	?	Huiet *et al.* (1992)
				v:19.8k	NCP	Disease-specific	
	5	1317	L13446	vc:44.2k	NS5	?	Huiet *et al.* (1993)
Rice hoja blanca virus (RHBV)	3	2299	L07940	vc:35k	N	Nucleocapsid protein	de Miranda *et al.* (1994)
				v:23k	NS3	?	
	4	1991	L14952	vc:32.5k	NSvc4	?	Ramirez *et al.* (1993)
				v:20.1k	NS4	Disease-specific?	

[a]vc, protein encoded in viral complementary-sense RNA; v, protein encoded in viral-sense RNA.
[b]A second complete sequence of RNA segment 2 from an unidentified isolate of RSV is in the database under accession number D13787; this sequence shows 96.7% nucleotide identity to the sequence D13176.

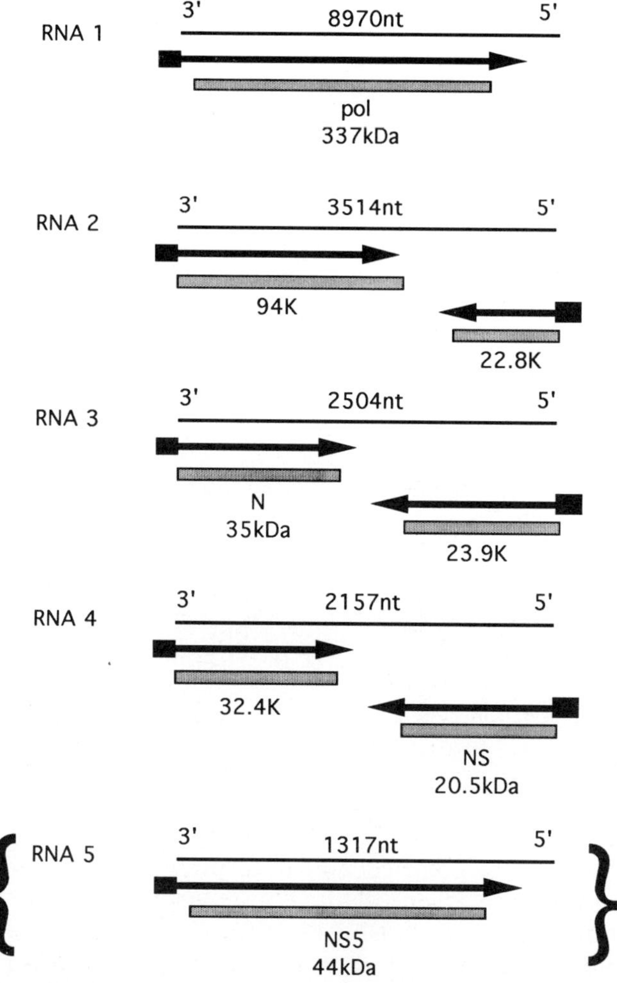

FIGURE 2. Coding strategy of the tenuivirus virus genome. Genomic RNAs are represented by thin lines (the number of nucleotides is given above) and mRNAs are shown as arrows (■ indicates host-derived primer sequence at 5' end, ▶ indicates 3' end). Gene products, with their size in kilodaltons, are represented by solid rectangles. The data for RNA segments 1 to 4 are based on rice stripe virus (RSV). Although one strain of RSV is reported to have five RNA segments, segment 5 of this strain has not been sequenced. Hence, the data for RNA segment 5 are for maize streak virus.

ilarities between the *Bunyaviridae* consensus sequences and those of other segmented genome viruses have been noted and these are also displayed in Fig. 3: de Haan *et al.* (1991) observed homology between the terminal sequences of tomato spotted wilt virus segments and RNA segment 3 of Thogoto virus (family *Orthomyxoviridae*), while the eight terminal nucleo-

Bunyaviridae

Bunyavirus	UCAUCACAU...
Hantavirus	AUCAUCAUCUG...
Nairovirus	AGAGUUUCU...
Phlebovirus	UGUGUUUCUG...
Tospovirus	UCUCGUUAG...

Orthomyxoviridae

Influenza A	UCGUUUUCGUCC...
Influenza B	UCGUCUUCGCUU...
Influenza C	UCGUUUUCGUCC...
Thogoto virus RNA 3	UCUCUUUAGUUU...
	|||| ||||
(*Tospovirus*)	UCUCGUUAG...

Arenaviridae

Tacaribe virus S RNA	GCGUGGCUCCUAGGA...
L RNA	GCGUGUCACCUAGGA...

Tenuivirus

Rice stripe virus	UGUGUUUCAG...
	|||||||| |
(*Phlebovirus*)	UGUGUUUCUG...

FIGURE 3. 3′ terminal consensus sequences of the genome RNAs of negative or ambisense segmented viruses. The similarities between the terminal sequences of tospoviruses and Thogoto virus, and between those of tenuiviruses and phleboviruses are indicated.

tides of tenuiviruses are identical to the terminal consensus sequences of phleboviruses (Kakutani *et al.*, 1990). These observations are suggested to reflect distant ancestral relationships of these different viral groups (Kakutani *et al.*, 1990; de Haan *et al.*, 1991); interestingly, by this criterion, the tenuiviruses seem more closely related to the animal-infecting phleboviruses than to the plant-infecting tospoviruses.

The 3′ terminal nucleotide of hanta- and nairovirus genome RNAs is a purine which implies that antigenome synthesis would be initiated with a pyrimidine. However, this would be unique as all viral RNA polymerases described so far initiate RNA synthesis with a purine (Banerjee, 1980). For Tacaribe arenavirus, which also has a 3′ terminal purine in the genome RNA, it has been shown that antigenomes contain a nontemplated G residue at their 5′ ends, apparently added by the polymerase via a slippage mechanism (Garcin and Kolakofsky, 1992). Recent data indicate that the 5′ residue of Hantaan virus RNA is in fact a monophosphorylated U residue (Garcin *et al.*, 1995). A model to explain this finding also employs a slippage event by the viral RNA polymerase: it is proposed that RNA synthesis is initiated by pppG opposite the third nucleotide (C) at the 3′ end of the RNA; after elongation of a few nucleotides the new RNA strand slips back on the template so that the second-incorporated base (a U residue) is opposite the 3′ terminal A of the template, and elongation then continues to the 5′ end of the template strand. The overhanging G residue is subsequently removed from the 5′ end of the newly synthesized RNA by the polymerase-associated endonuclease activity. Slippage of the viral polymerase may be a general feature of the *Bunyaviridae*, as slippage has also been proposed to account for the apparent reiteration of sequences in bunya- and nairovirus mRNAs during primer-dependent synthesis (Jin and Elliott, 1993b). The advantage to the virus of slippage during initiation of RNA synthesis probably rests with replication rather than transcription, however, as it provides a mechanism to repair damaged ends of RNA segments that may have lost one or two bases at their termini.

III. PHYLOGENETIC STUDIES ON *BUNYAVIRIDAE*

A. L Proteins

Comparison of the available sequence data indicates that there is little obvious similarity between the genes or gene products of viruses in different genera, perhaps not surprising as classification was originally based on serological relationships. The overall impression is that the current classification correlates with the observed phylogeny. Attempts to compare analogous structural proteins globally have not, so far, generated convincing alignments from which to draw phylogenetic trees for the family. A clearer picture of the evolutionary relationships within this family will be aided by identification of functional homologies between proteins from viruses in different genera, though such studies are limited at present. However, some insight can be obtained by analysis of the L protein, and in particular the putative polymerase domain. Poch *et al.* (1989) described four conserved motifs, dubbed the polymerase module, in all RNA-dependent RNA polymerases, which provides evidence for a common ancestral polymerase. Experimental evidence for the importance of the conserved amino acids in these motifs was provided by mutational analysis of the Bunyamwera virus L protein (Jin and Elliott, 1992). An alignment of these putative polymerase domains in the L proteins of the *Bunyaviridae, Arenaviridae,* and tenuivirus is shown in Fig. 4. In motif C the triplet SDD should be noted; this is characteristic of the polymerases of all negative and ambisense segmented genome viruses so far sequenced which suggests a distinct evolutionary pathway for these viral polymerases. Phylogenetic analysis of the complete L protein ORFs of nine members of the *Bunyaviridae,* representing four genera, was reported by Roberts *et al.* (1995), and showed the tospoviruses to be more related to bunyaviruses than to the other genera. An extension of these data, using the polymerase domain only, is given in Fig. 5. There are three major lineages,

FIGURE 4. Alignment of the polymerase domains of *Bunyaviridae*, *Arenaviridae*, and tenuivirus L or RNA polymerase proteins. The alignment was generated using the programs PILEUP and PRETTY in the UWGCG package (Devereux *et al.*, 1984) and CLUSTAL (Higgins and Sharp, 1989). The regions corresponding to the four motifs (A to D) identified by Poch *et al.* (1989) in all RNA polymerases are overlined and "Con" shows the residues conserved among all proteins analyzed. Virus abbreviations are given in Table I.

the first containing hanta-, bunya-, and tospoviruses, the second containing phlebo-, nairo-, and tenuiviruses, and the third containing the arenaviruses. Using this region of the polymerase the relationship between tospo- and bunyaviruses is apparent, as is that between tenui- and phleboviruses. Note that no correlation emerges between the use of an ambisense strategy and phylogeny of polymerase: phlebo-, tenui-, arena-, and tospoviruses all share the property of ambisense coding arrangements on some genome segments but the tospovirus polmerase domain is found in a lineage containing viruses that only use a negative-sense strategy. Furthermore, since the tenuivirus polymerase is more related to the phleboviruses than to the tospoviruses, this may suggest that the spread between animal and plant hosts occurred independently for the tenui- and tospoviruses. More detailed sequence alignments (but not phylogenetic analysis) of the polymerase proteins of segmented and unsegmented RNA viruses are given by Müller *et al.* (1994); these authors also suggest delineation of other putative functional domains which provide targets for experimental analysis.

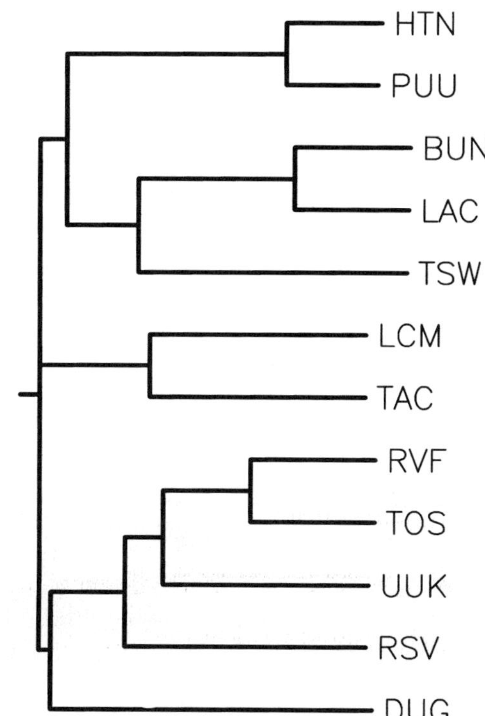

FIGURE 5. Preliminary phylogenetic tree of the polymerase domains of *Bunyaviridae, Arenaviridae,* and tenuivirus L or RNA polymerase proteins, based on the alignment shown in Fig. 4. The tree was constructed using CLUSTAL, and the programs DRAWTREE and DRAWGRAM in PHYLIP (Felsenstein, 1989). Virus abbreviations are given in Table I.

B. Relationships between Animal- and Plant-Infecting *Bunyaviridae* and Tenuiviruses

The apparent relatedness of bunya- and tospoviruses, and of phlebo- and tenuiviruses, is reinforced by analysis of other structural proteins. Kormelink *et al.* (1992) noted a small region of similarity in the G1 proteins of bunya- and tospoviruses (Fig. 6A). The functional significance, if any, of this region is unknown, though it seems more likely that this region would have a vector-associated function rather than a role associated with the plant or vertebrate host. An alignment of the tospo- and bunyavirus N proteins is presented in Fig. 6B; the proteins align reasonably well and there is conservation of certain bulky, aromatic residues (phenylalanine, F, and tyrosine, Y). One may speculate that these residues play a role in binding to viral RNA or interacting with the polymerase although experimental evidence for this is lacking.

Some regions of limited homology have also been noted between the 94k protein of rice stripe tenuivirus and the Punta Toro and Uukuniemi phlebovirus glycoproteins (Takahashi *et al.*, 1993), and an alignment of the RSV 94k protein with the three available phlebovirus glycoproteins is shown in Fig. 7A. Of note is the conservation of a number of cysteine residues, which

A

```
INS  YtTaPiqsthtdffstCTGkCsdcrkeqpitgYqdFcitpTSyWGCEEvwCLAIneGatCGfCrni
TSW  YtTaPiqsthtdfystCTGnCdtcrknqaltgFqdFcvtpTSyWGCEEawCFAIneGatCGfCrnv
SSH  YtTgPtsgintkhdelCTGpCpa..kinhqtgWltFakerTSsWGCEEfgCLAIsdGcvFGsCqdi
LAC  YsTgPtsgintkhdelCTGpCpa..ninhqvgWltFarerTSsWGCEEfgCLAVsdGcvFGsCqdi
BUN  YdTgPtisintkhdehCTGqCps..nieheanWltFsqerTSrWGCEEfgCLAVntGcvFGsCqdv
GER  YdTgPtininskhdelCTGqCpk..kipadpnWltFsqerTSrWGCEEfgCLAIntGcvYGsCqdv
Con  Y-T-P-----------CTG-C-----------W--F----TS-WGCEE--CLAI--G--FG-C---
```

B

```
        1                                                                                            100
BUN     ........mielefhdvaantsstFDpEvayanFkrvhttglsyDhIrifyikgreiktsLakRsewevtlnlggWkitvyntnfpgnrnnpVpddgLTL
GER     ........mlelefedvpnnigstFDpEsgytnFqrnylpgvtlDqIrifyikgreiknsLskRsewevtlnlggWkvpvlntnfpgnrnnaVpdygLTF
LAC     ........msdlvfydvastgangFDpDagymdFcvknaeslNlaaVrifflnaakakaaLsrkperkanpkfgeWqvevinnhfpgnrnnpIgnndLTi
AINO    .......manqfifqdvpqrnlatFnpEvgyvaFiakhgaqlNfDtVrffflnqkkakmvLsktaqpsvdltfggikftlvnnhfpqytanpVpdtaLTL
TSW     mskvk.ltkesivalltqg.kdleFEeDqnlvaFnfktfcleNiDqIkkmsvis..cltfLknRqsimkvikqsdFtfgkitikkt...sdrIggtdMTF
TSW-BL  ..mvk.ltkesivalltqg.kdleFEeDqnlvaFnfktfcleNlDqIkkmsiis..cltfLknRqsimkvikqsdFtfgkitikkt...sdrIgatdMTF
TSW-B   mskvk.ltkenivslltqs.adveFEeDqnqvaFnfktfcqeNlDlIkkmsits..cltfLknRqgimkvvnqsdFtfgkvtikkn...serVgakdMTF
INS     mnkak.itkenivklltqs.dsleFEetqnegsFnftdfftnNrEkIqnmttas..clsfLknRqsimrviksadFtfgsvtikktrnnserVgvndMTF
TSW-IV  msnvkqltekkikellaggtadveiEtEdstpGFsfkaffdnNkni..eitftn..clniLkcRkqifaacksgkYnfcgknivat...svdVgpedwTF
Con     ------------------------FE-------F--------N-D-I-------------L--R-----------F-----------------V-----TF

        101                                                                                          200
BUN     hRLsgFlaryllekm.lkvsepekliiksKIi.nPLaeknGitwn...DgeevyLsfFpGseM.................flgtfrFypLaIgiykvqrk
GER     hRisgYlaryllgky.laetepeklimrtKIv.nPLaeknGitwe...sgpevyLsfFpGaeM.................flgtfrFypLaIgiykvqrk
LAC     hRLsgYlarwvldqynenddesqhelirttIi.nPiaesnGvgwd...sgpeiyLsfFpGteM.................fletfkFypLtIgihrvkqg
AINO    hRLsgYlakwvadqcktnqiklaeam..eKIv.mPLaevkGctwt...EgltmyLgfapGaeM.................fletfeFypLvIdmhrvlkd
TSW     rRLdsLirvrlve...etgnsenlntiksKIashPLiqayGlpld...DaksvrLaiMlGgsLpliasvdsfemisvvlaiyqdakYkdLgIdpkkydtk
TSW-BL  rRLdsLirvrlve...etgnsenlntiksKIashPLsqayGlpld...DaksvrLaiMlGgsLpliasvdsfemisvvlaiyqdakYkdLgIdpkkydtr
TSW-B   rRLdsmirvklie...etannenlaiikaKIashPLvqayGlpla...DaksvrLaiMlGgsipliasvdsfemisvvlaiyqdakYkeLgIeptkyntk
INS     rRLdamvrvhlvgm..ikdngsalteainslpshPLiasyGlatt...DlkscvLgvLlGgsLpliasvlnfeiaalvpaiyqdakhveLgIdmskfstk
TSW-IV  kRteaFirtkivsmvekskneaarqemygKImelPLvaayGlnvpasyDscalrMmlciGgpLpllssirglapiifplayyqnikkekLgI..knfsty
Con     -RL--Y-----------------------KI---PL----G-------D-----L----G------------------------F--L-I--------

        201                                                                       285
BUN     EmepKylektmrqrYmgleaatwtvsklteVqsaltvvsslgwkktnvSaaardFlakFGInm*.........................
GER     EmdpKflektmrqrYlgidaqtwtttklgeVeaalkvvsglgwkktnvSsaareFlskFGIrm*.........................
LAC     mmdpqylkkalrqrYgtltadkwmsqkvaaIakslkdveqlkwgkggLSdtaktFlqkFGIrlp*........................
AINO    gmdvnfmrkvlrqrYgtltaeqwmtqkidaVraafnavgqlswaksgFSpaaraFlaqFGIni*.........................
TSW     EalgKvctvlkskaFemnedqvkkgkeyaaIlsssnpnakgsvamehYSetlnkFyemFGVkkq.aklaela*.............
TSW-BL  EalgKvctvlkrkaFemnedqvkkgkeyaaIlsssnpnakgsiamehYSetlnkFyemFGVkkq.akltela*.............
TSW-B   EalgKvctvlkskgFtmddaqinkgkeyakIlsscnpnakgsiamdyYSdnldkFyemFGVkke.akiagva*.............
INS     EavgKvctvlkskgYsmnsveigkakqyadIlkacspkakglaamdhYkegltsiysmFnatidfgkndsi*.............
TSW-IV  EqvcKvakvlsasqvefkddldvmfkqavkIlsesnpgtassislkkYedqvkymdrvFsanlsvddygdhskksskpstslev*
Con     E---K---------Y----------------I----------------YS-----F---FGI--------------------------
```

FIGURE 6. Similarity between bunyavirus and tospovirus proteins. (A) Alignment of a short region in G1 conserved between bunya- and tospoviruses. The residues compared are INS 652–717; TSW 672–737; SSH 1025–1088; LAC 1025–1088; BUN 1017–1080; GER 1021–1084. (B) Alignment of the N proteins of bunya- and tospoviruses. Virus abbreviations are given in Table I.

A

```
       401                                                                          500
RVF    ygltedcnfcrgmt.....gaslkkgsyplqdlfcqsseddgsklktkmkgvceVgvqahkkcdgqlstahevvpFavfknskkVYldkldlkteEnllp
PT     lemtedcafcrkikkkagqsvqvqktsvplqdaicqensdtysgpkipfkgvckIglikykeckfkts.syetvsFitlkekgkIYiehlmlkniEvvtn
UUK    qtlskdcasciekk.....ssimksehlvyddaicqsdysspeaiarpethlcrIgplhiqhctheakrv.qhvsWfwidgklrVY.ddfsvswtEgkfl
RSV94k ..............mhfksyfiyttifnmawgapipfpdthswmrnrerepseIvkvpcsarappckltyelngY..fienglICynrasvnyfEtcyt
Con    --------------------------------------------------I--------------------F----------Y---------E----

       501                                                                          600
RVF    dsFvCfehgqykgtmdsgqtkrelksfdisqcpkigghgskkctgdaafcsaYectaqyanaycshangsgivqiqvsgvwkkplCvgyervvVkrels
PT     vsFvCyehvgqdeqeve....hralkrvsvndckivdnskqkictgdhvfcekYdcstsypdvtcihapgsgplyinlmgswikpqCvgyervlVdrevk
UUK    slFdClnets...............kdhncnkavclegr....csgdlqfcteFtcsyakadcnckrnqvsgvavvhtkhgsfmpeCmgqslwsVrkpls
RSV94k gnYdY...........klplhpsfskfgghvylscddailqnvslvgiqqteYtss....pllitnsnsekisysnlktgflgivYavetracIqpdqa
Con    --F-C--------------------------------------------------Y----------------------------C--------V-----

       601                                                                          700
RVF    akpiqrvepcttcitkCephglvvrstgfkissavacasgvcvtgsqspsteitlkypgisqssgggdigVhmahddqsvsskivahCppqDpClvhgciv
PT     qpllapeqncdtcvseCldegvhikstgfeitsavacshgscisahqepstsvivpypgllasvggrigIhlshtsdsasvhmvvvCpprDsCaahncll
UUK    krsvtvqqpcmdcesdCkvdhilvivrhfypdhyqaclgstcltg.rakdkefkipfkmadrlsdshfeIriwdkersneyflesrCesvDaCaaitcwf
RSV94k kkpeeiinhgvaikpsCtdgvlyyinsacevn...............vsdqtfsipscesvklptyddtIevcdkggcqn....vtChpgEiCdkyermd
Con    ----------------C----------------------------------------------------I---------------C---D-C------

       701                                                                          800
RVF    cahgliNyqC..htalsafvvvfvfssiaiicLavlyrvLkclkiaprkvlnplmwitaFirWi....ykkmvarvahninqvnreigwmEggqlvlgnp
PT     cyhgilNyqC..hstlsailtsfllilfiytvFsvttniLyvlrlipkqlkspvgwlklFinWl....ltalriktrnvmrringrigwvDhhdv.....
UUK    cranwaNihC..fsk.eqvlilvavsslcillLasvlraLkviatftwkiikpfwwilsLlcRtcskrlnkraerlkesihsleeglnnvDegpreqnnp
RSV94k mimrikNyqCshiyryslysiilffvivivftLitimniLfflkpafwllkkvlysmvgLchRrpvvdevsvdmstvrvvdeaeegllvvEdsiapntnv
Con    ------N--C-----------------------L-------L-------------------------------------------------------

       801                                                                          900
RVF    apiprhapip.......rYstylmllLivsyas.aCse..liqassrittcstegvNtkCrlsgtaliragsvgaEaCLmlkgvkedqtkflkIktvsse
PT     erprhrepmr.......rFkttllltLimmtggnaCsn..tvvanskqtrcvqegsNtkCsitatitlragvigaEsCFiikgpmenqqktisIktisse
UUK    aravarpnvrqkmfnltrLspvvvgmLclacpvesCsdsisvtalsqrcstssdgvNs.CfvstssllqvspkgqEsCLilkgptgtavdsirIkttdik
RSV94k sdkvkrkg..rkvengliFipyvlmiLllvcsaesCqdlvssisnierct......NnsCdfiskmkltllntpqDfCF......ktstdvykIrfnsvr
Con    --------------------F--------L--------C----------------N--C----------------E-C----------------I------

       901                                                                         1000
RVF    lsCregqsYWtgsisp.kclssRrChlvgeChvnrclswrdnetSaeFsfvgesttmrenk.cfeqcggwgCgcFnvnpsclFvhtyLqsvrkeAlrVfn
PT     tvCregssFWtslyip.sclssRrChlvgdCvgnkcqswrddqlSreFsgvkdnhimmenk.cfeqcgaigCgcFninpsclYvhayLksarneAvrVfs
UUK    leCvrrdlYWvprvth.rcigtRrChlmgaCkgeacsefkindySpeWgheeelmaqlgwsycveqcggalCqcFnmrpscfYlrktFshlsqdAfnIye
RSV94k vmClsvplYYtnsfkrvisreeWkCfegegCrtdgthsiwgestSlsFdycvtdfhifsy...........CpaYhyn....WkrieYeptssrActImk
Con    --C------YW------------R-C-----C-------------S--F------------------------C--F--------Y----L------A-----

       1001                                                                        1100
RVF    CiDwvhkltleitdfdgsvstidLgasssrftnwgsvSlSldaegiSgsnsfsFiEspskgyaivdepFseiprq......gflGeIrCnsessvlsahe
PT     CsDwvhrvsfevkgpdgetelvtLgspgtkflnwgtlSlSldaegiSgtnsisFlEsskggfalydegYneipre......gflGeIrCssesaaisahk
UUK    CsEwsyrinvlvst.nsthsnltLklgvpdsiphgliSlS....svSqppaiaYsEcfged..lhgtkFhtvcnrrtdytlgriGeIqCptkadalavsk
RSV94k CmD..tkfeivgyiqknghvlkeLggitsky.dsplvSiSlsnyn.SarmpreYaEcdgkaylrtandLgsfdke.......llGnIqCptkedav....
Con    C-D--------------------L--------------S-S------S--------E------------F----------------G-I-C-----------

       1101                                                                        1200
RVF    sclrapnlisyKpmidqlecttnlIdpfvvfergsLpqtrndktFaaskgnrgVqafskgsvqadLtlmfdnfevdFvgaavscdaaflnltG.......
PT     scirapglikyKpmtdqiectaslVdpfaiflkgsLpqtrngqtFtstkdkktVqaftngaikalLsinlddheivFinkvkncdatflnvsG.......
UUK    rcissdsiifsKvhkdsvdcqssiIdpmtirnrnkLpstvgsvtFwptet..sVeaaipdlasatMlirldgytiqFrsdsnkcsprflslsG.......
RSV94k .......vlssKcktkilsnedlpViryierdgvdMlehvksepL......kdVlvsssgislstLdlfpvelnlqFkeaitsiitskislnGtsckitg
Con    -----------K--------------------L---------F---------V-------------L----------F----------------G-------

       1201                                                                        1300
RVF    ........cyscnagarvcls.ItstgtgslsahnkdgslhivlpsengtkdqcqilhFtvpeVeeefmYscdgderpllvkgtliaidpfddrreagge
PT     ........cyscdygahvcvk.VkssesadffaesedkttvlsfpiqsgthdycqvlhFqkplVderlsYscgsepklivikgtlvcmgvydfrnktggs
UUK    ........cynceagaklelehVtdfgtalgilecpslgyttyyevkntleksirtmhLngshVeakcyFrcpnsesqltirgeliylfnddirhhnqtl
RSV94k ierkfkkttvsiessnkvylsdIlaceglavcp..........milnnikkgtcitttYysvtVgsmikCkfiysgdtlmckydvspleitvispsldvs
Con    ------------------------------------------------------------F----V-----Y-----------------------------

       1301                                                            1382
RVF    Stvvnpksgs.WnffdwfsglmswFggplkl...yssfacmlhyqlgsfsslyileeqaslkcgllplrrphrsvrvkvic*
PT     Stvvnpsega.WsisnwfsglldwLggpmka...ilkilgfiaigivcfvlfmil.............iriavnsinikkkn*
UUK    SpglspksgsgWdpfgwfka..swLraiwai...lggtvsliigvviiymvftl..................clkvkks*.
RSV94k SfeavktsttnW........melLagivkdnpklslvasiipiglilktirsflddirqvd*..................
Con    S-----------W-------------L--------------------------------------------------------
```

FIGURE 7.

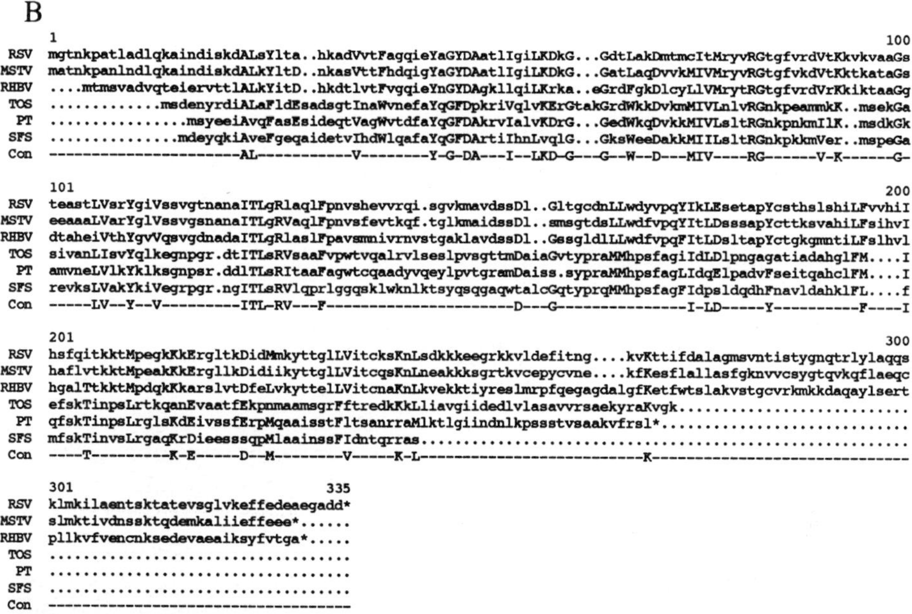

FIGURE 7. Similarity between phlebo- and tenuivirus proteins. (A) Alignment of the phlebo-virus M segment-encoded precursor protein with the RSV 94k protein encoded by RNA segment 2. Virus abbreviations are given in Table I. "Con" indicates conservation of amino acid in all four sequences compared. No overall similarity is observed in the N-terminal approximately 400 amino acids. (B) Alignment of the phlebo- and tenuivirus nucleocapsid proteins. "Con" indicates conservation of amino acid in four of the six sequences compared.

may suggest conservation of structural features as a result of disulfide bond formation in the mature proteins. Kakutani *et al.* (1991) reported sequence similarity between the nucleocapsid protein of RSV and Punta Toro N pro-tein; more extensive comparisons are shown in Fig. 7B. Again, however, the functional significance of the conserved residues requires experimental dem-onstration. Based on these data, together with the identity between the tenui- and phlebovirus terminal sequences, and the mode of tenuivirus tran-scription, it seems valid to conclude that the tenuiviruses share sufficient features with other members of the family for them to be included as a separate genus in the *Bunyaviridae*.

C. Evolutionary Pathway of Hantaviruses

Most complete genome sequences are available for hantaviruses (Table I), in part because of the interest generated by the emergence of hantavirus pulmonary syndrome, which has resulted in a number of phylogenetic anal-yses. Comparison of the structural protein sequences has revealed distinct evolutionary pathways which are influenced most strongly by the primary rodent host (Antic *et al.*, 1991a, 1992). Geographical factors do not appear to have been a major influence since Puumala-like and Hantaan-like viruses have been isolated from the same region in Europe but from different rodent hosts. Antic *et al.* (1992) suggest that the divergence of these pathways is not recent, citing as evidence the high homology between the M segments of Hantaan-like viruses isolated from *Apodemus* and humans over a 7-year period (Schmaljohn *et al.*, 1988). Phylogenetic trees and further discussion are provided in Chapters 3 and 11.

D. Phylogeny of Bunyavirus S Segments

As shown in Fig. 1, the bunyavirus S segment encodes two proteins in overlapping reading frames. Complete S segment sequences of 22 bunyaviruses representing three serogroups, Bunyamwera, California, and Simbu, have been determined (Table I). The segments vary from 850 to 1077 nucleotides in length, and show marked variation in the lengths of the untranslated regions. The N proteins are more uniform in size, being either 233 (Bunyamwera and Simbu serogroups) or 235 (California serogroup) amino acids; within a serogroup there is > 62% identity and between serogroups > 40% (Dunn *et al.*, 1994; Bowen *et al.*, 1995). However, identical residues are not distributed throughout the proteins but are clustered, particularly the region between residues 90 and 105 which shows almost total identity (Fig. 8A). The conserved amino acids, which presumably are involved in interacting with the viral RNA and the polymerase, do not match with the motifs described for other RNA-binding proteins (Burd and Dreyfuss, 1994). The NSs proteins are far more variable, ranging from 83 to 109 amino acids, show lower homology, and display few globally conserved residues (Fig. 8B).

Bowen *et al.* (1995) performed phylogenetic analyses on the nucleotide sequence of the N protein ORFs (Fig. 9) which showed that the viruses clustered into three monophyletic clades or lineages, corresponding to their respective serogroups. The California serogroup viruses split into three lineages which correspond to antigenic complexes assigned by Calisher and Karabatsos (1988). Trivittatus virus (TVT) was the sole member of one lineage, ancestral to the Melao and California complexes, in concordance with its serological classification by Calisher and Karabatsos (1988) into a separate monotypic complex within the California serogroup. The position of TVT suggests that it most closely resembles the progenitor virus from which the California serogroup viruses evolved (Bowen *et al.*, 1995). Since the serological assignments were based on neutralization and hemagglutination-inhibition assays, which measure antigens encoded by the M segment, the S and M segments of these viruses appear to have cosegregated during evolution. The Bunyamwera serogroup viruses occupy a distinct lineage, with Guaroa virus (GRO) occupying the most ancestral position. As noted previously, GRO is most probably the result of a reassortment event in nature between ancient California and Bunyamwera serogroup viruses (Dunn *et al.*, 1994; Chapter 8). Correlation of the phylogenetic tree and the length of the NSs ORF indicates that the trend in bunyavirus evolution is toward a smaller NSs protein. In the California serogroup, viruses occupying the more ancestral positions (MEL, CE, LUM, and TVT) have 97-residue NSs proteins whereas those viruses in lower descendant positions have 94-residue NSs proteins. Similarly in the Bunyamwera serogroup, GER and KRI have 109-residue NSs proteins whereas BAT, BUN, CV, MAG, MD, and NOR all have 101-residue NSs proteins. In all cases the shortening of the NSs is by truncation at the carboxy terminus. GRO does not fit this pattern, however, in having a very short NSs (83 residues) which shows low homology with the other members of the serogroup (Dunn *et al.*, 1994). Bowen *et al.* (1995) suggest that the shortening of the GRO NSs (which has occurred at both the amino and carboxy termini) represents a separate event in bunyavirus evolution. A shorter NSs protein may be of evoutionary advantage in freeing additional codons in the N ORF from constraints imposed by the overlapping NSs ORF.

E. Origins of the Ambisense Coding Strategy

In addition to the phlebovirus S segment and the tospovirus M and S segments, ambisense coding strategies have also been described for both RNA segments which comprise the arenavirus genome (Bishop, 1990), and for three of the genome RNA segments of the plant-infecting tenuiviruses (Ramirez and Haenni, 1994; Toriyama *et al.*, 1994). The use of an ambisense strategy may provide an additional, temporal, level of control in viral gene expression, in that some protein products are only expressed after genome replication has commenced. How might an ambisense strategy have arisen? The simplest and most likely explanation is that the viral polymerase jumped from one template to another of the opposite sense during replication, and the resulting hybrid RNA became fixed in the population. This RNA would thus encode two products in opposite orientations. Polymerase jumping is not unknown in RNA viruses as evidenced by the various forms of defective RNA molecules (copyback, mosaic, etc.) that have been identified (Holland, 1990). One template could originally have been another viral genome segment, and so the generation of ambisense RNAs represents a compaction of the genome into fewer segments. Alternatively, one template could have been RNA derived from a different coinfecting virus, from another intracellular parasite, or even from the cell's genome, thus expanding the genetic content of the virus. Future, more sensitive, computer-aided analyses of the ambisense encoded proteins may give some clues to their origins.

A

```
     1                                                                                                  100
LAC  .MsDLvFyDVastgangFDPDagYmdFcvknaeslnLaaVRIFFlnaakaKaaLsrkpErkanpkFGeWqVeViNnhFPGNRNnPIgnndLTiHRlSGYL
SSH  .MsDLvFyDVastgangFDPDagYmaFcvkyaesvnLaaVRIFFlnaakaKaaLsrkpErkanpkFGeWqVeVvNnhFPGNRNnPInsddLTiHRlSGYL
CE   .MsDLvFyDVastgangFDPDagYvdFcakhgesinLaaVRIFFlnaakaKaaLsrkpErkanpkFGeWqVeVvNnhFPaNRNnPIgnndLTiHRiSGYL
LUM  .MsDLvFyDVastgangFDPDagYvdFcvkhgesinLhsVRIFFlnaakaKaaLarkpErkaspkFGeWqVeIvNnhFPGNRNnPIdnndLTiHRiSGYL
JC   .MgDLvFyDVastgangFDPDagFvaFmadhgesinLsaVRIFFlnaakaKaaLarkpErkatpkFGeWqVeIvNnhFsGNRNnPIgnndLTiHRlSGYL
JS   .MgDLvFyDVastgangFDPDagFvaFmadhgesinLsaVRIFFlnaakaKaaLarkpErkatpkFGeWqVeIiNnhFPGNRNnPIgnndLTiHRlSGYL
KEY  .MgDLvFyDVastgangFDPDagYvaFmanhgesisLstVRIFFlnaakaKaaLtrkpErkatpkFGeWqVeIvNnhFPGNRNnPIgnndLTlHRlSGYL
MEL  .MgDLiFyDVastgangFDPDagYlaFtiahgeainLsaVRIFFlnaakaKaaLsrkpErkatpkFGdWqVeIvNnhFPGNRNnPIgnndLTiHRlSGYL
TVT  .MsELvFyDapstgangFDPDagYvaFiaahagsydLsaVRIFFlnaakaKnaLsrkpEgkvsikFGeWsVeVvNnhFPGNRNnPIgnndLTiHRiSGYL
CV   .MiELeFnDVaantsstFDPEiaYvnFkrihttglsYdhIRIFYikgreiKtsLtkrsEwevtlnLGgWkVtVfNtnFPGNRNsPVpddgLTlHRlSGYL
NOR  .MiELeFnDVaantsstFDPEvaYinFkrvyttglsYdhIRIFYikgreiKtsLtkrsEwevtlnLGgWkVtVfNtnFPGNRNsPVpddgLTlHRlSGFL
MAG  .MiELeFnDVaantsstFDPEiaYvnFkrihttglsYdhIRVLYikgreiKtsLtkrsEwevtlnLGgWkVaVfNtnFPGNRNsPVpddgLTlHRlSGFL
BAT  .MiELeFnDVaantsstFDPEvaYinFkriyttglsYdnIRIFYikgreiKtsLskrsEwevtlnLGgWkVtVfNtnFPGNRNsPVpddgLTlHRlSGFL
BUN  .MiELeFhDVaantsstFDPEvaYanFkrvhttglsYdhIRIFYikgreiKtsLakrsEwevtlnLGgWkItVyNtnFPGNRNnPVpddgLTlHRlSGFL
MD   .MiELeFhDVaanssstFDPEvaYasFkrvhttglsYdhIRIFYikgreiKtsLskrsEwevtlnLGgWkVaVfNtnFPGNRNsPVpddgLTlHRlSGFL
GER  .MlELeFeDVpnnigstFDPEsgYtnFqrnylpgvtLdqIRIFYikgreiKnsLskrsEwevtlnLGgWkVpVlNtnFPGNRNnaVpdygLTfHRiSGYL
GRO  .MaEieFfDVaqnatstFnPElqYatFkrtnttglnYdnIRIFYlngkrsKdtLskrsEqsvvlnFGgWrIpVvNthFPGNRNsPVlddsFTlHRvSGYL
KRI  .MsEieFhDVtantsstFDPEagYaaFkrrhttglnYdhIRIFFlngkkaKdtLskrsEttitlnFGgWkIpVvNthFleNRNmsVpddgLTlHRvSGYL
AINO manqFiFqDVpqrnlatFnPEvgYvaFiakhgaqlnFdtVRfFFlnqkkaKmvLsktaqpsvdltFGgikftlvNnhFPqytanPVpdtaLTlHRlSGYL
Con  -M-EL-F-DV-------FDPE--Y--F---------L--VRIFF------K--L----E------FG-W-V-V-N--FPGNRN-PV----LT-HR-SGYL

     101                                                                                                200
LAC  ARWvLDqynenddEsqhelIrttIINPiAEsNGVgWdsGpEIYLSFFPGtEMFLetFkFYPLtIGIhrVkqgmMDpqYLkKaLRQRYgtLtAdkWmsqKv
SSH  ARWvLEqykenedEsrrelIkttIINPiAEsNGVrWdsGaEIYLSFFPGtEMFLetFkFYPLtIGIYrVkqgmMDpqYLkKaLRQRYgsLtAdkWmsqKv
CE   ARWvLEqykenedEsqrelVkttVINPiAEsNGIrWenGaEIYLaFFPGtEMFLetFkFYPLtIGIYrVkngmMDsqYLkKaLRQRYgsLtAekWmsqKt
LUM  ARWvLEqfkenedaaqrelIkttVINPiAEsNGIrWdnGaEIYLaFFPGtEMFLetFnFYPLtIGIYrVkhgmMDpqYLkKaLRrRYgsLtAdkWmsqKt
JC   ARWvLEhfnsdddEsqrelIrstIINPiAEsNGIhWnnGpEIYLSFFPGtEMFLeiFkFYPLtIGIYrVkhgmMDpqYLkKaLRQRYgtLtAekWmaqKt
JS   ARWvLEhftadddEsqrelIrstIINPiAEsNGIhWnnGpEIYLSFFPGtEMFLeiFkFYPLtIGIYrVkhgmMDpqYLkKaLRQRYgtLtAekWmaqKt
KEY  ARWvLEhfgegedEsqkelIkstVINPiAEsNGIrWgnGvEIYLSFFPGtEMFLelFkFYPLtIGIYrVkhgmMDaqYLkKaLRQRYgtLtAdkWmaqKt
MEL  ARWvLDlfkenedEsqkelIqstIINPiAEsNGIhWanGvEIYLSFFPGtEMFLeaFrFYPLtIGIYrVkhglMDpqYLkKaLRQRYgtLtAdkWmaqKt
TVT  ARWvLEefkgqddEaqkdiIrstIVNPiAEsNGIhWdsGaDaYLSFFPGtEMFLesFdFlPLaIGIYrVkngmMDvqYLkKaLRQRYgtMtAdkWmstKt
CV   ARYlLEki.lkvsDpekviIkskIINPlAEkNGItWsdGeEVYLSFFPGsEMFLgtFkFYPLaIGIYkVqrkeMEpkYLeKtMRQRYmgLeAstWtisKv
NOR  ARYlLEki.lkvsEpekliIkskIINPlAEkNGItWtdGeEVYLSFFPGsEMFLgtFkFYPLaIGIYkVqrkeMEpkYLeKtMRQRYmgLeAstWtisKv
MAG  ARYlLEki.lkvsDpekliIkskIINPlAEkNGItWadGeEVYLSFFPGsEMFLgtFkFYPLaIGIYkVqkkeMEpkYLeKtMRQRYmgLeAatWtvsKv
BAT  ARYlLEki.lkvsDpekliIkskIVNPlAEkNGItWadGeEVYLSFFPGsEMFLgtFrFYPLaIGIYkVqrkeMEpkYLeKtMRQRYmgLeAstWtvsKl
BUN  ARYlLEkm.lkvsEpekliIkskIINPlAEkNGItWndGeEVYLSFFPGsEMFLgtFrFYPLaIGIYkVqrkeMEpkYLeKtMRQRYmgLeAatWtvsKl
MD   ARYlLEki.lkvsEpekllIkskIINPlAEkNGItWadGeEVYLSFFPGsEMFLgiFkFYPLaIGIYkVqrkeMEpkYLeKtMRQkYmnMdAatWtvtqv
GER  ARYlLgky.laetEpeklimrtkIVNPlAEkNGItWesGpEVYLSFFPGaEMFLgtFrFYPLaIGIYkVqrkeMDpkFLeKtMRQRYlgidAqtWtttKl
GRO  ARYlLEry.ltvsapeqaiIrskIINPiAasNGItWedGpEVYLSFFPGtEMFLetFkFYPLaIGIYkVqkkmMEakYLeKtMRQkYagLdAsqWtqqKy
KRI  ARYlLDrv.ysagEpeklkIkttIINPiAashGItWddGeEVYLSFFPGsEMYLttFkFYPLaIGIYkVqrklMDpkYLeKtMRQRYmnLdAsqWtqkhf
AINO AkWvaDqcktnqiklaeamek..IVmPlAEvkGctWteGltmYLgFaPGaEMFLetFeFYPLvIdmhrVlkdgMDvnFMrKvLRQRYgtLtAeqWmtqKi
Con  ARW-LE-------E-----I---IINP-AE-NGI-W--G-EVYLSFFPG-EMFL--F-FYPL-IGIY-V----MD--YL-K-LRQRY--L-A--W---K-
```

```
     201                                 237
LAC  aaIaksLkdVeqLkWgkgglSdtAktFLqKFGIrLp
SSH  taIaksLkeVeqLkWgrgglSdtARsFLqKFGIrLp
CE   gmIaksLkeVeqLkWgrgglSdtARtFLqKFGIkLp
LUM  vaIaksLkdVeqLkWgrgglSdtARvFLqKFGIkLp
JC   vlIaksLkdVeqLkWgrgglSdaARtFLiKFGVkLp
JS   vlIaksLkdVeqLkWgrgglSdaARtFLiKFGVkLp
KEY  smItksLkdVeqLkWgkgglSdtARaFLaKFGVrLp
MEL  tmIaksLkdVeqLkWgkgglSdaARtFLqKFGVrLp
TVT  tvIaktLkrVesFkWgkgglSeaARaFLsKFnVkip
CV   neVqaaLtvVsgLgWkktnvSaaAReFLaKFGIsM.
NOR  neVqaaLtvVsgLgWkktnvSaaAReFLaKFGInM.
MAG  neVqaaLtvVsgLgWkktnvSaaAReFLaKFGInM.
BAT  neVqsaLtvVsgLgWkktnvSsaAReFLaKFGIsM.
BUN  teVqsaLtvVssLgWkktnvSaaARdFLaKFGInM.
MD   seVqaaLtvVsgLgWkktnvSaaAReFLaKFGInM.
GER  geVeaaLkvVsgLgWkktnvSsaAReFLsKFGIrM.
GRO  neInaaLsvVssLgWkkanvSsaAReFLarFGIsL.
KRI  sdVnsaLtvVagLgWkkanvSiaAkdFLnKFGIni.
AINO daVraaFnaVgqLsWaksgfSpaARaFLaqFGIni.
Con  --I---L--V--L-W-----S--AR-FL-KFGI-L-


B
     1                                                                                                              112
CV   MMSlltpavlLtqrlhtltLsvstplglVmttFeSStlkdarlklvSqkevnGrLrLtLgaGRLlylIqIFLaTGTvQFqtmVLPStdsv..dtlpgtylrky.........
MAG  MMSlltpavlLtqrlhtltLsvstplglVmttFeSStlkdarlklvSqkevsGrLrLtLgaGRLlylIqIFLaTGTvQFqtmVLPStdsv..dslpgtylrkf..........
NOR  MMSlltpavlLtqrlhtltLsvstplglVmttFeSStlkdvrlklvSqrevnGrLrLtLgaGRLlylIqIFLaTGTvQFqtmVLPStdsv..dslpgtylrry.........
BAT  MMSlltpavlLtqrlhtltLsvstplglVmttLefStlkdarlklvSqkevnGrLrLtLgaGRLlylIqIFLgTGTvQFqtmVLPStdsv..dslpgtylrrf.........
MD   MMSlltpavlLtqrshtlvLsvstplglVtttFeSStlkderlklvSqrevnGrLrLtLgaGRLlylIqIFLaTGTvQFqtmVLPStdsv..dslpgtylrky..........
BUN  MMSlltpavlLtqrshtltLsvstplglVmttYeSStlkdarlklvSqkevnGkLhLtLgaGRLlyiIrIFLaTGTtQFltmVLPStasv..dslpgtylrrc.........
KRI  MMSlltpavlLiqkpgmlrLsvdtpqglImttYeSSflterrlkilSqrevrqqLrLtLgaGRYlwlIrIFLrTGTcQFqmmVLPStesv..dilpgtclteytllenqrn
GER  .MSlitsgvlLtqsqdtltFsvttcqglrltkFaSStlkdarlklvSqkevnGkLrLtLgaGRYlysIrIsLeTGTmQclttVLPStvsv..dtlpgtylestlqrqnqkss
GRO  .....mqqvrLtrssnmlhLnvqtqqglIttiLeSSismgrdpkilSlrevnnrLfLtLeaGeFlwlIhIFLeTGTvQssmihshfie.................
JC   MMShpqvqmdLiqmqglwhLwlttes1sIcqpLgSSslmqqkpkllSlvnrsGkLlLsLesGRWrssIiIFLeTGTtQLvttILPSigf..qdi.................
JS   MMShpqvqmdLiqmqglwhLwlttes1sIcqpLgSSslmqqkpkllSlvnrsGkLlLsLesGRWrssIiIFLeTGTtQLvttILPSigf..qdi.................
KEY  MMShpqvqmdLilmqgmwhLwltmgsrsVcqpLgSSslmpqkpkllSlvsrsGrLhLsLesGRWrssIiIFLeTGTtQLvttILPctgf..qdi.................
MEL  MMShqqvqmdLiqmqgiwhLqlrmgklsIcqpLgSSslmpqkpkllSlvnrrGkLlLnLetGRWklstiIFLeTGTtQLvttILPSigf..qdilpdgc...........
LAC  MMShqqvqmdLilmqgiwtsvlkmqnhstllqlgSSsssmlqrprllSrvsqrGrLtLnLesGRWrlsIiIFLeTGTtQLvttILPStdy..lgi................
SSH  MMShqqvqmdLilmqgiwhsvlnmqnsIllqLgSSsssmpqrprllSrvsqrGrqiLnLesGRWrlsIiIFLeTGTiQLtatILPStdc..qdi................
CE   MMShpqvqmdLilmqgmwtsvlnmqnqltllqLgSSsssmpqrprllSrvsqrGkLiLnLasGRWrlsIiIFqqTGTiQLvttILPStas..qdtlpdgs.........
LUM  MMShppvqmdLilmqgmwtFvlnmenqsIsipLgSfslmplrprllSlvsrrGrLvLnLesGRWrssIiIFLeTGTtQLittILPStgc..qgiwldgc.........
TVT  MMlhqqvqtdLipmqgmwhLllhmpdrtIflllgSSsssmlprprmlSrenqrGrLvLnLasGRWrwsIiIFLaTGTiQLvttILPStef..qaisqdgf.........
AINO .MflngislrLtrrsgmwhLllnmgpnsIsipLdSSssirrrprwySvrrhnqvLiLhLvassLhwlItIF..pnTqQilcqtLPSlsivsqdi..............
Con  MMS-------L--------L--------I---L-SS----------S-----G-L-L-L--GRW---I-IFL-TGT-QL---ILPS-----------------------
```

FIGURE 8. Alignments of bunyavirus S RNA segment gene products. (A) Alignment of N proteins. (B) Alignment of NSs proteins. Virus abbreviations are given in Table I. "Con" indicates conservation of amino acid in 17 of the 19 sequences compared.

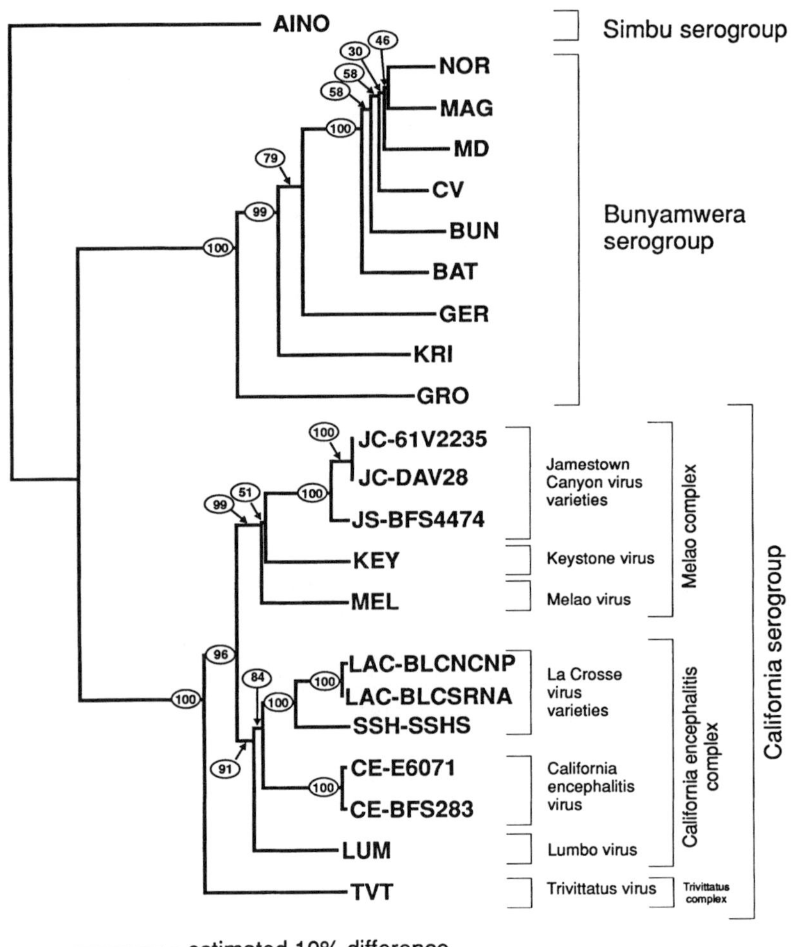

FIGURE 9. Phylogeny inferred from Fitch–Margoliash analysis of the nucleotide sequences encoding the N proteins of Bunyamwera and California serogroup bunyaviruses with Aino virus of the Simbu serogroup used as an outgroup. Numbers in ovals indicate percent of 1000 bootstrap sampling replicates that support the interior branch on which each oval is superimposed. Horizontal branch lengths are proportional to the scale bar. Vertical branches are for visual clarity only. Serological classification (Calisher and Karabatsos, 1988) is indicated by brackets. (Taken from Bowen *et al.*, 1995, with permission.)

IV. FUTURE RESEARCH: THE ADVENT OF REVERSE GENETICS

Study of the molecular biology of viruses with RNA genomes has been hampered by the inherent limitations imposed by the nature of their genetic material. For instance, site-directed mutagenesis of RNA to produce precisely defined nucleotide substitutions is not yet possible. A major advance in the study of positive-strand RNA viruses was the demonstration that cloned poliovirus cDNA was infectious (Racaniello and Baltimore, 1981), since in the DNA form viral sequences can be manipulated using recombinant DNA techniques and then reintroduced into viable virus. Subsequent studies have shown that RNA transcribed *in vitro* from cloned cDNA copies of a number of positive-strand genomes is a much more efficient initiator of infection, and these studies have revolutionized experimental analysis of replication processes of these viruses (Boyer and Haenni, 1994). Similar studies on negative-strand viruses have lagged but a "reverse genetics" system for influenza viruses was described in 1989 (Luytjes *et al.*, 1989). Since then, systems have been devised that allow or will allow investigation of *cis*-acting signals and protein requirements for transcription and replication of paramyxoviruses and rhabdoviruses as well. A common feature of these various systems is that the template RNA derives from a cDNA clone containing authentic viral terminal sequences. However, there are differences in the sources of viral proteins that recognize this template (helper virus infection or from recombinant sources), the delivery of the template (transfected naked RNA or ribonucleoprotein complex, or transcribed within the cell), and the sequences present in the template (reporter genes, defective-interfering RNA sequences, authentic viral transcripts, or mutated RNAs). A review of these approaches is given by Garcia-Sastre and Palese (1993).

With regard to genetically engineering a negative-strand RNA virus genome, again initial success was with influenza virus, whereby a cDNA-derived hemagglutinin gene was reassorted into the genome of a helper influenza virus (Enami *et al.*, 1990); subsequent studies have shown that other influenza virus segments may be manipulated and introduced by reassortment into the genome of a viable virus (Castrucci and Kawaoka, 1995, and references therein). More recently, infectious rhabdoviruses (rabies and vesicular stomatitis viruses) have been produced in a helper-independent manner: full-length cDNA clones of rhabdovirus genomes were transcribed intracellularly and subsequently replicated by the viral N, P, and L proteins produced by transfected plasmids. The amplified genomes were then the template for transcription of all five viral mRNAs, thus producing all of the viral proteins required for genome packaging and viral assembly, and infectious rhabdoviruses were recovered, albeit at low efficiency (Schnell *et al.*, 1994; Lawson *et al.*, 1995).

With regard to reverse genetics systems for the *Bunyaviridae*, most progress has been achieved with Bunyamwera and related bunyaviruses. Following the cloning and sequencing of the Bunyamwera virus L segment (Elliott, 1989b), a full-length cDNA was produced and cloned into a recombinant vaccinia virus. A protein the same size as authentic BUN L protein was made in infected cells, and a nucleocapsid transfection assay was established to demonstrate that the L protein was functional. BUN nucleocapsids were purified from BUN-infected cells by centrifugation on CsCl gradients. When transfected into cells it was expected that a low level of viral RNA synthesis would be observed (as detected by Northern blotting) but in fact no viral

RNA was produced, presumably because the purification procedure damaged the polymerase. However, when the cells expressed BUN L protein (by infection with the recombinant vaccinia virus) the synthesis of BUN S segment RNA was demonstrated (Jin and Elliott, 1991). By isolating nucleocapsids from purified virions, which contain only genome-sense RNA (as opposed to intracellular nucleocapsids which contain RNA of both polarities) it was possible to demonstrate that the recombinant L protein had both transcriptase and replicase activities (Jin and Elliott, 1993a).

The S positive-sense RNA produced was a functional mRNA as it was translated into N protein (Jin and Elliott, 1991) and contained nontemplated primer sequences at the 5' terminus (Jin and Elliott, 1993a). The only function that did not appear to be authentic was mRNA termination, which suggests that another bunyavirus protein may be required. A number of site-specific mutations in the L cDNA were made and their effect on RNA synthesis in the nucleocapsid transfection assay assessed; the results of these experiments allowed definition of the polymerase domain within the central part of the L protein (Jin and Elliott, 1992; see Fig. 4). However, this system has limitations in that authentic BUN nucleocapsids are used as the template for the recombinant L protein. At present there is no way of mutating the viral RNA or the encapsidating N protein in nucleocapsids extracted from infected cells, and thus it is not possible to define *cis*-acting signals in the RNA or examine the effects of specific mutations in the N protein.

Based on the influenza virus studies described previously (Luytjes *et al.*, 1989), a similar methodology was adopted to study transcription of a bunyavirus-like RNA containing a reporter gene (Dunn *et al.*, 1995). The plasmid pBUNSCAT (Fig. 10) directs the synthesis of a bunyavirus-like RNA containing a reporter gene. The antisense CAT coding sequence was cloned in place of the BUN S segment coding region in a negative-sense cDNA under control of a T7 promoter. The complete untranslated regions of the BUN S segment were included since these were presumed to contain all of the transcriptional control signals. Thus, T7 transcripts derived from pBUN-SCAT would have the authentic termini of BUN S segment RNA and would contain (in 5' to 3' order) the 174-nucleotide (nt) 5' noncoding region of BUN S segment, the entire CAT ORF in negative polarity, and the 85-nt BUN S segment 3' noncoding region. After transfection of this RNA into cells, CAT activity would only be detected if the RNA were recognized and transcribed into message-sense RNA by the bunyavirus polymerase. Hence, cells expressed the BUN L and S segment proteins, either by recombinant vaccinia viruses or using the vaccinia virus–T7 RNA polymerase system (Fuerst *et al.*, 1986) and measurable CAT activity was dependent on expression of the gene products of both BUN genes (Fig. 11). Further, the level of CAT activity produced following transfection of the chimeric RNA was also dependent on the amount and ratio of these proteins synthesized in the cells.

The bunyavirus S segment encodes two proteins, N and NSs, in overlapping reading frames which are translated from the same mRNA species as the

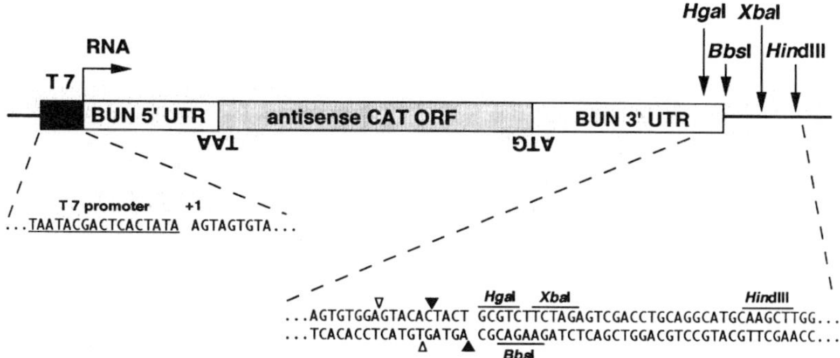

FIGURE 10. Essential features of pBUNSCAT. The truncated T7 promoter initiates transcription on the first base (A) of the BUN sequence. Run-off transcripts can be produced by linearizing the DNA at the indicated restriction sites; use of the *Bbs*I site, shown by solid triangles, produces RNA with the exact 3' terminal sequence of the BUN S segment.

result of initiation at alternate AUG codons (Fig. 1). To investigate whether both S segment gene products were needed, plasmids were constructed to express just one of the proteins: to make constructs expressing only N protein the third codon of NSs was converted to a stop codon by a single point mutation which did not alter the coding of the N protein ORF (Fig. 12), while to express just NSs the appropriate portion of the cDNA, omitting the N protein initiation codon, was subcloned downstream of a T7 promoter (Elliott and McGregor, 1989). Translation of T7 transcripts from these constructs in reticulocyte lysate confirmed that the plasmids expressed only the expected proteins. These plasmids were then used in conjunction with a BUN L-expressing plasmid to determine the S segment protein requirements for bunyavirus transcription and replication (Fig. 12). CAT activity was detected when just the BUN L and N proteins were expressed in the transfected cells. No CAT activity was observed when just the L and NSs proteins were expressed. Thus, from these results it was concluded that transcription of the BUNSCAT RNA required only the bunyavirus L and N proteins. These results are, therefore, in accord with the fact that transcriptase activity can be detected in detergent-disrupted bunyavirus particles (Bellocq and Kolakofsky, 1987; Bellocq *et al.*, 1987; Bouloy and Hannoun, 1976; Gerbaud *et al.*, 1987b; Patterson *et al.*, 1984; Vialat and Bouloy, 1992), assuming that NSs is a truly nonstructural protein and is not incorporated into virions. The role of NSs thus remains unknown. Vialat and Bouloy (1992) have suggested that NSs may be involved in mRNA transcription termination, but analysis of the 3' ends of message-sense chimeric CAT RNAs made in the presence and absence of NSs synthesis using the described system has not yet been reported.

Recently, Lopez *et al.* (1995) described a similar system for analyzing phlebovirus transcription. The template consisted of the Rift Valley fever

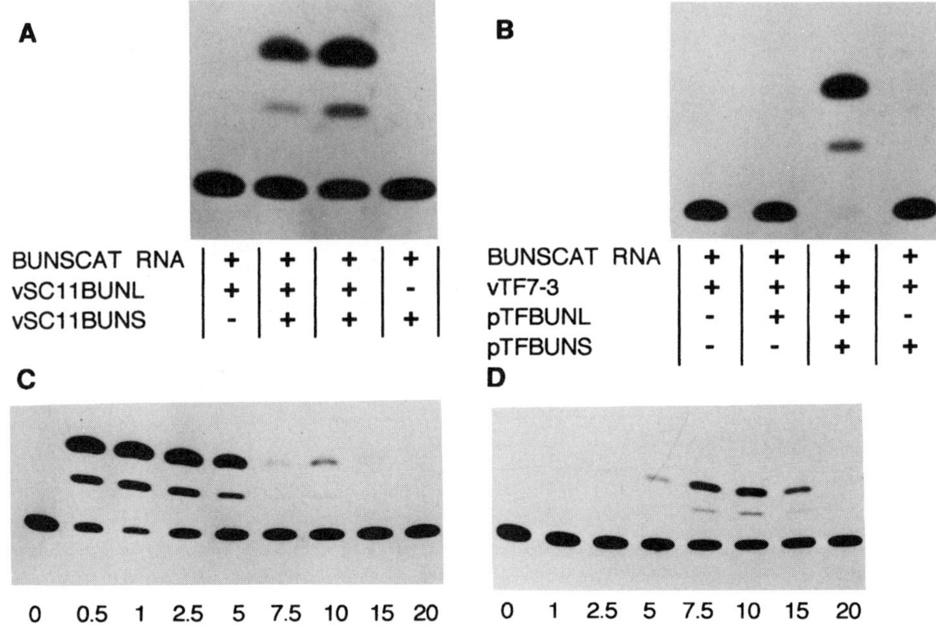

FIGURE 11. Transcription of BUNSCAT RNA by recombinant bunyavirus proteins. (A) CAT activity in recombinant vaccinia virus infected cells after transfection with BUNSCAT RNA. Cells were infected with vSC11BUNL (expressing BUN L protein) or vSC11BUNS (expressing BUN N and NSs proteins) or both viruses as indicated, and subsequently transfected with chimeric RNA. Cell extracts were prepared 20 hr postinfection, assayed for CAT activity, and the acetylated products separated by thin-layer chromatography. (B–D) CAT activity in cells transiently expressing BUN L and S segment gene products via the vaccinia virus–T7 system. Cells were infected with vTF7-3, then transfected with pTFBUNL (BUN L gene under T7 promoter control) and/or pTFBUNS (BUN S gene under T7 promoter control) and then transfected with BUNSCAT RNA. Cell extracts were prepared 16 hr later and assayed for CAT activity. (B) CAT activity is dependent on expression of both L and S segment gene products. (C) Titration of pTFBUNS. Cells were transfected with 5 μg of pTFBUNL DNA and varying amounts of pTFBUNS DNA as indicated (in μg). (D) Titration of pTFBUNL. Cells were transfected with 5 μg of pTFBUNS DNA and varying amounts of pTFBUNL DNA as indicated (in μg).

virus S RNA terminal untranslated regions flanking an antisense CAT gene (i.e., the template contained a single ORF rather than two ambisense ORFs, as in the authentic S segment), and RVF virus proteins were expressed from recombinant vaccinia viruses. Only the viral L and N proteins were required to transcribe the template after transfection into recombinant vaccinia virus infected cells, and additional expression of the RFV NSs protein apparently had no effect on reporter gene expression.

As discussed previously (Chapter 8), certain bunyaviruses can form reassortants by exchange of genome segments after mixed infection, and hence the above assay system provided a way to examine the molecular basis for restriction on reassortment by investigating whether the BUN L protein

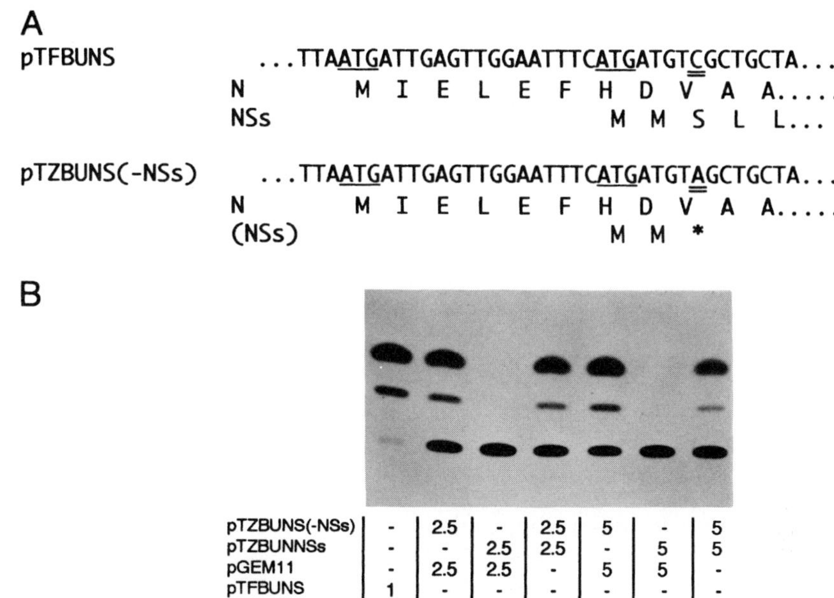

FIGURE 12. Only the BUN L and N proteins are required for transcription of BUNSCAT RNA. (A) Nucleotide and amino acid sequences of the wild-type and NSs ablation mutants of BUN S segment cDNA. The ATG initiation codons of N and NSs ORFs are underlined and the C-to-A mutation resulting in the introduction of a stop codon in the NSs ORF is double underlined. (B) CAT activity in transfected cell extracts. CV-1 cells were infected with vTF7-3, then transfected with 5 μg pTFBUNL and the indicated amounts (in μg) of BUN or MAG S segment cDNA-expressing plasmids. pGEM 11 DNA was added in some cases to maintain the overall concentration of DNA. The cells were subsequently transfected with aliquots of the same preparation of BUNSCAT RNA, and cell extracts prepared for CAT assay 16 hr later.

could function in concert with heterologous S segment gene products. Cells were infected with vTF7-3 and transfected with different combinations of BUN L and S segment cDNAs, followed by transfection with BUNSCAT RNA. As seen in Fig. 13, CAT activity was detected with the BUN L protein and the gene products of the S segments of BAT, CV, NOR, MAG, and MD viruses, but not with the S segments products of GRO, KRI, and LUM viruses. The compatible combinations are those with N proteins most closely related to BUN N (see Fig. 9). With one exception, compatibility in the CAT assay correlated with the ability of certain bunyaviruses to form reassortants by genome segment exchange in the course of a mixed infection (see Fig. 2 in Chapter 8). Thus, BUN could reassort with BAT, MAG, and NOR viruses, but not with KRI, GRO, or LUM viruses. However, MD virus was unable to form reassortants with BUN (or several other bunyaviruses) and was considered to be genetically isolated, although the MD N protein did function with BUN L protein to transcribe BUNSCAT RNA. Hence, restrictions on genome segment reassortment may be at more than one level, and

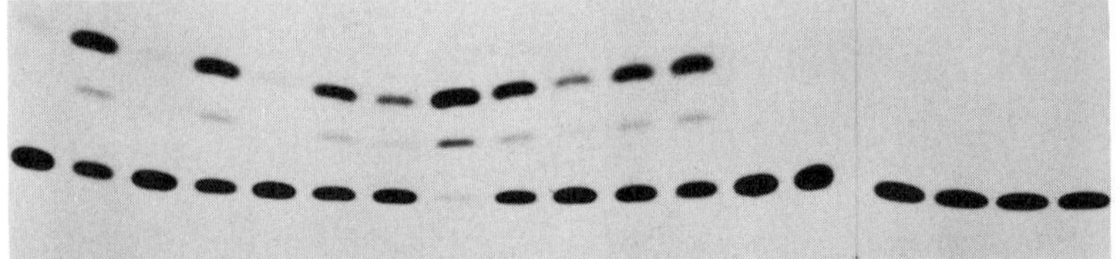

FIGURE 13. Transcription of BUNSCAT RNA by BUN L protein and the gene products of heterologous bunyavirus S segment. Cells were infected with vTF7-3, and then transfected with 5 µg pTFBUNL and 1 or 5 µg S segment plasmid as indicated. Cell extracts were assayed for CAT activity 16 hr after transfection with BUNSCAT RNA.

not just because of incompatibility between heterologous N and L proteins and the RNA template.

Reverse genetic approaches have now been described for representatives of most groups of negative-strand RNA viruses, though to date the production of infectious viruses derived from cDNAs has only been described for influenza viruses (Garcia-Sastre and Palese, 1993) and rhabdoviruses (Schnell *et al.*, 1994; Lawson *et al.*, 1995). The influenza virus system relies on reassortment of a cDNA-derived RNA segment into a helper influenza virus genome, whereas a helper-independent system utilizing transiently expressed proteins was described for the rhabdoviruses. The results described above with BUN virus cDNA clones pave the way to attempting a similar approach to recover infectious bunyavirus containing cDNA-derived genome segments. This will open up new opportunities to study all aspects of *Bunyaviridae* replication, pathogenesis, interaction with vectors, and possibly in the development of genetically engineered vaccine strains.

V. REFERENCES

Accardi, L., Gro, M. C., Bonito, P. D., and Giorgi, C., 1993, Toscana virus genomic L segment: Molecular cloning, coding strategy and amino acid sequence in comparison with other negative strand RNA viruses, *Virus Res.* **27**:119.

Akashi, H., and Bishop, D. H. L., 1983, Comparison of the sequences and coding of La Crosse and snowshoe hare bunyavirus S RNA species, *J. Virol.* **45**:1155.

Akashi, H., Gay, M., Ihara, T., and Bishop, D. H. L., 1984, Localized conserved regions of the S RNA gene products of bunyaviruses are revealed by sequence analyses of the Simbu serogroup Aino virus, *Virus Res.* **1**:51.

Antic, D., Lim, B.-U., and Kang, C. Y., 1991a, Molecular characterization of the M genomic segment of the Seoul 80-39 virus; nucleotide and amino acid sequence comparisons with other hantaviruses reveal the evolutionary pathway, *Virus Res.* **19**:47.

Antic, D., Lim, B.-U., and Kang, C. Y., 1991b, Nucleotide sequence and coding capacity of the large (L) genomic RNA segment of Seoul 80-39 virus, a member of the hantavirus genus, *Virus Res.* **19**:59.

Antic, D., Kang, C. Y., Spik, K., Schmaljohn, C. S., Vapalahti, O., and Vaheri, A., 1992, Comparison of the deduced gene products of the L, M and S genome segments of hantaviruses, *Virus Res.* **24:**35.

Arikawa, J., Lapenotiere, H. F., Iacono-Connors, L., Wang, M., and Schmaljohn, C. S., 1990, Coding properties of the S and M genome segments of Sapporo rat virus: Comparison to other causative agents of hemorrhagic fever with renal syndrome, *Virology* **176:**114.

Banerjee, A. K., 1980, 5' terminal cap structure in eukaryotic messenger ribonucleic acids, *Microbiol. Rev.* **44:**175.

Beaty, B. J., and Bishop, D. H. L., 1988, Bunyavirus–vector interaction, *Virus Res.* **10:**289.

Beaty, B. J., Rozhon, E. J., Gensemer, P., and Bishop, D. H. L., 1981, Formation of reassortant bunyaviruses in dually infected mosquitoes, *Virology* **111:**662.

Beaty, B. J., Sundin, D. R., Chandler, L. J., and Bishop, D. H. L., 1985, Evolution of bunyaviruses by genome reassortment in dually infected mosquitoes (*Aedes triseriatus*), *Science* **230:**548.

Bellocq, C., and Kolakofsky, D., 1987, Translational requirement of La Crosse virus S mRNA synthesis: A possible model, *J. Virol.* **61:**3960.

Bellocq, C., Raju, R., Patterson, J. L., and Kolakofsky, D., 1987, Translational requirement of La Crosse virus S mRNA synthesis, *J. Virol.* **61:**87.

Bishop, D. H. L., 1990, Arenaviridae and their replication, in: *Virology* (B. N. Fields and D. M. Knipe, eds.), pp. 1231–1243, Raven Press, New York.

Bishop, D. H. L., Gould, K. G., Akashi, H., and Clerx-van Haaster, C. M., 1982, The complete sequence and coding content of snowshoe hare bunyavirus (S) viral RNA species, *Nucleic Acids Res.* **10:**3703.

Bouloy, M., and Hannoun, C., 1976, Studies on Lumbo virus replication. I. RNA dependent RNA polymerase associated with virions, *Virology* **69:**258.

Bouloy, M., Krams-Ozden, S., Horodniceanu, F., and Hannoun, C., 1973/74, Three segment RNA genome of Lumbo virus (bunyavirus), *Intervirology* **2:**173.

Bowen, M. D., Jackson, A. O., Bruns, T. D., Hacker, D. L., and Hardy, J. L., 1995, Determination and comparative analysis of the small genomic sequences of California encephalitis, Jamestown Canyon, Jerry Slough, Melao, Keystone and Trivittatus viruses (*Bunyaviridae*, genus *Bunyavirus*, California serogroup), *J. Gen. Virol.* **76:**559.

Boyer, J.-C. and Haenni, A.-L., 1994, Infectious transcripts and cDNA clones of RNA viruses, *Virology* **198:**415.

Burd, C. G., and Dreyfuss, G., 1994, Conserved structures and diversity of functions of RNA-binding proteins, *Science* **265:**615.

Cabradilla, C. D., Holloway, B. P., and Obijeski, J. F., 1983, Molecular cloning and sequencing of the La Crosse virus S RNA, *Virology* **128:**463.

Calisher, C., and Karabatsos, N., 1988, Arbovirus serogroups: Definition and geographical distribution, in: *The Arboviruses: Epidemiology and Ecology* (T. P. Monath, ed.), pp. 19–57, CRC Press, Boca Raton.

Casals, J., and Whitman, L., 1961, Group C, a new serological group of hitherto undescribed arthropod-borne viruses: Immunological studies, *Am. J. Trop. Med. Hyg.* **10:**250.

Castrucci, M. R., and Kawaoka, Y., 1995, Reverse genetics system for generation of an influenza A virus mutant containing a deletion of the carboxyl-terminal residue of M2 protein, *Virology* **69:**2725.

Chandler, L. J., Beaty, B. J., Baldridge, G. D., Bishop, D. H. L., and Hewlett, M. J., 1990, Heterologous reassortment of bunyaviruses in *Aedes triseriatus* mosquitoes and transovarial and oral transmission of newly evolved genotypes, *J. Gen. Virol.* **71:**1045.

Collett, M. S., Purchio, A. F., Keegan, K., Frazier, S., Hays, W., Anderson, D. K., Parker, M. D., Schmaljohn, C. S., Schmidt, J., and Dalrymple, J. M., 1985, Complete nucleotide sequence of the M RNA segment of Rift Valley fever virus, *Virology* **144:**228.

de Haan, P., Wagemakers, L., Peters, D., and Goldbach, R., 1990, The S RNA segment of tomato spotted wilt virus has an ambisense character, *J. Gen. Virol.* **71:**1001.

de Haan, P., Kormelink, R., Resende, R. de O., Poelwijk, F. van, Peters, D., and Goldbach, R., 1991, Tomato spotted wilt L RNA encodes a putative RNA polymerase, *J. Gen. Virol.* **72:**2207.

de Haan, P., de Àvila, A. C., Kormelink, R., Westerbroek, A., Gielen, J. J., Peters, D., and Goldbach, R., 1992, The nucleotide sequence of the S RNA of Impatiens necrotic spot virus, a novel tospovirus, *FEBS Lett.* **306:**27.

de Miranda, J., Hernandez, M., Hull, R., and Espinoza, A. M., 1994, Sequence analysis of rice hoja blanca virus RNA 3, *J. Gen. Virol.* **75:**2127.

Devereux, J., Haeberli, P., and Smithies, O., 1984, A comprehensive set of sequence analysis programs for the VAX, *Nucleic Acids Res.* **12:**387.

Domingo, E., and Holland, J. J., 1988, High error rates, population equilibrium and evolution of RNA replicating systems, in: *RNA Genetics*, Vol. III (E. Domingo, J. J. Holland, and P. Ahlquist, eds.), pp. 3–36, CRC Press, Boca Raton.

Dunn, E. F., Pritlove, D. C., and Elliott, R. M., 1994, The S RNA segments of Batai, Cache Valley, Guaroa, Kairi, Lumbo, Main Drain and Northway bunyaviruses: Sequence determination and analysis, *J. Gen. Virol.* **75:**597.

Dunn, E. F., Pritlove, D. C., Jin, H. and Elliott, R. M., 1995, Transcription of a recombinant bunyavirus RNA template by transiently expressed bunyavirus proteins, *Virology* **211:**133–143.

Elliott, R. M., 1989a, Nucleotide sequence analysis of the S RNA segment of Bunyamwera virus, the prototype of the family Bunyaviridae, *J. Gen. Virol* **70:**1281.

Elliott, R. M., 1989b, Nucleotide sequence analysis of the large (L) genomic RNA segment of Bunyamwera virus, the prototype of the family Bunyaviridae, *Virology* **173:**426.

Elliott, R. M., 1990, Molecular biology of the Bunyaviridae, *J. Gen. Virol.* **71:**501.

Elliott, R. M., and McGregor, A., 1989, Nucleotide sequence and expression of the small (S) RNA segment of Maguari bunyavirus, *Virology* **171:**516.

Elliott, R. M., and Wilkie, M. L., 1986, Persistent infection of *Aedes albopictus* C6/36 cells by Bunyamwera virus, *Virology* **150:**21.

Elliott, R. M., Dunn, E., Simons, J. F., and Pettersson, R. F., 1992, Nucleotide sequence and coding strategy of the Uukuniemi virus L RNA segment, *J. Gen. Virol.* **73:**1745.

Enami, M., Luytjes, W., Krystal, M., and Palese, P., 1990, Introduction of site-specific mutations into the genome of influenza virus, *Proc. Natl. Acad. Sci. USA* **87:**3802.

Eshita, Y., and Bishop, D. H. L., 1984, The complete sequence of the M RNA of snowshoe hare bunyavirus reveals the presence of internal hydrophobic domains in the viral glycoprotein, *Virology* **137:**227.

Felsenstein, J., 1989, PHYLIP—Phylogeny inference package, *Cladistics* **5:**164.

Fuerst, T. R., Niles, E. G., Studier, F. W., and Moss, B., 1986, Eukaryotic transient-expression system based on recombinant vaccinia virus that synthesises bacteriophage T7 RNA polymerase, *Proc. Natl. Acad. Sci. USA* **83:**8122.

Garcia-Sastre, A., and Palese, P., 1993, Genetic manipulation of negative-strand RNA virus genomes, *Annu. Rev. Microbiol.* **47:**765.

Garcin, D., and Kolakofsky, D., 1992, Tacaribe arenavirus RNA synthesis *in vitro* is primer dependent and suggests an unusual mechanism for the initiation of genome replication, *J. Virol.* **66:**1370.

Garcin, D., Iezzi, M., Dobbs, M., Elliott, R. M., Schmaljohn, C. S., Kang, C. Y., and Kolakofsky, D., 1995, The 5′ ends of Hantaan virus (*Bunyaviridae*) RNAs suggest a prime and realign mechanism for the initiation of RNA synthesis, *J. Virol.:* **69:**5754.

Gerbaud, S., Vialat, P., Pardigon, N., Wychowski, C., Girard, M., and Bouloy, M., 1987a, The S segment of Germiston virus RNA genome can code for three proteins, *Virus Res.* **8:**1.

Gerbaud, S., Pardigon, N., Vialat, P., and Bouloy, M., 1987b, The S segment of the Germiston bunyavirus genome: Coding strategy and transcription, in: *The Biology of Negative-Strand Viruses* (B. W. J. Mahy and D. Kolakofsky, eds.), pp. 191–198, Elsevier, Amsterdam.

Giebel, L. B., Stohwasser, R., Zoller, L., Bautz, E. K. F., and Darai, G., 1989, Determination of the coding capacity of the M genome segment of nephropathia epidemica virus strain Hällnäs B1 by molecular cloning and nucleotide sequence analysis, *Virology* **172:**498.

Giorgi, C., Accardi, L., Nocoletti, L., Gro, M. C., Takehara, K., Hilditch, C., Morikawa, S., and Bishop, D. H. L., 1991, Sequences and coding strategies of the S RNAs of Toscana and Rift Valley fever viruses compared to those of Punta Toro, Sicilian sandfly fever and Uukuniemi viruses, *Virology* **180:**738.

Grady, L. J., Sanders, M. L., and Campbell, W. P., 1987, The sequence of the M RNA of an isolate of La Crosse virus, *J. Gen. Virol.* **68**:3057.

Heinze, C., Maiss, E., Adam, G., and Casper, R., 1995, The complete nucleotide sequence of the S RNA of a new Tospovirus species, representing serogroup IV, *Phytopathology* **85**:683.

Hewlett, M. J., Pettersson, R. F., and Baltimore, D., 1977, Circular forms of Uukuniemi virion RNA: An electron microscopic study, *J. Virol.* **21**:1085.

Higgins, D. G., and Sharp, P. M., 1989, Fast and sensitive multiple sequence alignment on a microcomputer, *CABIOS* **5**:151.

Hjelle, B., Chavez-Giles, F., Torrez-Martinez, N., Yates, T., Sarisky, J., Webb, J., and Ascher, M., 1994, Genetic identification of a novel hantavirus of a harvest mouse *Reithrodontomys megalotis*, *J. Virol.* **68**:6751.

Hjelle, B., Anderson, B., Torrez-Martinez, N., Song, W., Gannon, W. L., and Yates, T. L., 1995, Prevalence and geographic genetic variation of hantaviruses of New World harvest mice (*Reithrodontomys*): Identification of a divergent genotype from a Costa Rican *Reithrodontomys mexicanus*, *Virology* **207**:452.

Holland, J. J., 1990, Defective viral genomes, in: *Virology* (B. N. Fields and D. M. Knipe, eds.), pp. 151–165, Raven Press, New York.

Huang, C., Thompson, W. H., and Campbell, W. P., 1995, Comparison of the M RNA genome segments of two human isolates of La Crosse virus, *Virus Res.* **36**:177.

Huiet, L., Klaassen, V., Tsai, J. H., and Falk, B. W., 1991, Nucleotide sequence and RNA hybridization analyses reveal an ambisense coding strategy for maize stripe virus RNA3, *Virology* **182**:47.

Huiet, L., Tsai, J., and Falk, B. W., 1992, Complete sequence of maize stripe virus RNA4 and mapping of its subgenomic RNAs, *J. Gen. Virol.* **73**:1603.

Huiet, L., Feldstein, P. A., Tsai, J., and Falk, B. W., 1993, The maize stripe virus major noncapsid protein messenger RNA transcripts contain heterogeneous leader sequences at their 5′ termini, *Virology* **197**:808.

Ihara, T., Akashi, H., and Bishop, D. H. L., 1984, Novel coding strategy (ambisense genomic RNA) revealed by sequence analyses of Punta Toro phlebovirus S RNA, *Virology* **136**:293.

Ihara, T., Smith, J., Dalrymple, J. M., and Bishop, D. H. L., 1985, Complete sequences of the glycoproteins and M RNA of Punta Toro phlebovirus compared to those of Rift Valley fever virus, *Virology* **144**:246.

Isegawa, Y., Fujiwara, Y., Ohshima, A., Fukunaga, R., Murakami, H., Yamanishi, K., and Sokawa, Y., 1990, Nucleotide sequence of the M genome segment of hemorrhagic fever with renal syndrome virus strain B-1, *Nucleic Acids Res.* **18**:4936.

Isegawa, Y., Tanishita, O., Ueda, S., and Yamanishi, K., 1994, Association of serine in position 1124 of Hantaan virus glycoprotein with virulence in mice *J. Gen. Virol.* **75**:3273.

Jin, H., and Elliott, R. M., 1991, Expression of functional Bunyamwera virus L protein by recombinant vaccinia viruses, *J. Virol.* **65**:4182.

Jin, H., and Elliott, R. M., 1992, Mutagenesis of the L protein encoded by Bunyamwera virus and production of monospecific antibodies, *J. Gen. Virol.* **73**:2235.

Jin, H., and Elliott, R. M., 1993a, Characterization of Bunyamwera virus S RNA that is transcribed and replicated by the L protein expressed from recombinant vaccinia virus, *J. Virol.* **67**:1396.

Jin, H., and Elliott, R. M., 1993b, Non-viral sequences at the 5′ ends of Dugbe nairovirus S mRNAs, *J. Gen. Virol.* **74**:2293.

Kakutani, T., Hayano, Y., Hayashi, T., and Minobe, Y., 1990, Ambisense segment 4 of rice stripe virus: Possible evolutionary relationship with phleboviruses and uukuviruses (Bunyaviridae), *J. Gen. Virol.* **71**:1427.

Kakutani, T., Hayano, Y., Hayashi, T., and Minobe, Y., 1991, Ambisense segment 3 of rice stripe virus: The first instance of a virus containing two ambisense segments, *J. Gen. Virol.* **72**:465.

Kariwa, H., Isegawa, Y., Arikawa, J., Takashima, I., Ueda, S., Yamanishi, K., and Hashimoto, N., 1994, Comparison of nucleotide sequences of M genome segments among Seoul virus strains isolated from eastern Asia, *Virus Res.* **33**:27.

Kormelink, R., de Haan, P., Meurs, C., Peters, D., and Goldbach, R., 1992, The nucleotide sequence of the M RNA segment of tomato spotted wilt virus, a bunyavirus with two ambisense RNA segments, *J. Gen. Virol.* **73**:2795.

Lappin, D. F., Nakitare, G. W., Palfreyman, J. W., and Elliott, R. M., 1994, Localization of Bunyamwera bunyavirus G1 glycoprotein to the Golgi requires association with G2 but not with NSm, *J. Gen. Virol.* **75**:3441.

Law, M. D., Speck, J. V., and Moyer, J. W., 1992, The M RNA of Impatiens necrotic spot tospovirus (Bunyaviridae) has an ambisense genomic organisation, *Virology* **188**:732.

Lawson, N., Stillman, E. A., Whitt, M. A., and Rose, J. K., 1995, Recombinant vesicular stomatitis viruses from DNA, *Proc. Natl. Acad. Sci. USA* **92**:4477.

Lees, J. F., Pringle, C. R., and Elliott, R. M., 1986, Nucleotide sequence of the Bunyamwera virus M RNA segment: Conservation of structural features in the bunyavirus glycoprotein gene product, *Virology* **148**:1.

Li, D., Schmaljohn, C. S., Anderson, K., and Schmaljohn, C. S., 1995, Complete nucleotide sequences of the M and S segments of two hantavirus isolates from California: Evidence for reassortment in nature among viruses related to hantavirus pulmonary syndrome, *Virology* **206**:973.

Lopez, N., Muller, R., Prehaud, C. and Bouloy, M., 1995, The L protein of Rift Valley fever virus can rescue viral ribonucleoproteins and transcribe synthetic genome-like RNA molecules, *J. Virol.* **69**:3972–3979.

Luytjes, W., Krystal, M., Enami, M., Parvin, J. D., and Palese, P., 1989, Amplification, expression and packaging of a foreign gene by influenza virus, *Cell* **59**:1107.

Maiss, E., Ivanova, L., Breyel, E., and Adam, G., 1991, Cloning and sequencing of the S RNA from a Bulgarian isolate of tomato spotted wilt virus, *J. Gen. Virol.* **72**:461.

Marriott, A. C., and Nuttall, P. A., 1992, Comparison of the S RNA segments and nucleoprotein sequences of Crimean-Congo hemorrhagic fever, Hazara and Dugbe viruses, *Virology* **189**:795.

Marriott, A. C., Ward, V. K., and Nuttall, P. A., 1989, The S RNA segment of sandfly fever Sicilian virus: Evidence for an ambisense genome, *Virology* **169**:341.

Marriott, A. C., El Ghorr, A. A., and Nuttall, P. A., 1992, Dugbe nairovirus M RNA: Nucleotide sequence and coding strategy, *Virology* **190**:606.

Marriott, A. C., Polyzoni, T., Antonoades, A., and Nuttall, P. A., 1994, Detection of human antibodies to Crimean-Congo haemorrhagic fever virus using expressed viral nucleocapsid protein, *J. Gen. Virol.* **75**:2157.

Morzunov, S. P., Feldman, H., Spiropoulou, C. F., Semenova, V. A., Rollin, P. E., Ksiazek, T. G., Peters, C. J., and Nichol, S. T., 1995, A newly recognized virus associated with a fatal case of hantavirus pulmonary syndrome in Louisiana, *J. Virol.* **69**:1980.

Müller, R., Poch, O., Delarue, M., Bishop, D. H. L., and Bouloy, M., 1994, Rift Valley fever virus L segment: Correction of the sequence and possible functional role of newly identified regions conserved in RNA-dependent polymerases, *J. Gen. Virol.* **75**:1345.

Newton, S. E., Short, N. J., and Dalgarno, L., 1981, Bunyamwera virus replication in cultured *Aedes albopictus* (mosquito) cells: Establishment of a persistent viral infection, *J. Virol.* **38**:1015.

Obijeski, J. F., Bishop, D. H. L., Palmer, E. L., and Murphy, F. A., 1976, Segmented genome and nucleocapsid of La Crosse virus, *J. Virol.* **20**:644.

Pang, S. Z., Nagpala, P. G., Wang, M., Slightom, J. L., and Gonsalves, D., 1992, Resistance to heterologous isolates of tomato spotted wilt virus in transgenic tobacco expressing its nucleocapsid protein gene, *Phytopathology* **82**:1223.

Pang, S. Z., Slightom, J. L., and Gonsalves, D., 1993, The biological properties of a distinct tospovirus and sequence analysis of its RNA, *Phytopathology* **83**:728.

Pardigon, N., Vialat, P., Girard, M., and Bouloy, M., 1982, Panhandles and hairpin structures at the termini of Germiston virus RNAs (Bunyavirus), *Virology* **122**:191.

Pardigon, N., Vialat, P., Gerbaud, S., Girard, M., and Bouloy, M., 1988, Nucleotide sequence of the M segment of Germiston virus: Comparison of the M gene product of several bunyaviruses, *Virus Res.* **11**:73.

Parrington, M. A., and Kang, C. Y., 1990, Nucleotide sequence analysis of the S genomic segment of Prospect Hill virus: Comparison with the prototype hantavirus, *Virology* **175**:167.

Parrington, M. A., Lee, P. W., and Kang, C. Y., 1991, Molecular characterization of the Prospect Hill virus M RNA segment: A comparison with the M RNA segments of other hantaviruses, *J. Gen. Virol.* **72**:1845.

Patterson, J. L., Holloway, B., and Kolakofsky, D., 1984, La Crosse virions contain a primer-stimulated RNA polymerase and a cap-dependent endonuclease, *J. Virol.* **52**:215.

Plyusnin, A., Vapalahti, O., Lankinen, H., Lehväslaiho, H., Apekina, N., Myasnikov, Y., Kallio-Kokko, H., Henttonen, H., Lundkvist, Å., Brunner-Korvenkontio, M., Gavrilovskaya, I., and Vaheri, A., 1994a, Tula virus; a newly detected hantavirus carried by European common voles, *J. Virol.* **68**:7833.

Plyusnin, A., Vapalahti, O., Ulfves, K., Lehväslaiho, H., Apekina, N., Gavrilovskaya, I., Blinov, V., and Vaheri, A., 1994b, Sequences of wild Puumala virus genes show a correlation of genetic variation with geographic origin of the strains, *J. Gen. Virol.* **75**:405.

Poch, O., Suavaget, I., Delarue, M., and Tordo, N., 1989, Identification of four conserved motifs among the RNA-dependent polymerase encoding elements, *EMBO J.* **8**:3867.

Racaniello, V. R., and Baltimore, D., 1981, Cloned poliovirus cDNA is infectious in mammalian cells, *Science* **214**:916.

Raju, R., and Kolakofsky, D., 1987, Unusual transcripts in La Crosse virus-infected cells and the site for nucleocapsid assembly, *J. Virol.* **61**:667.

Raju, R., and Kolakofsky, D., 1989, The ends of La Crosse virus genome and antigenome RNAs within nucleocapsids are base paired, *J. Virol.* **63**:122.

Ramirez, B.-C., and Haenni, A.-L., 1994, Molecular biology of tenuiviruses, a remarkable group of plant viruses, *J. Gen. Virol.* **75**:467.

Ramirez, B.-C., Lozano, I., Constantini, L.-M., Haenni, A.-L., and Calvert, L. A., 1993, Complete nucleotide sequence and coding strategy of rice hoja blanca virus RNA 4, *J. Gen. Virol.* **74**:2463.

Ramirez, B.-C., Garcin, D., Calvert, L. A., Kolakofsky, D., and Haenni, A.-L., 1995, Capped nonviral sequences at the 5′ ends of the mRNAs of rice hoja blanca RNA4, *J. Virol.* **69**:1951.

Ravkov, E. V., Rollin, P. E., Ksiazek, T. G., Peters, C. J., and Nichol, S. T., 1995, Genetic and serological analysis of Black Creek Canal virus and its association with human disease and *Sigmodon hispidus* infection, *Virology* **210**:482.

Roberts, A., Rossier, C., Kolakofsky, D., Nathanson, N., and Gonzalez-Scarano, F., 1995, Completion of the La Crosse virus genome sequence and genetic comparison of the L proteins of the Bunyaviridae, *Virology* **206**:742.

Rönnholm, R., and Pettersson, R. F., 1987, Complete nucleotide sequence of the M RNA segment of Uukuniemi virus encoding the membrane glycoproteins G1 and G2, *Virology* **160**:191.

Samso, A., Bouloy, M., and Hannoun, C., 1975, Présence de ribonucléoproteins circulaires dans le virus Lumbo (Bunyavirus), *C. R. Acad. Sci. Ser. D* **280**:779.

Samso, A., Bouloy, M., and Hannoun, C., 1976, Mise en évidence de molécules d'acide ribonucleique circulaire dans le virus Lumbo (Bunyavirus), *C. R. Acad. Sci. Ser. D* **282**:1653.

Scallan, M. F., and Elliott, R. M., 1992, Defective RNAs in mosquito cells persistently infected with Bunyamwera virus, *J. Gen. Virol.* **73**:53.

Schmaljohn, C. S., 1990, Nucleotide sequence of the L genome segment of Hantaan virus, *Nucleic Acids Res.* **18**:6728.

Schmaljohn, C. S., Jennings, G. B., Hay, J., and Dalrymple, J. M., 1986, Coding strategy of the S-genome segment of Hantaan virus, *Virology* **155**:633.

Schmaljohn, C. S., Schmaljohn, A., and Dalrymple, J. M., 1987, Hantaan virus M RNA: Coding strategy, nucleotide sequence and gene order, *Virology* **157**:31.

Schmaljohn, C. S., Arikawa, J., Hasty, S. E., Rasmussen, L., Lee, P. W., and Dalrymple, J. M., 1988, Conservation of antigenic properties and sequences encoding the envelope proteins of prototype Hantaan virus and two virus isolates from Korean haemorrhagic fever patients, *J. Gen. Virol.* **69**:1949.

Schnell, M. J., Mebatsion, T., and Conzelmann, K. K., 1994, Infectious rabies virus from cloned cDNA, *EMBO J.* **13**:4195.

Shope, R. E., 1985, Bunyaviruses, in: *Virology* (B. N. Fields, ed.), pp. 1055–1082, Raven Press, New York.

Shope, R. E., and Causey, O. R., 1962, Further studies on the serological relationships of Group C arthropod-borne viruses and the application of these relationships to rapid identification of types, *Am. J. Trop. Med. Hyg.* **11**:283.

Simons, J. F., Hellman, U., and Pettersson, R. F., 1990, Uukuniemi virus S RNA segment: Ambisense coding strategy, packaging of complementary strands into virions, and homology to members of the genus *Phlebovirus*, *J. Virol.* **64**:247.

Spiropoulou, C. F., Morzunov, S., Feldman, H., Sanchez, A., Peters, C. J., and Nichol, S. T., 1994, Genome structure and variability of a virus causing hantavirus pulmonary syndrome, *Virology* **200**:715.

Stohwasser, R., Giebel, L. B., Zöller, L., Bautz, E. K. F., and Darai, G., 1990, Molecular characterization of the RNA S segment of nephropathia epidemica virus strain Hällnäs B 1, *Virology* **174**:79.

Stohwasser, R., Raab, K., Darai, G., and Bautz, E. K. F., 1991, Primary structure of the large (L) RNA segment of nephropathia epidemica virus strain Hällnäs B 1, *Virology* **174**:79.

Takahashi, M., Toriyama, S., Kikuchi, Y., Hayakawa, T., and Ishihama, A., 1990, Complementarity between the 5′ and 3′-terminal sequences of rice stripe virus RNAs, *J. Gen. Virol.* **71**:2817.

Takahashi, M., Toriyama, S., Hamamatsu, C., and Ishihama, A., 1993, Nucleotide sequence and possible ambisense coding strategy of rice stripe virus RNA segment 2, *J. Gen. Virol.* **74**:769.

Takehara, K., Min, M.-K., Battles, J. K., Sugiyama, K., Emery, V. C., and Dalrymple, J. M., 1989, Identification of mutations in the M RNA of a candidate vaccine strain of Rift Valley fever virus, *Virology* **169**:452.

Toriyama, S., Takahashi, M., Sano, Y., Shimizu, T., and Ishihama, A., 1994, Nucleotide sequence of RNA 1, the largest genomic segment of rice stripe virus, the prototype of the tenuiviruses, *J. Gen. Virol.* **75**:3569.

Vapalahti, O., Kallio-Kokko, H., Salonen, E. M., Brummer-Korvenkontio, M., and Vaheri, A., 1992, Cloning and sequencing of Puumala virus Sotkamo strains S and M RNA segments: Evidence of strain variation in hantaviruses and expression of the nucleocapsid protein, *J. Gen. Virol.* **73**:829.

Vialat, P., and Bouloy, M., 1992, Germiston virus transcriptase requires active 40S ribosomal subunits and utilizes capped cellular RNAs, *J. Virol.* **66**:685.

Ward, V. K., Marriott, A. C., El Ghorr, A. A., and Nuttall, P. A., 1990, Coding strategy of the S RNA segment of Dugbe virus (*Nairovirus;* Bunyaviridae), *Virology* **175**:518.

Xiao, S.-Y., Liang, M., and Schmaljohn, C. S., 1993a, Molecular and antigenic characterisation of HV114, a hantavirus isolated from a patient with haemorrhagic fever with renal syndrome, *J. Gen. Virol.* **74**:1657.

Xiao, S.-Y., Spik, K. W., Li, D., and Schmaljohn, C. S., 1993b, Nucleotide and deduced amino acid sequences of the M and S genome segments of two Puumala virus isolates from Russia, *Virus Res.* **30**:97.

Xiao, S.-Y., LeDuc, J. W., Chu, Y. K., and Schmaljohn, C. S., 1994, Phylogenetic analyses of virus isolates in the genus *Hantavirus*, family *Bunyaviridae*, *Virology* **198**:205.

Xu, X., Ruo, S. L., Tang, Y., Fisher-Hoch, S., and McCormick, J. B., 1991, Molecular characterization and expression of glycoprotein gene of hantavirus R22 strain isolated from *Rattus norvegicus* in China, *Virus Res.* **21**:35.

Yoo, D., and Kang, C. Y., 1987, Nucleotide sequence of the M segment of the genomic RNA of Hantaan virus 76-118, *Nucleic Acids Res.* **15**:6299.

Zhu, Y., Hayakawa, T., Toriyama, S., and Takahashi, M., 1991, Complete nucleotide sequence of RNA 3 of rice stripe virus: An ambisense coding strategy, *J. Gen. Virol.* **72**:763.

Zhu, Y., Hayakawa, T., and Toriyama, S., 1992, Complete nucleotide sequence of RNA 4 of rice stripe virus isolate T, and comparison with another isolate and with maize stripe virus, *J. Gen. Virol.* **73**:1309.

Index

THE VIRUSES
COMPLETE LISTING OF VOLUMES
Series Editors
HEINZ FRAENKEL-CONRAT, *University of California*
Berkeley, California

ROBERT R. WAGNER, *University of Virginia School of Medicine*
Charlottesville, Virginia

THE ADENOVIRUSES
Edited by Harold S. Ginsberg

THE ARENAVIRIDAE
Edited by Maria S. Salvato

THE BACTERIOPHAGES
Volumes 1 and 2 • Edited by Richard Calendar

THE BUNYAVIRIDAE
Edited by Richard M. Elliott

THE CORONAVIRIDAE
Edited by Stuart G. Siddell

THE HERPESVIRUSES
Volumes 1–3 • Edited by Bernard Roizman
Volume 4 • Edited by Bernard Roizman and Carlos Lopez

THE INFLUENZA VIRUSES
Edited by Robert M. Krug

THE PAPOVAVIRIDAE
Volume 1 • Edited by Norman P. Salzman
Volume 2 • Edited by Norman P. Salzman and Peter M. Howley

THE PARAMYXOVIRUSES
Edited by David W. Kingsbury

THE PARVOVIRUSES
Edited by Kenneth I. Berns

THE PLANT VIRUSES
Volume 1 • Edited by R. I. B. Francki
Volume 2 • Edited by M. H. V. Van Regenmortel and Heinz Fraenkel-Conrat
Volume 3 • Edited by Renate Koenig
Volume 4 • Edited by R. G. Milne

THE REOVIRIDAE
Edited by Wolfgang K. Joklik

THE RETROVIRIDAE
Volumes 1–4 • Edited by Jay A. Levy

THE RHABDOVIRUSES
Edited by Robert R. Wagner

THE TOGAVIRIDAE AND FLAVIVIRIDAE
Edited by Sondra Schlesinger and Milton J. Schlesinger

THE VIRUSES
COMPLETE LISTING OF VOLUMES

THE VIROIDS
Edited by T. O. Diener

THE VIRUSES: Catalogue, Characterization, and Classification
Heinz Fraenkel-Conrat